ELEMENTARY
QUANTUM
MECHANICS
Expanded Edition

E L E M E N T A R Y
QUANTUM
MECHANICS

Expanded Edition

PETER FONG
EMORY UNIVERSITY, USA

World Scientific

NEW JERSEY • LONDON • SINGAPORE • BEIJING • SHANGHAI • HONG KONG • TAIPEI • CHENNAI

Published by

World Scientific Publishing Co. Pte. Ltd.

5 Toh Tuck Link, Singapore 596224

USA office: 27 Warren Street, Suite 401-402, Hackensack, NJ 07601

UK office: 57 Shelton Street, Covent Garden, London WC2H 9HE

First published in 1962 by Addison-Wesley Publishing Company, Inc.

ELEMENTARY QUANTUM MECHANICS
(Expanded Edition)

ISBN 981-256-292-3

Printed in U.S.A.

PREFACE TO THE EXPANDED EDITION

Quantum mechanics is a difficult subject for students to learn after years of rigorous training in classical physics, which eliminated their sloppy thinking and unscientific notions. Then in quantum mechanics they have to abandon what they have laboriously learned and adopt a new system of thinking. Yet most amazing is the fact that after much trial and tribulation, the good old Newton's law came out intact unscathed.

In the old edition of this book published four decades ago the author reformulated Newtonian mechanics in a new form of wave motion in a classical theory with an undetermined constant H. When H approaches zero the theory reduces to Newton's exactly. When H is set equal to the Planck constant h, the theory reduces to the Schrödinger version of quantum mechanics. Thus the new theory, at least its mathematical format, can be learned without tortuous brain washing. The ideological changes on the fundamentals of physics have to be taken care of but can be done cleanly without ramification and complexity. Dr. Ralph Ciceroni, the Chancellor of UC, Irvine, has commented that it is more than a textbook. Over the years the book has shepherded the growth of a generation of physicists.

In this expanded edition the same trick is applied to introduce matrix mechanics. In the new formulation based on matrix, H and h are replaced by matrix $\mathbf{H}$ and $h\mathbf{I}$. We can expect the same benefit of introducing something alien under the security blanket of something familiar. One novel result is the derivation of the Sommerfeld quantum condition from the matrix commutation relation of $\mathbf{p}$ and $\mathbf{x}$. This wraps up the last loose end in the fabric of quantum mechanics, which was not known before, and is a gratifying concluding act to close the theoretical structure of the quantum theory.

It also clarifies the physical meaning of the enigmatic non-commutative algebra of $\mathbf{p}$ and $\mathbf{x}$ in quantum theory — it is an esoteric mathematical representation of the fundamentally inexplicable fact of the quantization of the action variable to the units of h. We can only marvel but not explain. The historical wonder is that the limiting case of h approaching zero was discovered earlier in the same algebra system but reduced to commutative when h is zero. This is the more direct connection of the two theories than obtainable from the Schrödinger theory. The matrix formulation also allows the quantum theory to be generalized to new physical systems such as the electron spin, which cannot be done by the Schrödinger approach.

P. F.

PREFACE TO THE FIRST EDITION

Quantum mechanics is usually a frustrating course for many physics students. Sometimes it is the stumbling block on their way to completing an education in physics. Many hard-working students may finish the course with an understanding of the mathematical manipulations but without a grasp of the underlying physical principles. They have seen that assumptions disconnected from reality and concepts alien to experience, when put together, may yield such familiar results as the Balmer formula and the Rutherford scattering equation. To them these results are amazing, but the physical principles that make the results possible seem beyond comprehension.

To simplify the presentation of a complicated subject, quantum mechanics may be developed by the deductive method, starting from a few axioms. However, the axioms are so highly mathematical in nature (e. g., observables are represented by noncommutative operators) that the student feels he is learning a branch of mathematics which, by some magic, predicts correctly the experimental results in physics. Physical meaning becomes obscure, in spite of the addition of piecemeal interpretations. To overcome these difficulties, it seems best to keep the development of quantum mechanics as close to classical mechanics as possible and to emphasize the fact that quantum mechanics is a natural extension of classical mechanics. The generalization of a well-understood theory is usually easier to accept than a completely new one; the physical picture is clearer and the logical relation simpler. It is not by good luck that quantum mechanics, starting with assumptions very different from classical mechanics, ends with results very similar. It is possible to trace out the intimate relations of the two at the beginning and at the different stages of development so that the similarity of their end results appears to be a matter of course rather than a surprise. To make clear the relations among all elements involved, one looks for a logical structure, not merely a few independent threads of thought but an interwoven texture embracing all logical connections through which we can foresee some of the results of quantum mechanics without having to await the laborious solution of a differential equation which seems so remote from the physical world.

In this introductory volume we try to develop as thoroughly as possible a few, but not all, of the new concepts introduced in quantum mechanics. Essentially, we concern ourselves with the many aspects of the Schrödinger equation (without spin) and its applications. In presenting our material we shall make rather extensive use of the concept of the wave packet. The wave packet is by no means of basic importance, but it serves as a useful scaffolding by means of which quantum mechanics may be built. The kinematics and dynamics of a wave packet are closely related to the classical mechanics of a particle; thus the mechanics of a wave packet may be considered as a formal extension of classical mechanics. On the other hand, a wave packet not only resembles a classical particle but also has an additional property, the uncertainty relation. Therefore, the results of the Schrödinger equation can either be reduced to the results of classical mechanics or be traced to the uncertainty relation (quantum phenomena). Take the linear harmonic oscillator, for example. It can be shown that the results of classical mechanics can be derived from a wave packet formed by superposition of eigenfunctions of large quantum numbers, and that a quantum effect, namely, the existence of the so-called zero-point motion, may be traced back to the uncertainty relation. The consequences of the extension from classical mechanics to quantum mechanics are thus made clear. Often a student, after learning quantum mechanics for several months, finds himself in possession of a large body of information concerning energy quantization, but fails to appreciate quantum mechanics as a dynamical system, i.e., a theoretical system capable of predicting the future development of physical systems from their initial conditions. The use of the wave packet helps bring out the dynamical aspect of the new mechanics. After quantum mechanics is thus introduced and made familiar to the student, a more general, abstract formulation is then presented in the later part of the book.

Following a historical introduction in Chapter 1, the mathematical formalism of quantum mechanics in a limited form will be presented in Chapter 2. The Schrödinger equation will be introduced. The meaning of its solutions will be discussed according to the two assumptions of Born. Schrödinger's equation and Born's assumptions form a self-consistent theoretical system. The similarity of and difference between this system and classical mechanics will be demonstrated by the derivation of Newton's second law and of the uncertainty relation.

In three following chapters this system of mechanics will be applied to three special cases: the free particle, the linear harmonic oscillator, and the potential barrier problems. To emphasize the similarity to classical mechanics, wave packet solutions will be obtained which reproduce the properties of the classical solutions. To emphasize the difference from classical mechanics, applications will be made to three quantum phenomena, i.e., the wave property of matter, the quantization of energy, and the penetration of potential barrier.

After having demonstrated the usefulness of this mathematical theory, we shall turn our attention in Chapter 6 to the physical meaning of quantum mechanics. Heisenberg's uncertainty principle will be discussed after a critical examination of the physical concepts used in classical mechanics. The causality law will be examined in connection with the quantum-mechanical description of physical processes. The inquiry into the physical meaning of quantum mechanics is usually expressed by students in the following questions: Is an electron a particle or a wave? Why is the energy quantized? How can a particle penetrate through a potential barrier? In anticipation of such inquiries, an explanation of the quantum phenomena will be given (in Section 6–3) in terms of the uncertainty principle and of the quantum potential of Eq. (2–65).

Chapters 7–11 will contain straightforward applications of quantum mechanics to various problems; they will be solved either exactly or approximately by the perturbation methods. When a problem may be solved in several ways, it is usually instructive to look into the relations among the different methods. Thus, the three-dimensional harmonic oscillator problem will be solved by using rectangular, cylindrical, and spherical coordinates respectively, and the relations among the solutions will be discussed. The Rutherford scattering problem will be treated by methods developed in the time-independent, as well as in the time-dependent, perturbation theory.

Before the book closes, the wave-packet scaffolding will be torn down and quantum mechanics will be reformulated in the operator form in Chapter 12. The operator formulation is the only one by which quantum mechanics may be adequately presented. But its abstractness makes it preferable, if not necessary, to delay its introduction until after the concepts and results of quantum mechanics have been made reasonably familiar. This chapter also serves the purpose of preparing students to read more advanced treatises.

In writing this book the author was inspired by the writings of Heisenberg, de Broglie, Schrödinger, Bohr, and Dirac. Nevertheless, he is solely responsible for the logical organization of the presentation. The subjects treated here, except for the derivation of a number of results, are found in published literature, although a few gaps must be filled to make possible a systematic presentation on an elementary level. Suggestions from users of the book will be appreciated. It is fitting to acknowledge the help the author has received in preparing this book. Dr. Melba Phillips made valuable suggestions in the earlier chapters. Two friends, Y. F. Bow and Robert Sanders, have read the manuscript and checked the equations carefully; they also made many important suggestions. The late Mrs. Ruth Hoople edited the manuscript. Mrs. Dorothy Sickels provided editorial advice and helped in proofreading. To them and to many others the author takes this opportunity to express his deep gratitude.

If the reader finds the assumptions involved in this presentation not like repulsive strangers, if the reader finds in the logical texture no loose thread, the author has achieved his aim.

P. F.

Utica, New York

TO THE INSTRUCTOR

The purpose of this book is to develop quantum mechanics on an elementary level with an emphasis on its physical meaning and to provide a textbook to satisfy the rising need for a one-semester introductory course in this subject. The requirements of these two objectives are not exactly the same, and therefore some compromises are unavoidable regarding the choice of subject material and the method of presentation.

It is not our intention in this book to present quantum mechanics with strict mathematical rigor, since a few other treatises have done so. The use of mathematics is dictated by the consideration that a senior physics student should be able to follow the text without mathematical difficulty. Many elegant methods of advanced mathematics will have to be left out; the prerequisite is limited to a course of advanced calculus.*

For our purposes, it seems best to limit the scope of this book to a discussion centered around the Schrödinger equation (without spin). Its mathematical implication, physical meaning, and various applications form a unit suitable for presentation in one semester. The *systematic* treatment of matrix mechanics, transformation theory, theory of angular momentum, theory of electron spin, identical particles, atomic structure, molecular structure, nuclear structure, theory of collisions and quantum statistics, usually the subject material of the second semester of a one-year course (and thus the contents of another book), will not be attempted here, though many individual problems from these topics will be included as illustrative examples. In a one-year course assuming no previous knowledge of quantum mechanics, this book may be used as the text for the first semester.

One major consideration in determining the order and method of presentation is that many of the students may not have completed a course in analytic mechanics. The subject material is so arranged and presented, at the risk of sacrificing elegance and conciseness, that this group of students may work with a minimum handicap. Starting from the very beginning of this book they may proceed without difficulty by omitting Section 2–7 (with no break of continuity) until they reach the last chapter, which may be omitted. It is solely for this reason that the operator formulation is left to the last chapter. For those who have had a course in analytic mechanics, or are taking it concurrently, it is advis-

---

* *All* mathematical prerequisites for reading this book, including an elementary knowledge of vector analysis, may be found in a textbook of advanced calculus such as Osgood's.

able to take up the operator formulation (Chapter 12) right after Chapter 8, about halfway through the book, or at a time when the Hamiltonian theory and the Poisson bracket have been introduced in the concurrent course.

A set of carefully planned problems is appended at the end of each chapter. The more difficult ones are marked by an asterisk(*). Problems denoted as *mathematical exercise* are purely mathematical problems and may be omitted for those students who are adequately prepared in mathematics. As the text is complete by itself, students are not required to solve any problem in order to be able to read the text. On the other hand the problems, usually appended with *answers, hints, notes* or *remarks*, supplement the text and thus form a part of the required reading. In many cases, to work out the problem in all its mathematical details is time consuming and may be unnecessary; the principles and methods of solving the problems should be emphasized, and the available time for working problems should be so allocated as to obtain the maximum benefit from these problems.

CONTENTS

CHAPTER 1

HISTORICAL INTRODUCTION

1-1 Planck's theory of blackbody radiation. In 1901 Planck*introduced in physics a drastically new idea: he postulated that an oscillator can take on only a discrete set of energy values instead of all continuous values as tacitly assumed in classical mechanics. Originally, this assumption was made to explain the law of blackbody radiation. As history unfolded, it turned out that the introduction of this idea marked the beginning of a new age in the development of physics, the age of quantum theory.

Planck's argument is roundabout and abstruse; it will not be discussed in detail here. Nor shall we go through the lengthy thermodynamical reasoning leading to the law of blackbody radiation.† It will be sufficient here to say that the Rayleigh-Jeans law and the Wien law are valid respectively in the low- and the high-frequency regions of blackbody radiation; that Planck found a formula which not only included these two laws as its special cases, but also proved to be the correct formula for the entire frequency range; and that Planck tried to find a theory to explain his empirical formula. Let us write the Rayleigh-Jeans formula:

$$u(\nu)\,d\nu = \frac{8\pi\nu^2 kT}{c^3}\,d\nu;\qquad(1\text{--}1)$$

the Wien formula (in convenient form):‡

$$u(\nu)\,d\nu \sim \nu^3 e^{-h\nu/kT}\,d\nu;$$

and the Planck formula:

$$u(\nu)\,d\nu = \frac{8\pi h\nu^3}{c^3}\,\frac{1}{e^{h\nu/kT}-1}\,d\nu,\qquad(1\text{--}2)$$

where $u(\nu)$ is the energy per unit volume of the blackbody radiation in the frequency range from ν to $\nu + d\nu$, k is the Boltzmann constant, T is the absolute temperature, c is the velocity of light, and h is an empirical

---

* M. Planck, *Ann. d. Phys.* **4**, 553 (1901).

† A discussion of blackbody radiation may be found, for example, in Richtmyer, Kennard, and Lauritsen, *Introduction to Modern Physics* (McGraw-Hill Book Co.).

‡ For convenience, the empirical constant Wien introduced is written h/k.

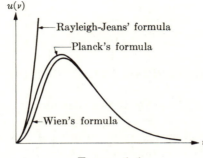

FIGURE 1-1

constant known as the Planck constant. The constant h may be determined by fitting Eq. (1–2) with the experimental data, but the following
best value of h is obtained later by other methods:

$$h = 6.625 \times 10^{-27} \text{ erg} \cdot \text{sec}.$$

It is easily verified that Planck's formula includes the other two as its
special cases.

All three formulas are represented graphically in Fig. 1–1. The Rayleigh-
Jeans formula consists of a product of $(8\pi\nu^2/c^3)\,d\nu$ and kT, the former
being the number of vibrational waves in the frequency range ν to $\nu + d\nu$
and the latter being the average energy of a mode of vibration at temperature T. Thus the Rayleigh-Jeans law is the result of a straightforward
application of classical statistical mechanics. Its difficulty lies in the fact
that it diverges at the high-frequency end. In Planck's formula, kT is
replaced by $h\nu/(e^{h\nu/kT} - 1)$, which approaches zero sufficiently rapidly
to make Eq. (1–2) converge at the high-frequency end. Thus the Planck
law seems to imply a modification of the principle of equipartition of energy, which may be specified by the following substitution:

$$kT \rightarrow \frac{h\nu}{e^{h\nu/kT} - 1}. \tag{1-3}$$

Such a modification may be realized if, according to the Planck theory,*
it is assumed that the energy of a mode of vibration of the radiation field

---

* We shall follow a simplified approach due to Debye (1910), according to
which a mode of vibration may be treated as an oscillator. Planck considered
that the energies given by Eq. (1–4) are those of the oscillators emitting blackbody radiation, and not the energies of the modes of vibration themselves. He
then derived his empirical formula by a consideration of the equilibrium between
the oscillators and the radiation field.

can assume only the following values:

$$E_n = nh\nu, \qquad n = 0, 1, 2, \ldots ; \qquad (1\text{--}4)$$

in other words, the energy must be an integral multiple of a *quantum* of energy, $h\nu$. If the oscillator can take on any continuous value from 0 to ∞, as the classical theory tacitly assumed, the average value of the energy will be

$$\overline{E} = \frac{\int_0^\infty E e^{-E/kT}\, dE}{\int_0^\infty e^{-E/kT}\, dE} = kT, \qquad (1\text{--}5)$$

where $e^{-E/kT}$, according to the Boltzmann distribution law in statistical mechanics, is the relative probability of finding the energy value to be E. According to the Planck hypothesis, Eq. (1–4), the average value becomes

$$\overline{E} = \frac{\sum_{n=0}^\infty nh\nu e^{-nh\nu/kT}}{\sum_{n=0}^\infty e^{-nh\nu/kT}}. \qquad (1\text{--}6)$$

These summations may be evaluated by using the following identities:

$$\frac{1}{1-x} = 1 + x + x^2 + \cdots,$$

$$\frac{x}{(1-x)^2} = x + 2x^2 + 3x^3 + \cdots. \qquad (1\text{--}7)$$

As a result, Eq. (1–6) becomes

$$\overline{E} = \frac{h\nu e^{-h\nu/kT}/(1 - e^{-h\nu/kT})^2}{1/(1 - e^{-h\nu/kT})} = \frac{h\nu}{e^{h\nu/kT} - 1}. \qquad (1\text{--}8)$$

Thus the modification specified by Eq. (1–3) is realized by the introduction of the quantum hypothesis, Eq. (1–4). The above mathematical result may be illustrated graphically as follows. In Fig. 1–2 we plot the energy of an oscillator or a mode of vibration of the radiation field on the vertical axis. The relative probability distribution among these energy values is plotted to the left of the energy axis. In Fig. 1–2(a) the energy is assumed to be continuous. From the shape of the exponential distribution curve the average value of energy may be roughly estimated to be near $E = kT$, where the probability has reduced to $1/e$. In Fig. 1–2(b) the energy of the oscillator is assumed to take on only a set of discontinuous values, each represented by a horizontal line. If the lines are very close to one another they may be regarded as forming a continuous band, as in Fig. 1–2(a), and the average energy will be very close to that shown

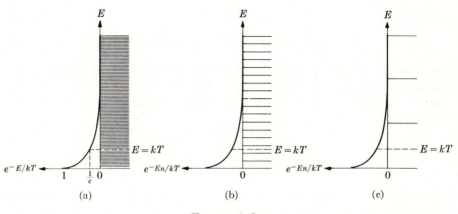

FIGURE 1–2

in Fig. 1–2(a). Therefore the continuous energy spectrum may be regarded as a limiting case of the discontinuous energy spectrum when the
spacing between two adjacent energy values approaches zero. On the
other hand, in Fig. 1–2(c), we have a situation where the discrete energy
values are so widely separated that no allowed energy values are to be
found near $E = kT$. In this case, all energy values except $E = 0$ have
negligible probabilities, so that the average energy of the oscillator is near
zero rather than kT. The average energy approaches zero when the spacing
approaches infinity. According to Eq. (1–4), the energy spacing is proportional to frequency ν. Thus in the low-frequency region the situation
will be like that in Fig. 1–2(b), while in the high-frequency region it will
be like that in Fig. 1–2(c). The divergence difficulty of the Rayleigh-
Jeans law is due to the fact that there are too many degrees of freedom
(proportional to ν^2) at the high-frequency end, each being assumed to
have a finite amount of energy kT. The introduction of the quantum
hypothesis does not change the situation in the low-frequency region as
demonstrated in Fig. 1–2(b). However, in the high-frequency region,
Fig. 1–2(c) shows that each degree of freedom, i.e., each mode of oscillation, shares a negligibly small amount of energy instead of a finite amount
kT, so that the total energy of all waves is finite instead of infinite. The
introduction of the discontinuous energy values thus renders the many
degrees of freedom at the high-frequency end ineffective so that the distribution function becomes convergent.

Although the formula of blackbody radiation may be "explained" by
the quantum hypothesis, the problem is not solved; it is merely shifted
from the radiation formula to the quantum hypothesis.

The classical theory of physics is formulated on the basis that dynamical
quantities, including energy, are all continuous. Restricting the energy

value of an oscillator to a selected set of numbers specified by Eq. (1–4) is a concept completely foreign to classical physics. Planck spent the rest of his life trying unsuccessfully to reconcile the quantum hypothesis with classical theory. With the advantage of retrospect, we now know that Planck's hypothesis is just a first step in a revolution in which many continuous physical quantities are to be replaced by discontinuous ones. In this revolution, the basic issue is discontinuity versus continuity, which is related to the issue of the finite versus the infinite. The divergence (infinity) of Eq. (1–1) originates from the fact that both the coordinates and the energy are continuous. The continuity of space makes it possible to conceive of waves of infinitely small wavelengths; the number of vibrational modes thus diverges to the square of ν. In Planck's theory the divergence is eliminated by making the energy variable discontinuous. Actually, the concepts of continuity and infinity are sophisticated mental products of mathematics. But we have become so accustomed to them by our training in mathematical analysis and its applications that we take for granted that all physical quantities are continuous without asking for experimental substantiation. In the following example we shall show that a theory based on the concept of continuity cannot always be continuous in every respect. Consider the specific heat of solids. Each atom in a solid is assumed to execute vibrational motion in three dimensions and thus is equivalent, in terms of energy, to three linear oscillators. If the energy variable is continuous, the average energy of one atom will be $3kT$. The average energy of one mole of material will be $3N_0kT$, where N_0 is the *Avogadro number*, and the molar specific heat will be $3N_0k$ or $3R$, where R is the *universal gas constant*. Now let the force constant of the oscillator increase. This will be accompanied by an increase in the frequency of vibration. Yet the specific heat remains $3R$ no matter how large the force constant may become. On the other hand, when the force constant reaches infinity, no vibration is possible and the specific heat should be zero. Therefore a discontinuous change of the specific heat from a constant value $3R$ to zero will take place when the force constant approaches infinity. This example shows that the assumption of continuity in energy leads to a result of discontinuity in specific heat. If we cannot avoid discontinuity in one physical quantity, there is no reason why we cannot accept discontinuity in another, say energy. In fact, if we accept Eq. (1–4) for the energy of an oscillator, the specific heat C of a solid will be given by

$$C = \frac{d}{dT}\left(N_0 \frac{3h\nu}{e^{h\nu/kT} - 1}\right). \tag{1–9}$$

When the force constant increases to infinity, this specific heat approaches zero gradually instead of discontinuously and there will be no discon-

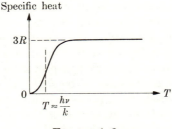

FIGURE 1–3

tinuity in specific heat. In a certain sense the theory based on discontinuous energy values, which may be reduced to the theory based on continuous values when the energy spacings are small, seems more flexible. It may well be that physical quantities are actually discontinuous and that the continuous theory of classical physics is but a macroscopic approximation.

The preceding discussion of specific heat serves to introduce another subject, namely, specific heat at low temperature, which provides further evidence that the energy of an oscillator is *quantized*, i.e., the oscillator can take on only a discrete set of energy values. The study of specific heat played an important part in the development of *quantum theory*, the theory dealing with quantized physical quantities. On the other hand, the quantum theory of specific heat is a straightforward extension of Planck's theory and no new idea is involved. It is thus appropriate to include a brief discussion of it in this section, disregarding that this work was done two years after another important step in the development of quantum theory, i.e., the theory of photoelectric effect which will be discussed in the next section.

As we have mentioned, classical theory sets forth that the molar specific heat of solids is a constant, $3R$, approximately 6 cal/mole/degree. This result has been verified for many solids at room temperature (Dulong-Petit's law). However, experimental evidence shows that when the temperature is reduced to near absolute zero, the specific heat decreases to zero, thus contradicting the prediction of classical theory (see Fig. 1–3). This deviation becomes noticeable at liquid air temperature, although a few materials such as the diamond reveal the beginning of this deviation even at room temperature. In attempting to explain this phenomenon, Einstein* in 1907 made the assumption that the oscillators in a solid behave just like the Planck oscillator. The specific heat of the solid, according to our discussion above, is then given by Eq. (1–9). It actually approaches zero when the temperature is reduced to zero. The reason is that in the low-temperature region kT is so small that the situation of

---

* A. Einstein, *Ann. d. Phys.* **22,** 180 (1907).

Fig. 1–2(c) prevails for all oscillators, and consequently the average energy of each oscillator becomes very small. In spite of the qualitative agreement, Einstein's theory is too simple to describe the complicated internal motion of a solid. Debye* in 1912 worked out a more complete theory which agreed well quantitatively with experimental results. Nevertheless, the basic idea involved is the same as Einstein's. The improvement lies in a more detailed analysis of the internal vibrational motion of a solid.

The specific heat of gases also furnishes evidence for the existence of quantized energy values. A diatomic molecule may execute oscillatory motion, the two atoms moving closer and farther apart alternately. The average energy of this oscillation, according to classical statistical mechanics, is again kT and thus contributes to the molar specific heat an amount equal to R. However, at room temperature experimental results do not show such a contribution in many diatomic gases. According to quantum theory, this may be explained by assuming that the frequency of oscillation is so high that even at room temperature the situation of Fig. 1–2(c) prevails and the average energy of this oscillation is nearly zero. However, at higher temperatures, when the situation of Fig. 1–2(b) prevails, the prediction of classical theory will be valid. This is also verified.

In Debye's theory the vibrational motion of a solid as a whole is analyzed. Such a theory is adequate for monatomic crystals. For polyatomic crystals, the vibrational motion of the molecule itself, like that of the diatomic molecule just mentioned, must also be considered. Accordingly, the vibrational frequencies may be divided into two parts: the *acoustical branch* for the vibration of the solid as a whole, and the *optical branch* for the vibration of the molecules. The first part may be treated by the Debye theory and the second by the Einstein theory. By such an analysis, which is rather complicated, the quantum theory of specific heat is able to account for many observed results.

1–2 Einstein's photon theory of light. In 1905 Einstein† took an important step in the development of quantum theory. Unlike Planck who, looking backward, tried to reconcile quantized energies with classical theory, Einstein, looking forward, tried to develop the new idea to its logical conclusion. His later application of quantum theory to specific heat (1907), discussed above, was another step in the same direction. If the quantization of energy in blackbody radiation is considered as a special instance indicating the existence of a general law of quantization, then the same situation may prevail for other types of oscillators, such

---

* P. Debye, *Ann. d. Phys.* **39,** 789 (1912).
† A. Einstein, *Ann. d. Phys.* **17,** 132 (1905).

as those contributing to the specific heat of solids and gases. The successful application of quantum theory to specific heat justifies Einstein's foresight. His work in 1905 on the photoelectric effect was carried out along the same line of thought. He assumed that the energy of the radiation field is quantized. Planck assumed that only the energy of an oscillator in equilibrium with blackbody radiation is quantized, and he derived the law of blackbody radiation by considering equilibrium between the radiation field and the oscillators. If the energy of an oscillator is assumed to take on only the values $nh\nu$, it can change only by an integral multiple of $h\nu$. In the emission and absorption processes, an oscillator thus can absorb or emit radiation energy only in units of $h\nu$. The logical conclusion seems to be that radiation energy itself exists only in units of $h\nu$, there being no fractions of one such unit. Since the frequency of an oscillator and that of the radiation it absorbs or emits are the same, the latter may also be denoted by ν, which was defined for the former. Thus the energy of a radiation field of frequency ν may be regarded as quantized to the unit of $h\nu$. This is what Einstein assumed. It may be noted parenthetically that Planck did not believe that the energy of a radiation field exists in discrete units.

By means of this assumption, Einstein was able to explain the photoelectric effect. This effect was first discovered by Hertz late in the nineteenth century in his experiments on electromagnetic waves to verify the Maxwell theory. It is found experimentally that light causes the emission of electrons from a metal surface. For a given metal there is a threshold frequency below which no electrons are ejected no matter what the strength of the light intensity. For a given frequency above the threshold, the electrons ejected, no matter how weak the light may be, have the same maximum energy, and the emission takes place within a short time interval, 10^{-8} to 10^{-9} sec. These facts are very difficult to understand from the point of view of the wave theory of light. According to the wave theory, the energy of a light wave is uniformly distributed over its wave front. Given enough intensity of light and enough time of exposure, the electron may absorb from the wave enough energy to break loose from the metal surface. Frequency has no role to play. Thus in wave theory there can be no threshold in frequency; the electron energy and the emission time are dependent on the light intensity. Experimental results contradict these predictions, and they indicate that frequency, rather than intensity, plays the dominant role. In some experiments in which light of very weak intensity is used, the time required to absorb the necessary amount of energy, according to the wave theory, is calculated to be many days, whereas actually the photoelectrons are ejected almost immediately. It seems that either the ejected electron is endowed with an unusual ability of collecting a large amount of light energy in a short time at the expense of the other electrons, or the necessary amount of light energy is delivered in a package and the ejected electron happens to be the lucky recipient. By the tradition of the nineteenth-century physics and scientific method, the second alternative seems more reasonable; the photoelectric effect may be explained by assuming the existence of discontinuous units of light energy. Generalizing Planck's idea and applying it to the radiation field, Einstein introduced the concept of the *unidirectional quantum*, i.e., a quantum of radiation energy $h\nu$ propagated in one fixed direction without spreading. Such a quantum has the additional attribute that it has a linear momentum and thus has all the earmarks of a particle. Einstein essentially revived the corpuscular theory of light, the light particle being called the *photon*. According to the Maxwell theory, the energy-to-momentum ratio of a plane electromagnetic wave is c, the velocity of light. The amount of momentum p associated with the energy $h\nu$ is thus $h\nu/c$ or h/λ. Therefore the energy and momentum of a photon are $h\nu$ and h/λ, respectively.

$$E = h\nu, \qquad p = \frac{h}{\lambda}. \tag{1–10}$$

These are known as the Einstein relations for a photon.

The photoelectric effect, according to the photon theory, may be considered as due to the absorption of a light particle by an electron. The electron takes over the energy and momentum of the photon. Conservation of energy leads immediately to the Einstein equation of photoelectric effect,

$$\tfrac{1}{2}Mv^2 = h\nu - \phi, \tag{1–11}$$

where $\tfrac{1}{2}Mv^2$ is the kinetic energy of the ejected electron and ϕ, the *work function*, is the energy necessary to liberate an electron from the metal.* From Eq. (1–11) the threshold frequency ν_0 may be determined, the result being $h\nu_0 = \phi$. Equation (1–11) also expresses the frequency dependence of the kinetic energy of the photoelectrons. Thus, by the Einstein equation the experimental results of the photoelectric effects are easily explained.

In spite of this success, the photon theory revived the old controversy between the wave and corpuscular theories of light. However, the point at issue now is not which of the two theories is correct but how to reconcile them. Equations (1–10) define the property of a photon in terms of ν and λ, which are quantities meaningful only in the wave theory. Thus the photon theory cannot be stated without reference to the wave theory;

---

* Although mass is conventionally represented by the lower-case letter m, we use M to distinguish mass from the magnetic quantum number (to be introduced later) which is always designated by m.

the two theories are obviously related instead of contradictory. Einstein himself considered the photon theory as heuristic. The unification of the two theories, however, was not achieved until a generation later when a more general theory, *the quantum theory of radiation*, was established which includes both the Maxwell theory and the Einstein theory as special cases, and thus explains both the wave properties (interference and diffraction phenomena) and the corpuscular properties (photoelectric effects) within a single theoretical framework. Since this book deals primarily with *quantum mechanics of particles*, only a very brief discussion of the quantum theory of radiation will be given, in Chapter 12.

Although Eq. (1–11) was proposed in 1905, its conclusive, quantitative verification was not made until several years later, the most careful experiment being that of Millikan in 1916. In the meantime, the photoelectric effect induced by x-rays was studied. For this case, the Einstein equation may be written

$$\tfrac{1}{2}Mv^2 = h\nu, \tag{1–12}$$

because the work function, being several *electron volts** in magnitude, is negligible compared with the x-ray energy. Also the reverse reaction, producing x-rays by stopping high-energy electrons, has been studied. The maximum frequency of the x-rays emitted was found to be related to the kinetic energy of the electron by Eq. (1–12). The process may thus be regarded as one in which a photon is emitted with energy $h\nu$ equal to the energy loss of the electron, $E_2 - E_1$:

$$E_2 - E_1 = h\nu.$$

The photon concept is thus useful in studying the emission process as well as the absorption process of a free electron.

While considering the corpuscular properties of light we may depart from historical order to cite another experimentally determined fact, namely, the *Compton effect*, as further evidence. When hard x-rays are scattered by electrons, an increase in wavelength is observed which cannot be accounted for by the wave theory. In the wave theory no wavelength change is predicted (*Thomson scattering*). In the photon theory, however, this *Compton scattering* may be regarded as a collision process between a photon and an electron. The outcome of this collision is determined by the laws of conservation of energy and momentum. The energy and momentum of a photon are given by Eqs. (1–10). The theory of Compton scattering may thus be worked out by analogy to the collision

---

* The electron volt (ev) is a unit of energy defined as the amount gained by an electron in passing through a potential difference of one volt, its numerical value being 1.6021×10^{-12} erg.

of two billiard balls. Qualitatively, the photon loses energy after collision and therefore the frequency decreases because of the relation $E = h\nu$. This corresponds to an increase in wavelength. Quantitatively, it can be shown (Problem 1–1 at end of chapter) that

$$\lambda' - \lambda = \frac{h}{Mc} (1 - \cos \theta), \qquad (1\text{–}13)$$

where $\lambda' - \lambda$ is the increase in wavelength, M is the electron mass, and θ is the angle of scattering. (The quantity h/Mc, numerically 2.42×10^{-10} cm, is called the *Compton wavelength of the electron.*) The agreement with experiment is another triumph of the photon theory. Another important feature of Compton scattering, determined experimentally, is that the scattered photon is *unidirectional,* contrary to the classical theory that the scattered x-ray forms a spherical wave.

1–3 Bohr's theory of the hydrogen atom. The next important step in the development of quantum theory was taken by Bohr* in 1913 when he proposed the quantum theory of the hydrogen atom and thus initiated the so-called *old quantum theory.* Our knowledge concerning the structure of the atom was fragmentary until 1911, when Rutherford performed his famous α-particle scattering experiments. This series of experiments established the atom as a planetary system with a small but heavy nucleus carrying a positive charge at the center of the atom and a number of electrons surrounding the nucleus. Once this is established, classical theory leads to a number of conclusions: First of all, according to electrostatics, the electrons cannot maintain a static equilibrium under the influence of the Coulomb forces (Earnshaw's theorem). Thus, they must be in motion like the planets revolving around the sun. Once granted that the electrons execute periodic motions, classical electrodynamics requires that the electrons radiate electromagnetic waves (just like the electrons in the antenna of a broadcasting station radiating radio waves), and thus lose their energy gradually. Eventually the energy will be dissipated and the electrons will collapse into the nucleus. Therefore, according to classical electrodynamics, the nuclear atom cannot be *stationary.* (A state is said to be stationary if it remains unchanged in the course of time.) Actually atoms usually do not radiate electromagnetic waves and they can be stationary. The theory of the nuclear atom thus faces a serious difficulty.

The origin of this difficulty seems to lie in the fact that we have presumed the continuity of energy and space so that the electron is given a chance to decrease its energy indefinitely and to curve in indefinitely. If the energy is assumed to be discontinuous, this process of losing energy

---

* N. Bohr, *Phil. Mag.* **26**, 1 (1913).

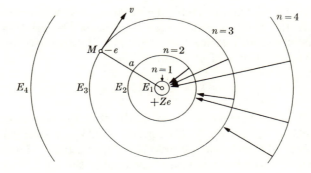

FIGURE 1–4

by radiation will come to an end when the lowest energy value is reached. We might have assumed the existence of things (such as the energy below the lowest energy value) which actually do not exist and thus have introduced a theoretical difficulty.

By 1913 a number of quantum phenomena were already known. The energy of the Planck oscillator was known to be quantized. Then energy in many other kinds of oscillators in solids and gases was known to be quantized. Not only mechanical systems but also the radiation field exhibited quantization. Photoelectric effects, the production of x-rays, and later the Compton effect provided ample evidence for the quantization of energy of the radiation field. Interactions between a mechanical system and a radiation field, i.e., the emission and absorption processes, also established the quantization of energy in these systems. Planck's oscillator absorbs and emits energy in the unit of $h\nu$. In photoelectric effects and production of x-rays, the electron (essentially a free electron) increases or decreases its energy from E_1 to E_2 upon the absorption or emission of light of frequency ν, the frequency being related to the energies by the relation $E_1 - E_2 = h\nu$. The interactions are such that the change of energy of a mechanical system is accompanied by the emission or absorption of a photon of the radiation field. With this background in mind it will be seen that the Bohr theory to be described presently is more or less a straightforward application of these ideas to the mechanics of the atom.

Bohr made two assumptions. The first is that the electron in the hydrogen atom can stay only on a series of selected circular orbits (see Fig. 1–4) specified by the condition that the angular momentum of the electron be an integral multiple of $h/2\pi$,

$$Mav = \frac{nh}{2\pi}, \qquad n = 1, 2, 3, \ldots, \tag{1–14}$$

where M is the mass of the electron, a is the radius of the orbit, and v is the linear velocity of the electron. These orbits are *assumed* to be stationary against radiation. The state of motion corresponding to each orbit is called a *stationary state* or a *quantum state*. The second assumption is that the electron may "jump" from one orbit with energy E_1 to another with energy E_2 and cause the emission or absorption of radiation of frequency ν given by

$$E_2 - E_1 = h\nu. \tag{1–15}$$

Since each orbit has a definite energy, the selection of a set of discrete orbits implies the quantization of energy, the quantized energy values being determined by Eq. (1–14). Equation (1–14) may thus be called the *quantum condition*. The second assumption determines the frequency of the radiation and Eq. (1–15) may thus be called the *frequency rule*. Actually, in the first assumption Bohr extended the idea of energy quantization from an oscillator to the hydrogen atom, although he introduced a rule of quantization different from that of Eq. (1–4). Once the energy is quantized, it can change, in the course of emission and absorption of radiation, only in discontinuous steps among the quantized energy values; the frequency rule, Eq. (1–15), thus appears to be a natural generalization of Eq. (1–13) from the photoelectric effect to the atomic system. At first look, the quantum condition, Eq. (1–14), does not seem to be related to that of the oscillator, Eq. (1–4). In the next section, however, we shall see that they are related.

Let us consider the consequences of these two assumptions. Denote the nuclear charge by Ze for generality where Z is the atomic number and e is the numerical value of the electronic charge, i.e.,

$$e = +4.8029 \times 10^{-10} \text{ esu.} \tag{1–16}$$

For an electron in a circular orbit the centripetal force must equal the Coulomb force between the nucleus and the electron,

$$\frac{Mv^2}{a} = \frac{Ze^2}{a^2}. \tag{1–17}$$

Combining Eqs. (1–14) and (1–17) we have

$$v = \frac{2\pi Ze^2}{nh} \tag{1–18}$$

and

$$a = \frac{n^2 h^2}{4\pi^2 M Ze^2}. \tag{1–19}$$

These equations give the velocity and radius of an orbit characterized by

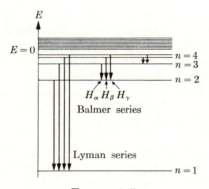

FIGURE 1–5

an integer n. The integer n is called the *quantum number*. The energy of this orbit is

$$E_n = \frac{1}{2} Mv^2 - \frac{Ze^2}{a}$$

$$= \frac{1}{2} \frac{Ze^2}{a} - \frac{Ze^2}{a}$$

$$= -\frac{2\pi^2 MZ^2 e^4}{n^2 h^2}, \qquad n = 1, 2, \ldots . \qquad (1\text{–}20)$$

These quantized energy values are plotted in Fig. 1–5. We note that the quantized energy values form a spectrum different from that of Eq. (1–4). The quantization of mechanical systems seems more complicated than that of the radiation field, which is always quantized to equal units $h\nu$. Each mechanical system has its own rule of quantization. The oscillator is quantized to equal intervals, whereas the hydrogen atom is quantized to a scheme shown in Fig. 1–5. For other mechanical systems, other rules of quantization are needed; and the task of finding the correct rule of quantization becomes the main concern of the old quantum theory.

After having obtained the quantized energy values, we now apply the second assumption to derive the frequencies of radiation that may be emitted or absorbed by the hydrogen atom. The frequency rule gives

$$\nu = \frac{2\pi^2 MZ^2 e^4}{h^3} \left(\frac{1}{n_f^2} - \frac{1}{n_i^2} \right),$$

for emission: $\quad n_f = 1, 2, \ldots$
$$n_i = n_f + 1, n_f + 2, \ldots \qquad (1\text{–}21)$$

for absorption: $\quad n_i = 1$
$$n_f = 2, 3, \ldots ,$$

where n_i and n_f are the initial and final values of n for a particular *quantum transition* (or "jump"). When $n_f = 1$, Eq. (1–21) reproduces the frequencies of the lines forming the Lyman series; $n_f = 2$ corresponds to the Balmer series; $n_f = 3$, the Paschen series, etc. The agreement between the predicted and observed values for the multitudes of spectral lines is so excellent that one feels the Bohr theory has made a breakthrough in the development of a theory of the atom. It is the first theory of line spectrum that agrees with experiments.

We shall not discuss other applications of the Bohr theory at this point. Some of the quantities derived here will be useful later, and we shall now express them in more convenient forms. From Eq. (1–20) we calculate the lowest of the quantized energy values by substituting $n = 1$ and $Z = 1$. The result is -13.6 ev (electron volt). The corresponding stationary state is called the *ground state*. Other states are called *excited states*. The energy values of the states may be represented by the *energy levels* in Fig. 1–5, which may be called the *energy level diagram* or the *level scheme*. Equation (1–20) may be more simply written as

$$E_n = -13.6 \frac{Z^2}{n^2} \text{ ev.} \qquad (1\text{–}22)$$

From Eq. (1–19) we calculate the *first Bohr radius* a_0, i.e., the radius of the first orbit, by substituting $n = 1$ and $Z = 1$,

$$a_0 = \frac{h^2}{4\pi^2 M e^2} = 0.529 \times 10^{-8} \text{ cm.} \qquad (1\text{–}23)$$

This radius of about half an angstrom agrees in order of magnitude with the radius of the atom estimated by the kinetic theory of gases—an interesting byproduct of the Bohr theory. The radii of the excited states may be expressed by

$$a = \frac{n^2}{Z} a_0. \qquad (1\text{–}24)$$

The velocity of the electron in the ground state may be calculated from Eq. (1–18), the result being 1/137 of the velocity of light. We may make use of the so-called *Sommerfeld fine structure constant* α, defined by

$$\alpha = \frac{2\pi e^2}{hc} = \frac{1}{137}. \qquad (1\text{–}25)$$

Equation (1–18) may thus be rewritten

$$v = \frac{Z}{n} \alpha c. \qquad (1\text{–}26)$$

The electron moving in a circular orbit gives rise to a circular current which has a magnetic moment μ, the magnitude of which in emu is given by the product of the current and the area of the circuit loop:

$$\mu = \pi a^2 \frac{ev}{2\pi ac} = n \frac{eh}{4\pi Mc}. \tag{1-27}$$

The magnetic moment is thus an integral multiple of a unit $eh/4\pi Mc$, this unit being known as the *Bohr magneton*.

The Bohr theory applies only to circular orbits. Bohr assumed that the hyperbolic orbits corresponding to positive energy values are not quantized. Thus the energies of the hyperbolic orbits form a continuous band above the $E = 0$ line in Fig. 1–5. Transitions to and from this band correspond to absorption and emission lines forming a continuous spectrum beyond the series limit of the discrete spectrum. The continuous spectrum is also observed.

According to the Bohr theory, the energy of an atom cannot take on any arbitrary continuous values. After Bohr propounded his theory in 1913, Franck and Hertz* immediately started experiments on electron collisions with atoms; and their results verified the existence of discrete energy levels. The quantization of energy of the atomic systems is thus an experimentally established fact.

In spite of the success of the Bohr theory, it must be kept in mind that this theory consists of incoherent parts and cannot be considered as complete and satisfactory. It retained classical mechanics but arbitrarily restricted it by the quantum condition; it rejected classical electrodynamics and replaced it by the frequency rule. However, its many successes make one feel that if a coherent and satisfactory theory is to be developed, it will have to embrace the salient points of the Bohr theory. As the quantization of energy is a universal phenomenon, any physical theory must adapt itself to accommodate it. Two problems thus arise. First, how is a general mechanical system quantized? In other words, what is the general rule of quantization for a mechanical system? We have seen that the oscillator and the hydrogen atom are quantized according to different rules. We want to know if there is a general rule which may be applied for the quantization of any mechanical system. Such a rule may be sought on an empirical basis at first, but eventually it has to be made a coherent part of the whole theory. Once the energy is quantized, any change of energy can take place only in discontinuous steps. Therefore, all physical processes involving energy exchange assume a discontinuous character. Now classical theory is based on the concept that

---

* Franck and Hertz, *Verhandlungen der Deutschen Physikalische Gesellschaft*, **15,** 613 (1913).

physical quantities and processes are all continuous, and this leads to the second problem: how to formulate a physical theory which describes physical processes in discontinuous steps. Furthermore, since classical theory is valid in macroscopic physics, the new theory, if it is a general theory, must include classical theory as a limiting case. The continuum of energy in classical theory most likely is an approximation of the quantized energy levels when the latter are so densely spaced that a continuous approximation may be valid. The classical theory may be just a continuous approximation of the quantized theory. In developing such a quantum theory, the rules of quantization for the oscillators and the hydrogen atom are important clues for the first problem. The frequency rule, which is essentially equivalent to the Einstein equation for the photoelectric effect, is important for the second problem. In the next section we shall discuss the work in this direction between 1913 and 1925 which constitutes the old quantum theory. The satisfactory solution of these problems, however, has to be found later in quantum mechanics.

1–4 Sommerfeld's generalization and the old quantum theory. Sommerfeld took the next important step after Bohr in developing quantum theory. He succeeded in introducing a general rule of quantization* which, in spite of its limitations, gave good results in a number of cases and thus partially solved the first problem mentioned at the end of the last section. For a multiply periodic system described by the generalized coordinates $q_1, q_2, \ldots, q_f$ and the generalized momenta $p_1, p_2, \ldots, p_f$, Sommerfeld proposed the following quantum conditions that determine the stationary states and quantized energy levels,

$$\oint p_k \, dq_k = n_k h, \qquad \begin{aligned} &n_k: \text{ integers,} \\ &k = 1, 2, \ldots, f. \end{aligned} \qquad (1\text{–}28)$$

In this equation the integral is to be performed over one cycle of the variable q_k, and f is the number of degrees of freedom. If we apply this condition to the circular orbit, the result obtained is the same as the Bohr quantum condition,

$$\oint p_\theta \, d\theta = \oint Mva \, d\theta = 2\pi Mva = n_\theta h; \qquad (1\text{–}29)$$

$$\therefore Mva = \frac{n_\theta h}{2\pi}, \qquad n_\theta = 1, 2, \ldots. \qquad (1\text{–}30)$$

---

* A. Sommerfeld, *München Sitz.*, pp. 425, 459 (1915); *Ann. d. Phys.* **51**, 1 (1916). The rule was proposed independently and almost simultaneously by W. Wilson, *Phil. Mag.* **29**, 795 (1915); **31**, 156 (1916). It is thus known as the Wilson-Sommerfeld quantum condition.

Applying this condition to an oscillator the coordinate of which is given by $x = A \sin \omega t$, with the period of oscillation denoted by T and energy by E, we obtain the Planck quantum condition,

$$\oint p_x \, dx = \oint M A^2 \omega^2 \cos^2 \omega t \, dt = M A^2 \omega^2 \tfrac{1}{2} T; \qquad (1\text{-}31)$$

$$\therefore \oint p_x \, dx = ET = nh; \qquad (1\text{-}32)$$

$$\therefore E = nh\nu, \qquad n = 0, 1, 2, \ldots. \qquad (1\text{-}33)^*$$

That the Wilson-Sommerfeld quantum condition includes the Planck and the Bohr quantum conditions as its special cases gives us some confidence in its general applicability. Sommerfeld applied his quantum conditions to the elliptical orbits of the hydrogen atom and thereby generalized the Bohr theory. For elliptical orbits there are two generalized momenta: the radial momentum p_r and the angular momentum p_θ. The two quantum conditions are

$$\oint p_r \, dr = n_r h, \qquad n_r = 0, 1, 2, \ldots, \qquad (1\text{-}34)$$

$$\oint p_\theta \, d\theta = n_\theta h, \qquad n_\theta = 1, 2, \ldots. \qquad (1\text{-}35)$$

Here, n_θ starts from 1 instead of 0 because of the following consideration: $n_\theta = 0$ means no angular momentum; the elliptical orbit thus collapses into a line passing through the nucleus, which does not seem physically possible. Whereas the Bohr quantum condition is satisfied by a set of discrete circular orbits, the Sommerfeld quantum conditions are obeyed by a set of discrete elliptical orbits. We shall omit the mathematical derivation (Problem 1–2) and simply state the results. The energies of the elliptical orbits are

$$E = -\frac{2\pi^2 M Z^2 e^4}{(n_r + n_\theta)^2 h^2}. \qquad (1\text{-}36)$$

Let

$$n = n_r + n_\theta, \qquad n = 1, 2, \ldots. \qquad (1\text{-}37)$$

Respectively, n_r, n_θ, and n are called the *radial*, *azimuthal*, and *principal quantum number*. Equation (1–36) may be rewritten as follows:

$$E = -\frac{2\pi^2 M Z^2 e^4}{n^2 h^2}, \qquad n = 1, 2, \ldots, \qquad (1\text{-}38)$$

---

* There is some confusion as to where the integers start. In Eq. (1–30) they start at unity, while in Eq. (1–33) they start at zero. This point is not clearly settled in the Sommerfeld theory, and we shall see other examples of it later. This is an unsatisfactory feature of the old quantum theory.

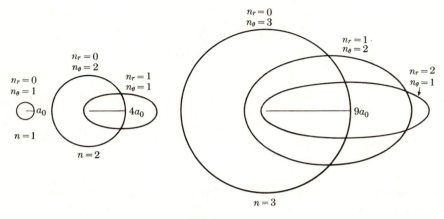

FIGURE 1–6

which is identical with Bohr's result, Eq. (1–20). Thus the success of the Bohr theory is retained. The elliptical motion has two degrees of freedom and thus has two constants of motion: one is the energy E and the other is the angular momentum p_θ. The energy is quantized according to Eq. (1–31). On the other hand, Eq. (1–35) gives rise to the equation

$$p_\theta = n_\theta \frac{h}{2\pi}, \qquad n_\theta = 1, 2, \ldots, \tag{1–39}$$

so that the angular momentum is also quantized (a result also implied in Bohr's theory). According to the Wilson-Sommerfeld quantum conditions, a system of f degrees of freedom has f integral quantum numbers and therefore the f constants of motion are all quantized. Thus the concept of quantization is extended from energy to many other quantities. Having described the dynamical properties of the elliptical orbits, we now turn to their kinematical properties. The semimajor axis a and semiminor axis b may be shown to be given by the following expressions:

$$a = \frac{n^2}{Z} a_0, \tag{1–40}$$

$$b = \frac{n n_\theta}{Z} a_0, \tag{1–41}$$

where a_0 is the first Bohr radius. Both the energy and the semimajor axis depend on the principal quantum number n only. For a given n, n_θ can be any number from 1 to n corresponding to n ellipses having the same energy and semimajor axis, but differing in their angular momenta and semiminor axes. The ellipses may thus be classified according to the principal quantum number n as in Fig. 1–6. The energy level diagram is

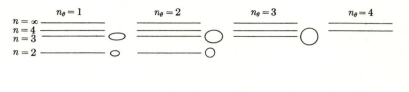

FIGURE 1-7

the same as in the Bohr theory. However, each level designated by n actually represents the energy of n elliptical orbits that happen to have the same energy. As these n orbits are different in other respects, it seems appropriate to show them separately in the energy level diagram. We adopt a scheme shown in Fig. 1-7 where the energy of each ellipse is represented by a line and energy levels of the same n_θ are grouped together in a column. In Fig. 1-7 we show some of the ellipses for easy identification. The vertical scale represents the energy and the columns specify the angular momentum.

Sommerfeld's next step was to introduce relativity into the quantum theory. According to the special theory of relativity, the mass of a particle increases with its velocity; the change becomes appreciable when the velocity approaches the velocity of light. The dependence of mass on velocity is given by

$$M = \frac{M_0}{\sqrt{1 - (v^2/c^2)}}.\qquad (1\text{-}42)$$

The velocity of the electron in the first Bohr orbit is $c/137$. The corresponding change of mass is small but detectable. This change of mass gives rise to small changes of the orbit; a detailed mathematical analysis shows that the orbit is no longer a closed curve but may be approximated by a precessing ellipse, similar to the advance of perihelion of the planetary orbits. The relativistic effect also gives rise to small changes in energy. Sommerfeld carried out the calculation in detail and showed that the energies of the n "ellipses" with the same principal quantum number n are no longer the same. Thus each level in Fig. 1-5 splits into a number of closely spaced levels: the ground state does not split, the first excited state ($n = 2$) splits into two, the nth into n levels. As a result each line in the spectrum splits into a number of lines of nearly the same frequency. The splitting of a line into a number of components was actually observed in spectroscopy and this fact was known as the *fine structure* of spectral lines. The success of the Sommerfeld theory in deriving the correct fine

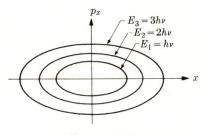

FIGURE 1-8

structure formula of the hydrogen atom is a major triumph of the old quantum theory. Although Sommerfeld's success, as we shall see later, is partly accidental, the formula bearing his name survived many changes of the theory until the discovery of the *Lamb shift* in 1947.

The meaning of the Wilson-Sommerfeld quantum condition may be brought out clearly by considering the motion in *phase space*. Let us take the oscillator as an example. The phase space for the oscillator is a two-dimensional space with x and p_x as its coordinates (see Fig. 1-8). As the oscillator moves, both x and p_x change with respect to time. Thus the representative point (x, p_x) moves in phase space. Because the motion of the oscillator is periodic, the representative point will return to the starting point after one period of oscillation, and its path will be a closed curve. For a linear harmonic oscillator, x and p_x are trigonometric functions with a phase difference of 90°, the closed curve being an ellipse. The integral appearing in the quantum condition is actually the *phase integral*, and its value represents the area inside the closed curve. The quantum condition thus means that the area inside the closed curve is an integral multiple of h. Figure 1-8 shows that the series of curves representing motions obeying the quantum condition cut the phase space into equal areas of h. In other words, each stationary state corresponds to an area of h in the phase space, and the quantum condition thus implies the quantization of the phase space. In classical statistical mechanics the phase space is assumed to be continuous, capable of being subdivided into parts as small as we wish without limit. The quantum theory thus requires a modification of classical statistical mechanics. In fact the failure of the latter in the theory of blackbody radiation is rooted on the presumption that phase space may be indefinitely subdivided. Planck was led to his theory by considering the energy distribution from the point of view of phase space. He noted that Eq. (1-3) requires that at low temperature the average energy of an oscillator be much lower than kT. This may be the case only if we forbid the energy values immediately above zero to appear, i.e., the energy must be discontinuous from zero up.

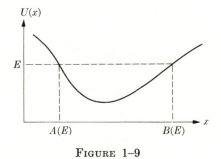

FIGURE 1-9

The significance of the phase integral may be demonstrated by a theorem in classical mechanics described below. Consider a particle of mass M moving in a potential $U(x)$ shown in Fig. 1-9. Given energy E, the particle is confined in a region $A < x < B$, where A and B are two points dependent on E. Let the phase integral be I. Then

$$I = \oint p_x \, dx$$

$$= 2 \int_{A(E)}^{B(E)} M\sqrt{(2/M)[E - U(x)]} \, dx. \tag{1-43}$$

Differentiation with respect to E leads to the following result (remember that the integrand vanishes at both limits):

$$\frac{dI}{dE} = M\sqrt{\frac{2}{M}} \int_{A(E)}^{B(E)} \frac{dx}{\sqrt{E - U(x)}} + 0 + 0$$

$$= 2 \int_{A(E)}^{B(E)} \frac{dx}{v}$$

$$= T,$$

where T is the period. In terms of frequency ν, we have

$$\frac{dE}{dI} = \nu. \tag{1-44}$$

We now consider this relation in connection with the quantum theory, taking the Bohr orbits of the hydrogen atom as an example. When the quantum number n is large, say 1,000,000, a small change of n, say 1 or 2 units, will not change the energy and radius of the orbit much and we may consider the two orbits of $n = 1,000,000$ and $n = 1,000,001$ as approximately the same. In other words, in the high quantum number region, the quantized energies and orbits are so close that they may be treated as being continuous. As a result, classical mechanics may be valid, and we may invoke Eq. (1–44),

$$\nu = \frac{dE}{dI} = \frac{E_{n'} - E_n}{I_{n'} - I_n}. \tag{1–45}$$

Since I is an integral multiple of h, the value of $I_{n'} - I_n$ will be h if $n' - n = 1$. Therefore

$$E_{n+1} - E_n = h\nu. \tag{1–46}$$

Note that the quantity ν here is the frequency of the mechanical oscillation of the nth orbit (or the $(n + 1)$th, as these two are nearly indistinguishable). Comparing Eq. (1–46) with the Bohr frequency condition, we conclude that the frequency of the radiation emitted when transition $(n + 1) \to n$ takes place is the same as the instantaneous frequency of the mechanical oscillation. This is exactly what the classical electrodynamics asserts.* In other words, Bohr's frequency rule agrees with classical electrodynamics in the high quantum number region. This is a very important result. When we first stated the frequency rule, it appeared to have no connection of any kind with classical electrodynamics. This result is a strong suggestion that classical theory, mechanical as well as electrical, is a continuous approximation of a discontinuous theory and is valid in the high quantum number region where the discrimination of an orbit of $n = 1,000,001$ from that of $n = 1,000,000$ is unnecessary. From this point of view, the classical and quantum laws must be the same in a region where n is large enough to be regarded as a continuous variable. There then exists a correspondence between these two, and the quantum laws in the high quantum number region may be guessed at from the classical laws. This is the essence of the so-called *correspondence principle* of Bohr (1923). Once the quantum laws in the high quantum number region are obtained in this way, they are assumed to apply equally well in the low quantum number region; a complete quantum theory may thus be established. The correspondence principle is thus the answer by the old quantum theory to the second problem stated at the end of the last section: the formulation of a theory for discontinuous physical processes. As an illustration of the application of the correspondence principle we discuss the *selection rules* of optical transitions which have some bearing in later discussions.

---

* Classical electrodynamics also predicts the emission of the harmonics 2ν, $3\nu, \ldots, n\nu$, in addition to the fundamental frequency ν. These correspond to quantum transitions in which $I_{n'} - I_n$ equals $2h$, $3h, \ldots, nh$, respectively, since the Bohr frequencies of these transitions are 2ν, $3\nu, \ldots, n\nu$ by virtue of Eq. (1–45).

In spectroscopy it was found that not all transitions among the levels in Fig. 1–7 are possible. The correspondence principle leads to the result that in a transition the quantum number n_r may change by any units, whereas n_θ may change only by one unit,* i.e.,

$$\Delta n_\theta = \pm 1. \qquad (1\text{–}47)$$

This means that transitions can take place only between levels which belong to adjacent columns in Fig. 1–7. Selection rules of this kind have been derived for a number of cases and they agree with experimental results.

The old quantum theory has been extensively developed, based on the Wilson-Sommerfeld quantum condition and the Bohr correspondence principle. By generalization and elaboration it was made applicable to a number of atomic systems and physical processes. It succeeded in explaining a large body of experimental data in spectroscopy. Most importantly, it explains not only the *Zeeman effect* (changes in spectra due to a magnetic field) but also the *Stark effect* (changes in spectra due to an electric field). While the Zeeman effect can be explained by classical theory, the Stark effect cannot. The success in explaining the Stark effect was considered a major triumph of the old quantum theory. It may be mentioned parenthetically that the Zeeman effect is accounted for by introducing space quantization, i.e., the orientation of an elliptical orbit in space is quantized and can take on only a finite number of specified orientations with respect to an external field. In spite of many successes, the old quantum theory is unable to account for the half-integer quantum numbers, the existence of which is forced on us empirically. It is unable to give even a qualitative account of the spectra of the helium atom and the hydrogen molecule. It soon becomes evident that the difficulties of the old quantum theory cannot be removed by minor modifications. A new theory is required. We add here in passing that some of the results of the Sommerfeld theory, such as the assignment of the angular momentum to the states represented in Fig. 1–7, are not correct and have to be revised according to quantum mechanics.

---

* The orbital motion of a charged particle when the orbit is not a closed curve may be considered as a closed-curve motion with angular frequency ω_r superimposed with a precessional motion of angular frequency $\omega_\theta - \omega_r$. The coordinates x, y, z expressed in Fourier series thus contain all harmonics of ω_r but only the fundamental of ω_θ. These are the frequencies, according to classical electrodynamics, that appear in the radiation emitted by the particle. In the high quantum number region these frequencies correspond to quantum transitions with arbitrary Δn_r and $\Delta n_\theta = \pm 1$ by virtue of Eq. (1–45). According to the correspondence principle, we conclude that Δn_r may be arbitrary and Δn_θ must be ± 1. Once these rules are established in the high quantum number region, they are assumed to hold for all quantum numbers.

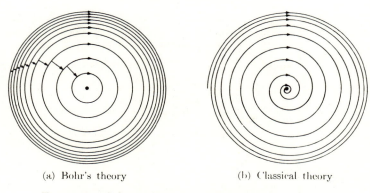

(a) Bohr's theory   (b) Classical theory

FIG. 1–10. Schematic diagram of transition of states.

The first appearance of quantum effects puzzled and shocked physicists. It seemed as if nature had unexpectedly betrayed us. After the first impact is over, we come to realize that the discontinuous nature of physical quantities and natural processes is a fact of universal occurrence, ignored previously because of the inadequacy of experimental observation. Our physical theory must then be generalized in such a way that dynamical quantities are allowed to be discontinuous and natural processes are described in terms of discrete transitions rather than continuous evolution. Such a theory is called a *quantized theory.*

In the region of high quantum numbers (corresponding to macroscopic phenomena) the discreteness of the dynamical quantities and natural processes are so inconspicuous that a continuous approximation may be valid. The classical theory of physics may be just such an approximation of the quantized theory. On the other hand, in the low quantum number region (corresponding to atomic phenomena) the difference between the classical theory and the quantized theory becomes apparent. The quantized theory thus appears to be a natural extension of classical theory into a region where classical theory, verified only by macroscopic observation, has no right to claim validity. This observation may be illustrated with the Bohr theory as an example. The Bohr theory provides a series of discrete orbits over which the radiating electron jumps down in cascades to the lowest energy level. Having reached the lowest level it stops jumping; there is no other level to jump into [see Fig. 1–10(a)]. In the region of high quantum numbers the orbits are so close together that the cascading process through a number of orbits may be approximated by a continuous spiral [see Fig. 1–10(b)]. Actually, according to classical electrodynamics the electron should follow such a spiral path. The frequencies of radiation emitted during the cascading transitions are represented by the slopes $(E_2 - E_1)/h$ of the broken line in Fig. 1–11. Again,

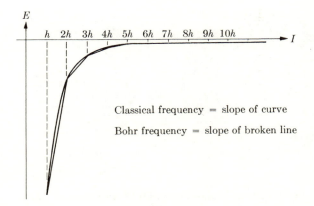

FIGURE 1–11

in the high quantum number region a continuous approximation may be valid which takes the form of the curve in Fig. 1–11. In fact the slope of this curve gives the classical frequency dE/dI. Both continuous approximations agree with the classical theory in the high quantum number region. On the other hand, both cease to be valid in the low quantum number region. There the curve of Fig. 1–11 is no longer a good approximation of the broken line and the spiral in Fig. 1–10(b) curves into the nucleus instead of terminating at the innermost quantized orbit.

Although the general features of the quantized theory are borne out in the old quantum theory, the latter is not a coherent theory. Efforts to formulate a satisfactory quantized theory culminated in 1925 in the establishment of *quantum mechanics*, which will be the main concern of this book. The remainder of this chapter will be devoted to the historical development of this theory.

1–5 De Broglie's wave theory of matter. A new page was turned in 1923–1924 when de Broglie* introduced the wave theory of matter. Inasmuch as many elements of his theory will be discussed mathematically in detail in Chapter 2, we shall not present it here as fully as we did the Bohr theory. We shall concern ourselves only with the historical aspect.

The similarity of the equations of motion in corpuscular mechanics and wave propagation (the principle of least action and the Fermat principle) strongly suggested to de Broglie that there exists a close relation between the concepts of particle and wave. Since radiation is known to have a dual property in that it behaves like a wave in interference and

---

* L. de Broglie, Thèse, Paris (1924); *Phil. Mag.* **47,** 446 (1924); *Ann. d. Phys.* **3,** 22 (1925).

diffraction phenomena and like a particle in photoelectric and Compton effects, de Broglie reasoned that matter might also have a dual property. A wave of frequency ν and wavelength λ might be assumed to be associated with a moving particle of energy E and momentum p. In strict analogy to the Einstein relations, Eqs. (1–10), he assumed the relations between ν, λ and E, p to be

$$E = h\nu, \qquad p = \frac{h}{\lambda}. \tag{1-48}$$

These are known as the *de Broglie relations*. In arriving at these, he was guided by a consideration based on the special theory of relativity. The phase of a wave, $2\pi(\sigma_x x + \sigma_y y + \sigma_z z - \nu t)$, where σ_x, σ_y, σ_z are the three components of the wave number vector, is a relativistic invariant. As a result the four quantities σ_x, σ_y, σ_z, ν transform like a four-vector when a space-time transformation (Lorentz transformation) is carried out. On the other hand, the four quantities p_x, p_y, p_z, E transform in exactly the same way. It seems reasonable to assume that the two sets of quantities are proportional to each other. If the proportionality constant is identified with the Planck constant, we have the Einstein or de Broglie relations.

Once the frequency and wavelength are given, the phase velocity V follows immediately. Thus

$$V = \lambda\nu$$

$$= \frac{E}{p}$$

$$= \frac{M_0 c^2}{\sqrt{1 - (v^2/c^2)}} \div \frac{M_0 v}{\sqrt{1 - (v^2/c^2)}}$$

$$= \frac{c^2}{v}, \tag{1-49}$$

where v is the velocity of the moving particle. Although the phase velocity is greater than the velocity of light, no contradiction to relativity will ensue, for the phase velocity is not the velocity of energy propagation. In wave motion, it is usually the *group velocity* (to be discussed more in detail in Section 2–1) that represents the velocity of energy propagation. The group velocity v_g of the *de Broglie waves* may be calculated according to a formula to be derived in Chapter 2 (Eq. 2–14),

$$v_g = \frac{d\nu}{d(1/\lambda)} = \frac{dE}{dp} = v. \tag{1-50}$$

The group velocity thus turns out to be the same as the velocity of the particle. Therefore a moving particle may be represented by a *wave packet* (to be discussed in detail in Section 2–1) of de Broglie waves. The propagation of a wave packet is determined by the Fermat principle which, by the substitution of the de Broglie relation $p = h/\lambda$, leads to the principle of least action (also see Section 2–1). Thus the equation of motion of the particle is implied in the wave theory. The kinematical and dynamical attributes of a particle may therefore be regarded as manifestations of de Broglie waves and the wave theory may include the corpuscular mechanics as a special case. Furthermore, the wave properties so introduced may be invoked in an attempt to explain quantum phenomena. In fact, de Broglie succeeded in *deriving* the quantum condition from such a consideration. In a circular orbit of the hydrogen atom, the length of the circle, according to the wave theory, must be an integral multiple of the wavelength. Otherwise the wave, after having traveled once around, will be out of phase with the original wave; the waves, having traveled 1, 2, 3, . . . , times around, will have random phase relations to one another at any point so that the resultant wave will be zero (due to interference). The condition that a nonvanishing wave may be established along a circular orbit of length s is thus

$$\oint \frac{ds}{\lambda} = n, \qquad n = 1, 2, \ldots . \tag{1–51}$$

The de Broglie relation then leads to the following equation:

$$\oint p_s \, ds = nh, \qquad n = 1, 2, \ldots , \tag{1–52}$$

which is the quantum condition. In the old quantum theory, the quantum condition is introduced without any explanation. In the de Broglie theory the origin of the quantum condition is traced to an assumed wave property of matter. Incidentally, another pleasing point is that light may be regarded as a special kind of de Broglie wave and the dual property of light may thus be accounted for.

While the dual property of light had been well established by 1924, the dual property of matter was merely speculation by de Broglie at that time. This speculation was experimentally verified by Davisson and Germer* in 1927 and later by many others. Diffraction of electron beams was achieved by using crystals as the grating, and the de Broglie relation $p = h/\lambda$ was verified for electrons of various energies. Not only electron beams but also atomic beams of helium and molecular beams of hydrogen were diffracted on a crystal surface. The wave property of matter is thus an experimentally established fact.

---

* Davisson and Germer, *Phys. Rev.* **30**, 705 (1927); *Proceedings of the National Academy*, **14**, 317, (1928).

The success of de Broglie's theory only leads to a more intriguing problem concerning the dual property of matter and radiation. However, we shall not concern ourselves with the question: Is the electron a particle or a wave? Electron diffraction experiments tell us merely that under certain conditions an electron beam behaves like a wave. They do not identify it as a wave. At best we may say that the electron has a wave-like property. Similarly in view of the photoelectric effect we may merely say that light has a particle-like property. Experimental evidence tells us only that both matter and radiation have some particle-like properties and some wavelike properties. The task before us is to construct a logically consistent and coherent theory, not merely a collection of *ad hoc* assumptions, which accounts for all the observed properties. Classical corpuscular theory explains the particle-like properties but not the wavelike properties. Classical wave theory does just the opposite. Both are incomplete. Quantum mechanics, as we shall see in later chapters, accounts for both kinds of properties. Once such a theory is established, the dual nature of matter and radiation is no longer a mystery and the question of whether the electron is a particle or a wave disappears.

We have discussed two classes of quantum phenomena, namely, the quantization of energy and the wave property of matter. The latter cannot be accounted for in the old quantum theory, a serious defect of this theory. Both may be accounted for in the de Broglie theory without introducing any *ad hoc* assumption regarding quantization. This important achievement is retained in quantum mechanics, as will be seen in later chapters.

To emphasize that a wavelike property does not necessarily guarantee the existence of a physical wave, we shall discuss here a quantum theory of diffraction by Duane* (1923). The diffraction of light was considered as evidence proving that light is a wave. Nevertheless, we shall see in this theory that the diffraction phenomena may be explained by a quantum theory of the light particle. Consider that a photon impinges on a grating of spacing d with an incident angle i and emerges with an angle r (see Fig. 1–12). The momentum transferred to the grating along the surface of the grating is

$$p_y = \frac{h}{\lambda} (\sin i - \sin r), \qquad (1\text{–}53)$$

according to the law of conservation of momentum. On the other hand, an infinite grating may be considered as a periodic system, and a vertical motion over a distance d

FIGURE 1–12

------

* W. Duane, *Proc. Nat. Acad. Sci.* **9**, 158 (1923).

completes one cycle of a periodic motion. The Wilson-Sommerfeld quantum condition thus gives rise to the following equation:

$$\oint p_y \, dy = p_y d = nh \tag{1-54}$$

or

$$p_y = \frac{nh}{d}. \tag{1-55}$$

The quantum condition thus requires that the momentum transferred to the grating be quantized to the unit of h/d. Combining Eqs. (1-53) and (1-55), we obtain

$$n\lambda = d\,(\sin i - \sin r). \tag{1-56}$$

This is identical with the equation derived in the wave theory specifying the directions of the diffracted rays caused by a grating. Thus the same result may be derived by either the wave theory or the corpuscular theory. This example shows that the observation of a diffraction phenomenon does not necessarily prove that light is a wave. The wave theory is but one of the theories that explain the diffraction phenomenon. From such an experiment we can say no more than that light has a wavelike property.

1-6 The development of quantum mechanics. In 1925 Heisenberg* introduced a system of mechanics, later known as *matrix mechanics*, in which classical concepts of mechanics were drastically revised. He considered that the atomic theory should emphasize the observable quantities, namely, the frequencies and intensities of the spectral lines, rather than those not directly observable, such as the shape of the electronic orbit. This theory was rapidly developed by Heisenberg, Born, and Jordan, making use of matrix algebra. We shall not discuss it in detail. It may be considered as a calculus of the observable quantities. If the mathematical problem involved can be solved completely, this theory can predict the frequencies and intensities of the spectral lines.

Parallel to the development of matrix mechanics, Schrödinger† in 1926 initiated a new line of study, inspired by de Broglie's wave theory of matter, which was then developed into a system of *wave mechanics*. He introduced an equation, now bearing his name, as the "equation of motion" of the de Broglie waves. From this equation, quantization follows automatically; wave mechanics thus becomes a powerful tool for the study of quantization. Its mathematical apparatus, involving the solution of a partial differential equation, is more convenient to handle than that of matrix mechanics and thus it is more widely used in practical applications.

---

* W. Heisenberg, *Zeits. f. Phys.* **33,** 879 (1925).

† E. Schrödinger, *Ann. d. Phys.* **79,** 361, 489; **80,** 437; **81,** 109 (1926).

Not long after, Schrödinger proved that wave mechanics was mathematically equivalent to matrix mechanics. On the other hand, the physical meaning of the new theory was not clear at first. Schrödinger first considered the de Broglie wave as a physical entity, i.e., the electron is actually a wave. But this soon led to difficulty. A wave may be partially reflected and partially transmitted at a boundary, but an electron cannot be split into two parts for transmission and reflection. The difficulty was removed by Born,* who proposed a statistical interpretation of de Broglie waves which is now generally accepted. The introduction of the statistical interpretation results in a drastic change in scientific thought, for it replaces the deterministic classical theory by a probabilistic theory. De Broglie, conceding that kinematics may be probabilistic because of the wave property, once tried to retain a deterministic dynamics (the pilot wave theory) but was not successful. The new theory based on the statistical interpretation was very rapidly developed into a general, coherent system of mechanics which now bears the name *quantum mechanics*. To this, Dirac, Jordan, Heisenberg, and Pauli made important contributions. Applications of the theory, made in many branches of physics, met with remarkable success.

So far we have confined ourselves to the development of quantum mechanics as applied to atomic systems. This subdivision may be called *quantum mechanics of particles*. However, atomic systems interact with the radiation field, and a complete theory cannot leave out the latter. Furthermore, the quantum theory originated from a study of the quantum effects of the radiation field (the blackbody radiation, the photoelectric effects); and to account for these effects, a quantum theory of radiation is necessary. On the other hand, the development of the general theory of quantum mechanics leads to the establishment of a general technique of quantization which may be applied to many other physical systems. The application of the general methods of quantization to the radiation field resulted in the development of the *quantum theory of radiation*.† In spite of many difficulties encountered, the quantum theory of radiation is able to make successful predictions for many observable processes. It is able to account for both the particle-like properties and wavelike properties of radiation. No previous theory was able to explain all of these properties at the same time.

The application of the general methods of quantization to the radiation field is just one of many. Extensive work has been done in extending

---

* M. Born, *Zeits. f. Phys.* **37**, 863; **38**, 803 (1926).

† P. A. M. Dirac, *Proc. Roy. Soc.* **114**, 243, 710 (1927); P. Jordan and W. Pauli, *Zeits. f. Phys.* **47**, 151 (1928); E. Fermi, *Rev. Mod. Phys.* **4**, 131 (1932); W. Heisenberg and W. Pauli, *Zeits. f. Phys.* **56**, 1 (1929); **59**, 169 (1930).

quantum theory to many other fields. In this introductory volume we shall limit ourselves to an introduction of quantum mechanics of particles, primarily quantum mechanics of one particle without spin. Nevertheless, general methods of quantum mechanics and its various applications will be discussed briefly in the last chapter.

Let us summarize the difficulties of the classical theories which quantum mechanics is supposed to resolve. We know as fact that physical quantities of mechanical systems and radiation fields are quantized and, as a result, physical processes take place by discontinuous transitions. We also know that both the radiation field and the material particle have the particle-wave dual property. Classical theories consider the physical quantities of both material particles and radiation fields as continuous, and the two pictures of wave and particle are mutually exclusive.

Quantum mechanics will have to be formulated on the basis of discrete physical quantities, and its equation of motion must deal with discrete transitions in physical processes. Under certain circumstances, it should include classical theories as special cases. At the same time, quantum mechanics should be such that the wave and the particle pictures become compatible. In the next chapter we shall formulate a theory which allows discontinuous physical quantities and reduces to classical mechanics of a particle as a special case. Although this theory is based on the particle picture, it nevertheless accounts for the wavelike properties. In the following chapters this theory will be applied to study the quantization of physical quantities in various problems and to describe the physical processes in terms of discrete transitions.

From classical theory to quantum theory a basic change is that the concept of continuity is given up and replaced by discontinuity. This change necessitates the abandonment of *determinism* in classical theory and the adoption of *indeterminism*. Scientific thought since the Greek times has been dominated by determinism (the *law of causality*). But a quantized theory may be shown to be necessarily probabilistic. In classical theory all physical quantities are continuous and thus all rates of change are continuous. Once the equation of motion specifies the rate of change, the initial condition determines completely and uniquely the future development of the system. On the other hand, when physical quantities are quantized, a precise specification of the rate of change and the initial condition does not necessarily dictate a unique course of development of the system. To illustrate this we consider the radiating electron in the hydrogen atom. In classical theory the electron circles the nucleus in a uniquely determined spiral path, since the rate of energy loss is precisely specified by the theory of radiation. Once the energy is quantized, energy loss takes place by quantum transitions. The specified rate of energy loss may be realized in a number of alternative courses of transitions (an electron in the state $n = 3$ may go to the state $n = 1$ directly or first to the state $n = 2$ and then to the state $n = 1$), and the electron is given a chance to choose its future course among a number of alternatives without violating the equation of motion. The future course of the electron is thus not uniquely determined by the equation of motion and the initial condition. Therefore, only probabilistic predictions of its future, not deterministic predictions, may be made. We thus see that the introduction of probability in quantum mechanics, which has profound consequences in many respects, is forced on us because of the experimental fact of quantization. This important point will be elaborated later on a number of occasions.

PROBLEMS

1-1. Derive Eq. (1–13), the equation giving the wavelength change in Compton scattering, the process being considered as a collision of a photon with an electron. [*Hint:* Set up three equations by considering conservation of energy and momentum in the collision process. There are four variables involved: the scattering angle θ of the photon, the recoil angle φ of the electron, the wavelength after scattering λ', and the velocity of recoil of the electron v. One of the four, say θ, is to be left undetermined.]

1-2. Derive Eq. (1–36) of the Sommerfeld theory. [*Hint:* In applying Eq. (1–34) remember that

$$p_r = \sqrt{2M[E + (ze^2/r)] - (p_\theta^2/r^2)}$$

and carry out the integration over one cycle of the elliptical orbit.]

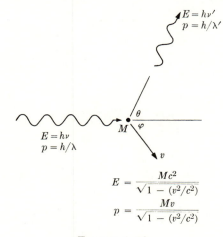

FIGURE 1–13

*1–3. By the method of the old quantum theory, determine the quantized energy levels of a three-dimensional harmonic oscillator the potential function of which is specified by

$$U(r, \theta, \varphi) = \tfrac{1}{2}kr^2. \tag{1–57}$$

Note that in classical mechanics any central force problem is two-dimensional and angular momentum is a constant of motion. The problem may thus be conveniently solved by using the plane polar coordinates. As an orientation the student may determine the quantized energy values of the circular orbits first.

*1–4. Solve Problem 1–3 by using rectangular coordinates. The method of separation of variables may be used (the treatment of the Stark effect by the old quantum theory, to be given in Section 10–4, may be consulted for this purpose).

---

* Indicates more difficult problems.

CHAPTER 2

THE SCHRÖDINGER EQUATION AND
ITS MATHEMATICAL IMPLICATION

In classical physics, the mathematical formalisms of the mechanics of a particle and of the theory of wave motion exhibit a striking similarity which suggests that one may be represented by the other to a certain degree of approximation. In this chapter we shall first formulate a classical wave theory that may be regarded as a mathematical substitute for classical mechanics. This theory contains an undetermined parameter which is required to be small. By setting this parameter equal to Planck's constant h, we obtain the mathematical formalism of quantum mechanics in the form of the Schrödinger equation. An interpretation of the Schrödinger equation will be set forth, based on two assumptions of Born. The similarity between classical and quantum mechanics will be further demonstrated by showing that Newton's second law of motion may be derived from Schrödinger's equation, and the difference between them will be brought out by establishing the so-called uncertainty relation. At the end of the chapter the mathematical implication of the Schrödinger equation will be analyzed within the framework of the Hamiltonian theory of classical mechanics, and the similarity and difference between classical and quantum mechanics will be discussed again in the light of this analysis.

2–1 Relations between the law of mechanics of a particle and the law of wave propagation. We have seen the success of classical mechanics in many applications and we have seen its failure in atomic physics. This dilemma seems to suggest that classical mechanics is not exactly correct but is an approximation of a more general theory. In the applications where the approximation is good (or bad), classical mechanics is good (or bad). The task before us is thus to formulate a more general theory of mechanics which includes classical mechanics as a special case and, at the same time, explains the atomic phenomena which classical mechanics fails to explain. Among such phenomena are the wave property of matter, the quantization of energy, and the penetration of potential barrier. The basic laws on which this more general mechanics—quantum mechanics—is to be founded cannot be derived from something else. As the starting point of a theoretical system, they have to be taken as postulates, like the axioms of geometry, and are justified only by their consequences.

Since quantum mechanics is expected to be able to account for the wave property of matter, we start by examining the classical laws governing the motion of a particle and the propagation of a wave. From these laws we hope to find a clue leading to the establishment of the new mechanics. These two subjects have been studied thoroughly in classical physics. A striking similarity between them is found if their equations of motion are written in the variational form.* The motion of a particle of mass M in a force field described by a potential $U(x, y, z)$ may be determined by the *principle of least action:*

$$\delta \int_A^B \sqrt{2M[E - U(x, y, z)]}\, ds = 0 \tag{2-1}$$

or

$$\delta \int_A^B p\, ds = 0, \tag{2-2}$$

where p and E are the momentum and energy of the particle. The principle of least action is equivalent to Newton's second law of motion. We digress for a moment to show this equivalence. The left-hand side of Eq. (2-1), dropping the constant factor $\sqrt{2M}$, is calculated as follows:

$$\delta \int_A^B \sqrt{E - U}\, ds$$

$$= \int_A^B \frac{-[(\partial U/\partial x)\, \delta x] - [(\partial U/\partial y)\, \delta y] - [(\partial U/\partial z)\, \delta z]}{2\sqrt{E - U}}\, ds + \int_A^B \sqrt{E - U}\, \delta\, ds.$$

Since

$$ds^2 = dx^2 + dy^2 + dz^2,$$

we have

$$d\,\delta s = \delta\, ds = \frac{dx}{ds}\, \delta\, dx + \frac{dy}{ds}\, \delta\, dy + \frac{dz}{ds}\, \delta\, dz.$$

The second integral may be evaluated as follows by integration by parts:

$$\int_A^B \sqrt{E - U}\, \delta\, ds$$

$$= \int_A^B \sqrt{E - U} \left(\frac{dx}{ds}\, \delta\, dx + \frac{dy}{ds}\, \delta\, dy + \frac{dz}{ds}\, \delta\, dz \right)$$

$$= \int_A^B \sqrt{E - U} \left(\frac{dx}{ds}\, d\, \delta x + \frac{dy}{ds}\, d\, \delta y + \frac{dz}{ds}\, d\, \delta z \right)$$

$$= -\int_A^B \left\{ \delta x\, d\left(\sqrt{E - U}\, \frac{dx}{ds} \right) + \delta y\, d\left(\sqrt{E - U}\, \frac{dy}{ds} \right) + \delta z\, d\left(\sqrt{E - U}\, \frac{dz}{ds} \right) \right\}.$$

---

* The rudimentary knowledge of the calculus of variations necessary for reading this section may be found in many textbooks of advanced calculus, for example, Osgood's *Advanced Calculus*.

(The integrated part vanishes at the endpoints.)   Equation (2–1) thus requires

$$-\frac{1}{2}\frac{1}{\sqrt{E-U}}\frac{\partial U}{\partial x} = \frac{d}{ds}\left(\sqrt{E-U}\,\frac{dx}{ds}\right),$$

$$-\frac{1}{2}\frac{1}{\sqrt{E-U}}\frac{\partial U}{\partial y} = \frac{d}{ds}\left(\sqrt{E-U}\,\frac{dy}{ds}\right),$$

$$-\frac{1}{2}\frac{1}{\sqrt{E-U}}\frac{\partial U}{\partial z} = \frac{d}{ds}\left(\sqrt{E-U}\,\frac{dz}{ds}\right).$$

These equations may be written in the following form:

$$-\frac{\partial U}{\partial x} = Mv\frac{d}{ds}\left(v\frac{dx}{ds}\right) = M\frac{d^2x}{dt^2},$$

$$-\frac{\partial U}{\partial y} = Mv\frac{d}{ds}\left(v\frac{dy}{ds}\right) = M\frac{d^2y}{dt^2},$$

$$-\frac{\partial U}{\partial z} = Mv\frac{d}{ds}\left(v\frac{dz}{ds}\right) = M\frac{d^2z}{dt^2},$$

which are the equations of motion according to Newton's second law of motion. Return to Eq. (2–1). For a given energy E, the trajectory of the particle will be completely determined by Eq. (2–1) once the endpoints A and B are specified.

We now turn to the wave motion. The propagation of a wave may be determined by the *principle of least time*.* We note that the propagation of a wave is governed by the laws of reflection and refraction. In the case of reflection we can show that the ray AOB in Fig. 2–1, obeying the law of reflection, is the quickest route for light to travel from A to B by way

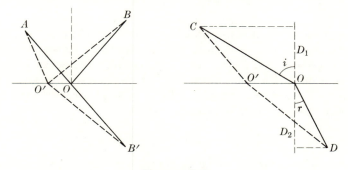

FIGURE 2–1

_____

* Also known as the Fermat principle.

of the reflecting surface. This is because the distance AOB' is the shortest compared with any other route $AO'B'$. In the case of refraction it can also be shown that the ray COD, obeying Snell's law of refraction, is the quickest route for light to travel from C to D. Let the velocities of light in the two media be v_1 and v_2. The total time required from C to D is $(D_1/v_1 \cos i) + (D_2/v_2 \cos r)$. The condition that the time is a minimum is, by differentiation,

$$\frac{D_1 \sin i}{v_1 \cos^2 i} \, di + \frac{D_2 \sin r}{v_2 \cos^2 r} \, dr = 0.$$

Since

$$D_1 \tan i + D_2 \tan r = \text{constant},$$

we have

$$D_1 \sec^2 i \, di + D_2 \sec^2 r \, dr = 0.$$

Therefore the condition of minimum time is

$$\frac{\sin i}{\sin r} = \frac{v_1}{v_2},$$

which is Snell's law of refraction. In a continuous medium the propagation of a wave may be regarded as consisting of a series of refractions, and the total time of travel must be a minimum. Hence the principle of least time.

The mathematical expression of the principle of least time is

$$\delta \int_A^B \frac{1}{v} \, ds = 0. \tag{2-3}$$

For a monochromatic wave of a given frequency ν, it may be written as

$$\delta \int_A^B \frac{1}{\lambda} \, ds = 0, \tag{2-4}$$

since $v = \lambda\nu$ and ν is a constant. The similarity between Eq. (2–2) and Eq. (2–4), though purely formal, suggests that the motion of a particle may be described, in a formal way, by a wave. This wave is used here merely as a mathematical artifice without any physical meaning for the time being. For a particle having a fixed value of energy E, the momentum at any point (x, y, z) is given by

$$p(x, y, z) = \sqrt{2M[E - U(x, y, z)]}. \tag{2-5}$$

Let us create an *artificial* monochromatic wave in a medium the index of refraction of which varies according to $U(x, y, z)$ in such a way that the reciprocal of the wavelength λ of this wave at any point (x, y, z) is always proportional to the momentum of the particle, i.e.,

$$p(x, y, z) = \frac{H}{\lambda(x, y, z)}, \tag{2-6}$$

where H is a proportionality constant. Equation (2-4) thus becomes identical with Eq. (2-2) and the rays of the artificial monochromatic wave determined by Eq. (2-4) coincide exactly with the trajectories of the particle with fixed energy E, determined by Eq. (2-2). Thus, by assuming Eq. (2-6), a monochromatic wave may be set up so that its rays represent a set of particle trajectories all corresponding to the same energy E. Let it be emphasized that *a single wave represents a set of infinitely many trajectories, whereas a single frequency ν corresponds to a single energy E.* Furthermore, the representation is purely *geometrical*, not *kinematical*. In other words, the shapes of the trajectories and rays are identical but the time rates at which these trajectories or rays are traversed by the particle or wave are not yet related to each other.

In order to establish a complete description of the motion of a particle by the motion of a wave, we have to do two things: (1), find a suitable wave representation of a single particle (not a group of them); and (2), establish the kinematical equivalence of a ray and a trajectory. In wave motion, we are familiar with a kind of localized wave, the amplitude of which is zero everywhere except in a small region in space. Such a localized wave, known as a *wave packet,* may move in space as a whole for some time without dispersing itself into a nonlocalized wave. When a wave packet is localized in a very small region, it may be regarded as a point. Its locus in space is essentially a point trajectory. Thus we may try to describe the motion of a single particle by the motion of a wave packet. If we can establish the kinematical equivalence of the two kinds of motion, then the particle and the wave packet are mathematically identical so far as the motion in space is concerned. In order to establish this equivalence, we have to know how a wave packet moves in space.

A wave packet may be represented mathematically by a superposition of infinitely many monochromatic plane waves which have nearly the same wavelengths and frequencies. This is a mathematical result of the Fourier analysis,* but we shall illustrate it in terms of simple concepts of wave motion. We limit our discussion to the one-dimensional wave, although the result may be shown to be true for three-dimensional waves also. A monochromatic plane wave may be represented by a function $\Psi(x, t)$ in a number of forms:

---

* Fourier analysis is of fundamental importance in quantum mechanics. An introduction to it is usually found in textbooks of advanced calculus, such as Kaplan's and Osgood's.

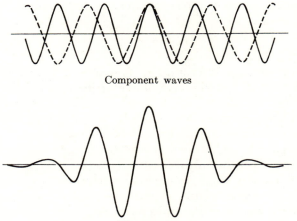

Component waves

Wave packet

FIGURE 2–2

$$\Psi(x, t) = A e^{2\pi i(\sigma x - \nu t)} = A e^{2\pi i[(x/\lambda) - (t/T)]} = A e^{i(kx - \omega t)}, \qquad (2\text{–}7)$$

where σ, ν, λ, T, k, ω are the wave number, frequency, wavelength, period, propagation vector, and angular frequency, respectively. These notations will be used throughout the book. σ and ν are related, the equation relating them being called the dispersion equation; thus we may write

$$\nu = f(\sigma). \qquad (2\text{–}8)$$

The amplitude of a wave packet is expressed analytically by

$$\Psi(x, t) = \sum_{\sigma_0}^{\sigma_0 + \Delta\sigma} A_\sigma e^{2\pi i(\sigma x - \nu t)}. \qquad (2\text{–}9)^*$$

In Fig. 2–2 we plot two of the component waves of the superposition. Because of their slightly different wavelengths, their phase relation cannot remain the same everywhere. At the point where their phases are the same, the resultant amplitude is a maximum. Away from this point of phase agreement, the resultant amplitude becomes smaller. Eventually the phases will be π radians apart and the resultant amplitude will be zero. Farther out, the phases of these two waves will come closer. However, in a wave packet, we have infinitely many waves of all wavelengths between these two. If all waves agree in phase at one point, they will never agree

---

* The summation sign $\sum$ is understood to indicate an integral when the summation index becomes a continuous variable. This convention will be adopted throughout the book.

again at any other points. Usually, a few wavelengths away from the point
of phase agreement the phase relations of all waves are so random that
the resultant amplitude is nearly zero. Thus the waves of the packet
interfere with one another and result in a localized disturbance. The motion of the wave packet as a whole is determined by the motion of the
point of phase agreement, since the whereabouts of the packet is determined by this point. The condition that the phases of the two waves
specified by (σ_0, ν_0) and $(\sigma_0 + d\sigma, \nu_0 + d\nu)$, respectively, agree at $t = t_1$
and $x = x_1$ is

$$2\pi i(\sigma_0 x_1 - \nu_0 t_1) = 2\pi i[(\sigma_0 + d\sigma)x_1 - (\nu_0 + d\nu)t_1]. \qquad (2\text{–}10)$$

The condition that their phases agree at a later time $t = t_2$ and at $x = x_2$ is

$$2\pi i(\sigma_0 x_2 - \nu_0 t_2) = 2\pi i[(\sigma_0 + d\sigma)x_2 - (\nu_0 + d\nu)t_2]. \qquad (2\text{–}11)$$

Taking the difference between the two equations, we have

$$0 = d\sigma(x_2 - x_1) - d\nu(t_2 - t_1), \qquad (2\text{–}12)$$

and therefore

$$\frac{x_2 - x_1}{t_2 - t_1} = \frac{d\nu}{d\sigma}. \qquad (2\text{–}13)$$

The left-hand side of the above equation is actually the velocity of the
point of phase agreement, which may be considered as the velocity of the
wave packet. It is called the *group velocity* of the wave packet v_g. Thus
we have

$$v_g = \frac{d\nu}{d\sigma}. \qquad (2\text{–}14)$$

(Additional discussion on group velocity is to be found in Problem 2–3.)

Knowing the kinematics of a wave packet, we now return to the
problem of creating a wave packet in such a manner that it will describe
the motion of a particle. The velocity of a particle, in both Newtonian
and relativistic mechanics, is related to energy and momentum by

$$v_p = \frac{dE}{dp}. \qquad (2\text{–}15)$$

In Eq. (2–6), where we established the relation between a particle and
a wave, we made use of the wavelength only. We have not made use of
the frequency except to say that a single frequency (monochromatic wave)
corresponds to a single energy value. This implies that E and ν are related, but the function relating them is not specified and remains arbitrary.
Therefore, we are free to choose one particular relation between E and ν;

this will be chosen in such a manner that the group velocity of the packet may be made to agree with the velocity of the particle to be described. This may be accomplished if we assume that the frequency of the artificial wave is related to the energy of the particle by

$$E = H\nu, \tag{2-16}$$

because

$$v_g = \frac{d\nu}{d\sigma} = \frac{d\nu}{d(1/\lambda)} = \frac{dE}{dp} = v_p, \tag{2-17}$$

where v_p is the velocity of a particle having energy E and momentum p. Thus the velocity of the wave packet is identical with that of the particle. As a result, the wave packet moves in exactly the same way as the particle moves and may be used as a mathematical apparatus to describe the motion of the particle. For a particle of mass M, momentum p_0, and energy E_0 at time t_0, we need to construct a wave packet by superposition of monochromatic waves having frequencies in the neighborhood of E_0/H and wavelengths in the neighborhood of H/p_0. Let the phases of all waves be adjusted equal at the time t_0 at the point x_0 where the particle is situated and we have the required wave packet. The position of the packet (considered as a point) at any later time t will be exactly the same as that of the particle by virtue of Eq. (2–17). This relation, though proved in the one-dimensional case only, may be shown to be valid for three-dimensional waves.

This mathematical artifice is made possible by the similarity of the equations of motion of a particle and a wave, by virtue of which we are able to describe the motion of a particle *in a formal way* by the propagation of an artificial wave. The wavelength is chosen according to Eq. (2–6) so that the rays of a monochromatic wave serve as a geometrical representation of *a set of* particle trajectories of the same energy. In addition, the frequency of the wave is chosen according to Eq. (2–16) so that the motion of a wave packet serves as a kinematical representation of the motion of a *single* particle. The proportionality constant H may take any fixed value. These results are direct consequences of classical physics and we have not mentioned quantum mechanics in deriving them. The motion of a particle and the motion of a wave packet are equivalent with only one difference, i.e., the wave packet has a spatial dimension whereas a particle (a mass-point) has not. If the wave packet can be made small enough, the distinction becomes insignificant. This will be discussed more in detail in Section 2–3.

2–2 The equation of motion of a wave packet. We proceed to formulate mathematically the law of propagation of the artificial wave, the properties of which have been specified above. We wish to obtain a wave equation having the following properties:

1. It admits solutions in the form of monochromatic waves which have wavelengths and frequencies specified by Eq. (2–6) and Eq. (2–16).

2. It admits solutions in the form of a linear superposition of monochromatic waves so that wave packets may be included among its solutions.

Such a wave equation will have solutions possessing all the properties we require of the artificial wave to describe the motion of a particle of mass M in a potential field $U(x, y, z)$; it will be the equation of motion of the artificial wave.

The general equation of wave propagation is

$$\nabla^2 \Psi = \frac{1}{V^2} \frac{\partial^2 \Psi}{\partial t^2}, \tag{2–18}$$

where V is the phase velocity of the wave (the product of wavelength and frequency). A monochromatic wave may be represented by

$$\Psi(x, y, z, t) = \psi(x, y, z)e^{-2\pi i \nu t}. \tag{2–19}$$

The equation which $\psi(x, y, z)$ has to satisfy may be obtained by substituting Eq. (2–19) in Eq. (2–18):

$$\nabla^2 \psi = -\frac{4\pi^2 \nu^2}{V^2} \psi. \tag{2–20}$$

Since

$$V = \nu\lambda, \tag{2–21}$$

we have

$$\nabla^2 \psi = -\frac{4\pi^2}{\lambda^2} \psi. \tag{2–22}$$

According to the properties ascribed to the wave, λ must be related to the momentum p of the particle to be described by Eq. (2–6). Since the rays of this monochromatic wave of frequency ν represents the trajectories of a particle with a given energy E equal to $H\nu$, the momentum p may thus be determined from the energy E and the potential $U(x, y, z)$, its magnitude being $\sqrt{2M(E - U)}$. Thus, we have, by Eq. (2–6),

$$\nabla^2 \psi = -\frac{4\pi^2}{H^2} 2M(E - U)\psi \tag{2–23}$$

or

$$\nabla^2 \psi + \frac{8\pi^2 M}{H^2} (E - U)\psi = 0. \tag{2–24}$$

The meaning of the equation is this: *Its solution, containing the parameter E, represents the space-dependent part of a monochromatic wave the rays of which represent the trajectories of a classical particle of mass M and energy E, moving in the field of a potential $U(x, y, z)$.* The frequency ν of this monochromatic wave, of course, is related to the energy E by $E = H\nu$.

For a given energy E, we may obtain a solution $\psi_E(x, y, z)$ from the above differential equation. As a result a monochromatic wave,

$$\psi_E(x, y, z)e^{-2\pi i(E/H)t},$$

is obtained satisfying the requirements of Eqs. (2–6) and (2–16). Different values of E lead to waves of different frequencies satisfying different differential equations of the form Eq. (2–24). However, we want one single differential equation which all monochromatic waves and also their superposition

$$\sum_E a_E \psi_E(x, y, z)e^{-2\pi i(E/H)t}$$

(where the a_E's are constants of superposition) may satisfy.

Obviously, we want a differential equation more general than Eq. (2–24). Such an equation cannot be obtained by a process of derivation, since a derivation usually leads from a general equation to a special one. The procedure to be followed here necessarily involves some kind of speculation and is justified only by the usefulness of the final result obtained. Let us write Eq. (2–24) as follows:

$$\nabla^2\left(\psi_E e^{-2\pi i(E/H)t}\right) + \frac{8\pi^2 M}{H^2}(E - U)\psi_E e^{-2\pi i(E/H)t} = 0. \quad (2\text{–}25)$$

Since

$$\frac{\partial}{\partial t}\left(\psi_E e^{-2\pi i(E/H)t}\right) = -2\pi i\frac{E}{H}\psi_E e^{-2\pi i(E/H)t}, \quad (2\text{–}26)$$

we have

$$\nabla^2\left(\psi_E e^{-2\pi i(E/H)t}\right) - \frac{8\pi^2 M}{H^2}U\left(\psi_E e^{-2\pi i(E/H)t}\right)$$

$$+ \frac{4\pi M i}{H}\frac{\partial}{\partial t}\left(\psi_E e^{-2\pi i(E/H)t}\right) = 0. \quad (2\text{–}27)$$

In this form, the parameter E does not appear in the *differential equation* (it appears only in the *solution* $\psi_E e^{-2\pi i(E/H)t}$). Thus monochromatic waves of all frequencies satisfy the same differential equation. Furthermore, this differential equation is a linear one, so that it admits solutions in the form of a superposition of monochromatic waves, i.e.,

$$\nabla^2 \left(\sum_E a_E \psi_E e^{-2\pi i (E/H)t} \right) - \frac{8\pi^2 M}{H^2} U \left(\sum_E a_E \psi_E e^{-2\pi i (E/H)t} \right)$$

$$+ \frac{4\pi M i}{H} \frac{\partial}{\partial t} \left(\sum_E a_E \psi_E e^{-2\pi i (E/H)t} \right) = 0. \quad (2\text{–}28)$$

From these results we may state that the differential equation

$$\nabla^2 \Psi - \frac{8\pi^2 M}{H^2} U \Psi + \frac{4\pi M i}{H} \frac{\partial \Psi}{\partial t} = 0 \qquad (2\text{–}29)$$

is the one which satisfies all of our requirements. It has solutions in the form of a linear superposition of monochromatic waves by virtue of Eq. (2–28). Each of the monochromatic waves satisfies the conditions imposed on the wavelength and frequency by virtue of Eq. (2–27) and Eq. (2–24). Thus our task of searching for a wave equation, the solution of which may be used to describe the motion of a particle, is accomplished. It may be added that the procedure by which we arrived at Eq. (2–29) from Eq. (2–24) is actually an elimination procedure by which a set of infinitely many differential equations in the form of Eq. (2–24), with different values of the parameter E, are reduced to one single differential equation. This is accomplished by eliminating the parameter E at the expense of raising the order of the differential equation with respect to the time variable t.

Since Eq. (2–29) contains the imaginary number i, its solution in general is a complex quantity. Actually, the superposition of waves given by Eq. (2–9) representing a wave packet is a complex quantity. The use of the imaginary number to represent a real physical entity is by no means mystic, because we merely want some mathematical quantity which may be made appreciably different from zero at the position where the particle is located, but equal to zero elsewhere. Any mathematical quantity capable of being made equal to zero as well as appreciably different from zero may be used for this purpose. This category includes the real number, complex number, hypercomplex numbers, and many other algebraic systems such as vectors and spinors.

2–3 Mathematical formulation of quantum mechanics—the Schrödinger equation. Up to this moment we have not introduced quantum mechanics. We have succeeded in obtaining a mathematical substitute for the classical mechanics of a particle. Instead of solving the Newtonian equations of motion, we may solve the wave equation (2–29) and obtain a wave packet which moves in space in exactly the same way as a particle moves according to classical mechanics. However, there is one difference, i.e., the wave packet has a dimension whereas the classical particle is assumed to be dimensionless. This point will now be discussed more in detail.

We note that the proportionality constant H introduced in Eq. (2–6) is still undetermined. All previous statements remain true, whatever value the constant H takes. However, it is obvious that the dimension of the wave packet depends on the magnitude of H. Since p equals H/λ, a smaller value of H results in smaller values of λ; the dimension of a wave packet, being of the same order of magnitude of λ, as may be seen in Fig. 2–2, will be small. When H approaches zero the dimension of the wave packet approaches zero, and for all practical purposes the wave packet may be treated as a point. Its motion, described by Eq. (2–29), thus becomes exactly the same as the motion of a classical mass-point. In fact, we may say that classical mechanics is a special case of the above-mentioned wave mechanics when H approaches zero.

We are now ready to introduce quantum mechanics. Quantum mechanics is a system of mechanics the mathematical expression of which (in the Schrödinger formalism) may be obtained by setting the constant H equal to the Planck constant h, which has a value of 6.6252×10^{-27} erg·sec. In other words, quantum mechanics may be formulated by introducing the following quantum condition:*

$$H = h. \qquad (2\text{–}30)$$

Equations (2–6) and (2–16) then take the following familiar forms identical with the de Broglie relations:

$$E = h\nu, \qquad p = \frac{h}{\lambda}. \qquad (2\text{–}31)$$

Equation (2–29) takes the form known as the *time-dependent Schrödinger equation*,

$$\nabla^2 \Psi - \frac{8\pi^2 M}{h^2} U\Psi + \frac{4\pi M i}{h} \frac{\partial \Psi}{\partial t} = 0, \qquad (2\text{–}32)$$

and Eq. (2–24) takes the form known as the *time-independent Schrödinger equation*,

$$\nabla^2 \psi + \frac{8\pi^2 M}{h^2} (E - U)\psi = 0. \qquad (2\text{–}33)$$

---

* The Wilson-Sommerfeld quantum condition does two things at one time—it introduces not only the Planck constant h but also the quantum number n. We shall use the term "quantum condition" in quantum mechanics in a broader and more refined sense: to denote the equation that introduces the Planck constant [see also Eq. (12–94)]. The quantum numbers will be derived in quantum mechanics, not introduced.

The time-dependent Schrödinger equation, Eq. (2–32), thus becomes the equation of motion in quantum mechanics. It expresses the fundamental law of the new mechanics. The quantities $\Psi(x, y, z, t)$ and $\psi(x, y, z)$ are called the *time-dependent* and *time-independent wave functions* or simply wave functions; the notations Ψ and ψ will be used for them throughout this book. The Schrödinger equation is also called the *wave equation*.

That the value of H is fixed at h has to be taken as an assumption—a basic one in quantum mechanics. Like other basic assumptions in physics, it cannot be derived and is justified only by its consequences. We shall judge the validity of quantum mechanics according to the standard set up in the very beginning: it should reduce to classical mechanics as a special case and it should be able to account for such quantum phenomena as the wave property of matter, the quantization of energy, and the penetration of potential barrier.

That quantum mechanics includes classical mechanics as a special case may be deduced from the fact that Eq. (2–32) contains a special solution representing a wave packet the motion of which resembles that of a classical particle. The dimension of the wave packet is now determined by Planck's constant which, in macroscopic physics, is an extremely small quantity. Thus in macroscopic physics the wave packet of quantum mechanics reduces to a point and quantum mechanics reduces to classical mechanics.

On the other hand, in dealing with atomic phenomena, when h cannot be regarded as small, quantum mechanics will be very different from classical mechanics. Here lies the potentiality of quantum mechanics to account for the quantum phenomena, where classical mechanics has failed. Since the de Broglie relations are implied in quantum mechanics by virtue of Eqs. (2–31), we may expect quantum mechanics to account for the wave properties of matter; this will be discussed in Chapter 3. The quantization of energy will be discussed in Chapter 4; and in Chapter 5, the problem of barrier penetration. It is the successes of quantum mechanics in these and other applications that justify its basic assumptions and establish its validity. Revolutions and modifications of physical theories may happen in the future, but the new theories will almost certainly have to include quantum mechanics as a special case, and many quantum-mechanical results will last as established parts of our scientific knowledge.

In most textbooks, Planck's constant is introduced at the very beginning. The purpose of introducing it at a later stage, as has been done here, is to show that many elements of this discussion are essentially classical and may be understood without any association with the quantum-mechanical concepts.

2–4 The statistical interpretation of the Schrödinger equation. We have based our discussion on the simple idea that a particle may be described by a wave packet. In the classical limit the dimension of the wave packet is so small that there is no ambiguity in using it to describe a particle. However, when h cannot be regarded as small, the dimension of the wave packet is no longer negligible and the identification of a wave packet with a particle becomes more difficult. Thus, we are required to specify more exactly the relation between a particle and a wave packet. Again a basic assumption has to be introduced here; it cannot be derived and is justified only by its consequences. However, the assumption is required to agree with classical mechanics in the limiting case where h may be regarded as negligibly small.

It seems natural that historically the first attempt to establish this relation (due to Schrödinger) was based on an assumption that a particle is in fact not a point but a physical wave, more or less localized. This interpretation meets difficulties in a number of situations. For example, when a wave packet impinges on a boundary surface it may be partly transmitted and partly reflected like a light wave. This interpretation would imply that an electron might be subdivided into two parts for transmission and reflection. Since no electron may be subdivided, this interpretation has to be abandoned.

Born offered a more satisfactory interpretation which is conventionally accepted at the present and is contained in the statement known as *Born's first assumption*. According to this assumption, a particle such as an electron is still a point indivisible and without dimension. But its position may not always be exactly known. In general, only its probable positions may be known. Born's first assumption states that:

"$|\Psi(x, y, z, t)|^2 \, dx \, dy \, dz$ *represents the relative probability at time t of finding the particle, by an act of observation, located at (x, y, z) in the volume element $dx \, dy \, dz$.*"

It establishes the relation between a particle and a wave function, the latter including the wave packet as a special case. Since $\Psi(x, y, z, t)$ is in general a complex quantity, $|\Psi(x, y, z, t)|^2$ may be written as $\Psi^*(x, y, z, t)\Psi(x, y, z, t)$, where Ψ^* is the complex conjugate of Ψ. The *square* of $|\Psi|$, instead of Ψ itself, is assumed to determine the probability, so that the probability will always be positive as it should. (That the probability is identified with the square, instead of the first power, of $|\Psi|$ is designed to bring out the interference phenomena. See Chapter 3.) $\Psi^*\Psi$ is called the *probability density* of the particle described by the wave function, and Ψ is called the *probability amplitude*. According to this assumption, a wave packet represents a dimensionless particle the position of which is somewhere in a limited region defined by the packet. When the packet reduces to a point, it represents a particle having a definite position. This interpretation is therefore reducible to the classical description as a special case. Since any act of observation to determine a physical quantity involves experimental error, and the precise position of a particle in classical mechanics is but a mathematical abstraction, the use of probability does not seem particularly foreign. According to this interpretation we now consider an electron impinging on a boundary surface to have a definite probability of being transmitted as a whole and a definite probability of being reflected as a whole. The difficulty of splitting the electron is avoided, though at the expense of giving up a deterministic description of the physical processes. The behavior of an electron at a boundary surface will no longer be dictated by a deterministic law which allows no freedom of choice. It will now be controlled by chance. On the other hand, when a large number of electrons impinge on a boundary surface, the transmission coefficient, i.e., the fraction of electrons transmitted, always has a definite value. In cases like this, Born's interpretation and Schrödinger's original interpretation give the same result. The difference appears when the behavior of an *individual* particle is considered. A probabilistic point of view has to be introduced in order to preserve the individuality of the particle.

In classical mechanics the position of a particle at any time is exactly given. Its velocity, being the time derivative of the position, is thus also exactly known. From these values, the momentum and energy may also be known exactly. If a wave packet is interpreted according to a probabilistic point of view, then the velocity, momentum, and energy all become indeterminate. Consider a wave packet of a very small dimension. Such a wave packet may be obtained by superposing a group of monochromatic waves of small wavelengths λ in a fairly large range $\Delta\lambda$. Such a packet, as may be shown from Fig. 2–2, has a small dimension, and may be made to approach a point when $\lambda \to 0$. Let us consider a finite wave packet consisting of waves the wavelengths of which lie in the range from λ_1 to λ_2. According to Eq. (2–6), the two waves of λ_1 and λ_2 correspond to two momentum values p_1 and p_2. These two momentum values correspond to two velocities v_1 and v_2. If the range $\lambda_2 - \lambda_1$ is not infinitesimal, as we require in order to form a narrow packet, we may divide the range $\lambda_2 - \lambda_1$ into infinitesimal segments. Consider the segment $\delta\lambda$ near λ_1. The point of phase agreement of all waves in this segment will move forward according to a group velocity which, by Eq. (2–17), is equal to a particle velocity v_1 related to λ_1 by the de Broglie relation. Similarly, waves within the segment $\delta\lambda$ near λ_2 will have their point of phase agreement moving with velocity v_2. For any intermediate segment, the point of phase agreement will move with an intermediate velocity between v_1 and v_2. Thus the points of phase agreement do not move with the same

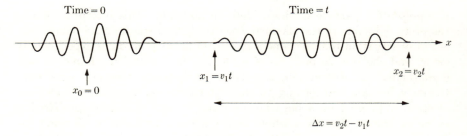

FIGURE 2–3

velocity and will spread out more and more as time goes on. Hence a wave packet, no matter how small to start with, will become dispersed. Its dimension at any time t may be given roughly by (see Fig. 2–3)

$$\Delta x = v_2 t - v_1 t. \qquad (2\text{--}34)$$

Under such circumstances we ask what velocity may be assigned to the particle represented by this dispersing wave packet. If the packet remained compact, then at times T_1 and T_2 we would have no great difficulty in assigning positions X_1 and X_2 for the packet, and the velocity would be $(X_2 - X_1)/(T_2 - T_1)$. (The small dimension of the wave packet would not prevent us from determining this velocity if we made the time interval $T_2 - T_1$ long enough.) However, in the present situation the spreading of the wave packet increases with time, the dimension of the packet Δx being $v_2 t - v_1 t$. According to Born's first assumption, the particle at time t has a certain probability of being observed at any point between x_1 and x_2, where $x_1 = v_1 t$ and $x_2 = v_2 t$. If the particle is observed at x_2 at time t, the velocity of the particle should be $v_2 t/t$ or v_2. If the particle is observed at x_1, the velocity of the particle should be $v_1 t/t$ or v_1. If the particle at time t has a probability P_1 (or P_2) of being observed at x_1 (or x_2), then it has the same probability P_1 (or P_2) of having velocity value v_1 (or v_2). Therefore the probabilistic point of view of the position forces us to take a probabilistic point of view of the velocity of the particle. Thus the particle represented by a wave packet does not have a fixed value of velocity; instead, it has a certain probability of being observed with a velocity v_1 or v_2 or any velocity between v_1 and v_2. Since there is a one-to-one correspondence between velocity and momentum, we are forced to accept a probabilistic point of view of the momentum of the particle. The particle will thus have a certain probability of having momentum equal to p_1 or p_2 or to any value between p_1 and p_2. Furthermore, since the momentum value determines the energy value, we have to accept a probabilistic point of view of the energy of the particle. Therefore, all kinematical and dynamical information of a particle will have to be considered from a probabilistic standpoint.

We note that the probability distribution of momentum in the range between p_1 and p_2 arises from the fact that the wave packet is formed of waves with wavelengths within a range between λ_1 and λ_2, the values of λ_1 and λ_2 being related to p_1 and p_2 by Eq. (2–6). If we choose a different range of λ, say from λ_1' to λ_2', we will have a wave packet with a probability distribution of momentum in a range from p_1' to p_2' where p_1' equals h/λ_1' and p_2' equals h/λ_2'. It necessarily follows that the mixing of waves of different wavelengths results in the mixing of probabilities of the corresponding momentum values. This example makes it easy for us to understand and accept the so-called *Born's second assumption* which, for the present special case, may be stated as follows:

"In a superposition of plane waves $\Psi = \sum_i a_i e^{2\pi i[(x/\lambda_i)-\nu_i t]}$, the square of the absolute value of the coefficient of superposition $|a_i|^2$, or $a_i^* a_i$, represents the relative probability of finding the particle, upon an act of observation, to have a momentum value p_i equal to h/λ_i."

The above argument should not be considered as a proof of this assumption, because we have not shown why the probability has to be proportional to the *square* of $|a_i|$, instead of the fourth or sixth power of $|a_i|$ (the odd powers are ruled out because they give rise to negative probability). It does show that this assumption is consistent with the fact that classical mechanics may be made a special case of quantum mechanics. This assumption establishes the quantitative relation between the coefficients of superposition and the probability distribution of momentum. As a corollary, $|a_i|^2$ in the plane wave expansion also gives the relative probability of finding the particle to have energy E_i equal to $h\nu_i$. (The meaning of Born's second assumption will be further discussed in Problem 2–2 at the end of this chapter.)

When the time t is fixed, the wave function is a function of the position only, and a_i is the Fourier coefficient* of this function. Born's two assumptions may thus be stated as follows:

"The physical meaning of a wave function at a given time $\psi(x)$ is that $\psi^*(x)\psi(x)$ represents the probability distribution of the position of the particle and $a_i^* a_i$ represents the probability distribution of the momentum value where a_i is the Fourier coefficient of $\psi(x)$."

With respect to the second assumption, this statement is more general but is equivalent to the previous one for the special case of a wave packet

---

* A statement of the Fourier theorem may be found in a mathematical note in Section 2–9.

made of plane waves; we *assume* by generalization that it holds for any wave function.

Born's two assumptions establish a probabilistic kinematics. In the classical limit, when the wave packet dimension approaches zero, this statistical interpretation reduces to the deterministic description of classical mechanics, as we required of the new theory. In the field of atomic physics, where h cannot be regarded as small, these assumptions lead to results quite different from those of classical mechanics. The experimental verification of these results is the final judgment of the validity of the Born assumptions.

2–5 The uncertainty relation. The statistical interpretation of quantum mechanics, when h cannot be regarded as small, leads to a conclusion very much at variance with classical mechanics. This result will be discussed below.

When h is not negligible, the dimension of a wave packet is no longer negligible. Turning to Fig. 2–2, we note that the two points where the two waves differ in phase by π radians may be regarded as representing the "boundary" of the wave packet. Let the distance from one of the points to the center of the wave packet be d. We have

$$d = n(\lambda + \Delta\lambda) = (n + \tfrac{1}{2})\lambda \qquad (2\text{--}35)$$

and thus

$$n\,\Delta\lambda = \tfrac{1}{2}\lambda, \qquad (2\text{--}36)$$

where n is the number of wave crests within the distance d. The dimension of the wave packet Δx is thus

$$\Delta x \cong 2d \cong 2n\lambda = \frac{\lambda^2}{\Delta\lambda} = \left|\frac{1}{\Delta(1/\lambda)}\right| = \frac{h}{\Delta p}, \qquad (2\text{--}37)$$

where Δp is the momentum range corresponding to the wavelength range $\Delta\lambda$. We thus have

$$\Delta x\,\Delta p \cong h. \qquad (2\text{--}38)$$

This equation expresses an *uncertainty relation* between x and p. According to the statistical assumptions, the particle represented by the wave packet has an uncertainty in position Δx and an uncertainty in momentum Δp. Equation (2–38) states that the product of these two uncertainties is a fixed quantity of the order of h. If Δx (or Δp) is made to approach zero, the other quantity Δp (or Δx) becomes very large because h is finite. The uncertainties of both x and p cannot be made equal to zero simultaneously. If we let h approach zero as we did in going over to the classical limit, both Δx and Δp may then be made equal to zero, as tacitly assumed in classical mechanics. That the value of h is finite, no matter how small, necessarily introduces an intrinsic uncertainty, i.e., an exact knowledge of both x and p at the same time is not obtainable. The experimental error inherent in the determination of x or p in classical mechanics is assumed to be reducible to zero when the precision of the measurement increases. Therefore the existence of experimental errors does not prevent us, in classical mechanics, from describing a particle by a theory in which all quantities assume exactly specified values without any uncertainty. The intrinsic uncertainty in quantum mechanics differs from the experimental errors in classical mechanics in that we cannot conceive a situation in which both Δx and Δp may vanish. Thus quantum mechanics is intrinsically probabilistic, not merely a system of deterministic mechanics with a calculus of errors built in; and it will not be surprising to find it leading to results quite contrary to the expectations of the classical theory.

The uncertainty relation was first derived by Heisenberg.* Another derivation, mathematically more elaborate than the one given here but physically equivalent, will be given in Chapter 12; it shows specifically that Eq. (2–38) holds for any solution of the time-dependent Schrödinger equation. As described above, this relation brings out a point of sharp contrast between classical mechanics and quantum mechanics which may be taken as a point of departure to exploit the physical meaning of the new theory. Let us consider the assumption in classical mechanics that the errors of both x and p may be reduced indefinitely and thus may be ignored in theoretical discussion. Actually this assumption is unwarranted. It seems to have its root in the supposition that the errors, like any other physical quantities, are continuous quantities and thus allow themselves to be reduced indefinitely. The quantization of physical quantities upsets this situation; the feasibility of reducing errors indefinitely thus cannot be taken for granted. We shall leave the detailed discussion to Section 6–1, where the conclusion will be reached that if we take into consideration the quantization of physical quantities involved in a process of measurement, it will be impossible to reduce the errors of measurement of both x and p to zero; furthermore the limit by which the errors may be reduced is given by an expression which turns out to be identical with Eq. (2–38). In the light of this discussion, the above-mentioned assumption in classical mechanics cannot be substantiated. Moreover, quantum mechanics which includes Eq. (2–38) as an integral part of its theoretical structure seems to be a theory that corroborates the actual physical situation.

When we make use of the classical concepts to describe any atomic phenomena, the values of x and p must not be specified more accurately

---

* W. Heisenberg, *Zeitschrift für Physik*, **43**, 172 (1927).

than permitted by Eq. (2–38), so that no contradiction to quantum mechanics is thus initiated. *The uncertainty relation thus sets the limit within which the classical concepts may be valid.* It will also be pointed out in Section 6–3 that if the classical concepts are so restricted, the description of the quantum phenomena based on them will cease to be paradoxical; in other words, the quantum phenomena may be "explained" by the uncertainty relation.

The statistical interpretation of the wave function applies to all kinds of wave packets. For a wave packet consisting of waves in a *narrow* band of wave numbers, $\Delta\sigma$, its dimension in space will be large. The statistical interpretation asserts that the position of the particle has a large uncertainty, whereas the momentum uncertainty is limited to a narrow range corresponding to $\Delta\sigma$. Consider the limiting case $\Delta\sigma \to 0$. Here we have a "wave packet" consisting of a single monochromatic plane wave. The statistical interpretation of this "wave packet" will give us the physical meaning of a monochromatic plane wave. According to Born's second assumption, the momentum uncertainty of the particle is zero, i.e., the particle represented by a monochromatic plane wave has a definite value of momentum. On the other hand, Born's first assumption asserts that the probability distribution of the position is given by $|e^{2\pi i(\sigma x - \nu t)}|^2$, which is a constant. This means that the probability of finding the particle is everywhere the same. Thus the position is completely uncertain. Therefore $\Delta p = 0$ is accompanied by $\Delta x \to \infty$, in agreement with the uncertainty relation. In cases like this we have to think of a particle whose position is not limited to a small locality but is completely uncertain. Such a situation never arises in classical mechanics; its physical meaning will be discussed in Chapter 6.

In de Broglie's wave theory of matter (Section 1–5), a monochromatic plane wave is associated with a particle of momentum p. The same mathematical association appears here, but in the present case we have a definite physical interpretation of the wave.

2–6 Newton's second law of motion as a special case of quantum mechanics. We now proceed to show that the basic law of classical mechanics, i.e., Newton's second law of motion, may be derived, with certain approximations, from the basic law of quantum mechanics, namely, the time-dependent Schrödinger equation (the Ehrenfest* theorem). Considering the origin of the Schrödinger equation, this is an expected result. The derivation will not give us anything new physically. Nevertheless, it is instructive to see how the same result may be obtained in a different manner. As the mathematical operations involved in the derivation are

---

* P. Ehrenfest, *Z. Physik* **45,** 455 (1927).

typical ones in quantum mechanics, an acquaintance with them at an early stage will be most helpful.

For a particle described by a wave packet $\Psi(x, y, z, t)$ we define a point, $(\bar{x}, \bar{y}, \bar{z})$ called the center of gravity of the wave packet, by the following equations:

$$\bar{x} = \iiint_{-\infty}^{\infty} x\Psi^*\Psi \, dx \, dy \, dz,$$

$$\bar{y} = \iiint_{-\infty}^{\infty} y\Psi^*\Psi \, dx \, dy \, dz, \qquad (2\text{–}39)$$

$$\bar{z} = \iiint_{-\infty}^{\infty} z\Psi^*\Psi \, dx \, dy \, dz,$$

where Ψ^* is the complex conjugate of Ψ. According to Born's first assumption, $\bar{x}$, $\bar{y}$, and $\bar{z}$ may be considered as representing the average values of the three coordinates of the particle. When the size of the wave packet is made smaller and smaller, the whole packet reduces to a point identical with $(\bar{x}, \bar{y}, \bar{z})$. Thus the point $(\bar{x}, \bar{y}, \bar{z})$ is the classical limit of a wave packet. The position of this point changes in time because Ψ^* and Ψ are time-dependent. The behavior of Ψ^* and Ψ in the course of time is determined by the time-dependent Schrödinger equation. We shall show that the motion of this point $(\bar{x}, \bar{y}, \bar{z})$ is identical with the motion of the particle determined by Newton's second law of motion.

For simplicity, we introduce a new notation

$$\hbar = \frac{h}{2\pi}. \qquad (2\text{–}40)$$

The time-dependent Schrödinger equation thus takes the following form:

$$\nabla^2\Psi - \frac{2M}{\hbar^2} U\Psi = -\frac{2Mi}{\hbar} \frac{\partial \Psi}{\partial t}. \qquad (2\text{–}41)$$

As Ψ is a complex number, both sides of the above equation are complex numbers. The complex conjugate quantities of both sides must also be equal. Thus

$$\nabla^2\Psi^* - \frac{2M}{\hbar^2} U\Psi^* = \frac{2Mi}{\hbar} \frac{\partial \Psi^*}{\partial t}. \qquad (2\text{–}42)$$

Equations (2–41) and (2–42) are completely equivalent, one following the other automatically. The velocity of the point $(\bar{x}, \bar{y}, \bar{z})$ may be obtained by differentiation with respect to time. As the dependence of $\bar{x}$, $\bar{y}$, and $\bar{z}$ on time comes in through the time dependence of Ψ^* and Ψ *only*, x, y, z will be held as constants in differentiation with respect to time:

$$\frac{d\bar{x}}{dt} = \iiint\limits_{-\infty}^{\infty} \frac{\partial \Psi^*}{\partial t}\, x\Psi\, dx\, dy\, dz$$

$$+ \iiint\limits_{-\infty}^{\infty} \Psi^* x\, \frac{\partial \Psi}{\partial t}\, dx\, dy\, dz$$

$$= \iiint\limits_{-\infty}^{\infty} \frac{\hbar}{2Mi} \left(\nabla^2 \Psi^* - \frac{2M}{\hbar^2}\, U\Psi^* \right) x\Psi\, dx\, dy\, dz$$

$$- \iiint\limits_{-\infty}^{\infty} \Psi^* x\, \frac{\hbar}{2Mi} \left(\nabla^2 \Psi - \frac{2M}{\hbar^2}\, U\Psi \right) dx\, dy\, dz$$

$$= \frac{\hbar}{2Mi} \iiint\limits_{-\infty}^{\infty} [(\nabla^2 \Psi^*)x\Psi - \Psi^* x(\nabla^2 \Psi)]\, dx\, dy\, dz. \tag{2-43}$$

The second equality results from substituting $\partial \Psi/\partial t$ and $\partial \Psi^*/\partial t$ according to the Schrödinger equation and its complex conjugate equation. We next transform the first term of the last expression by the well-known procedure of integration by parts in three dimensions:

$$\iiint\limits_{-\infty}^{\infty} (\nabla^2 \Psi^*)(x\Psi)\, dx\, dy\, dz$$

$$= \iiint\limits_{-\infty}^{\infty} \{\nabla \cdot [(\nabla \Psi^*)(x\Psi)] - (\nabla \Psi^*) \cdot \nabla(x\Psi)\}\, dx\, dy\, dz. \tag{2-44}$$

The first term may be transformed to an integral over a surface at infinity. Since the $\Psi(x, y, z, t)$ of a wave packet, by definition, vanishes at infinity, the surface integral drops out. The second term of Eq. (2-44) is symmetric with respect to Ψ^* and $x\Psi$. Thus we have

$$\iiint\limits_{-\infty}^{\infty} (\nabla^2 \Psi^*)(x\Psi)\, dx\, dy\, dz = \iiint\limits_{-\infty}^{\infty} \Psi^* \nabla^2 (x\Psi)\, dx\, dy\, dz. \tag{2-45}*$$

---

* The differential operator ∇^2, satisfying Eq. (2-45), is said to be *Hermitian*. Hermitian operators, important in quantum mechanics, will be discussed in Chapter 12.

It follows that

$$\frac{d\bar{x}}{dt} = \frac{\hbar}{2Mi} \int\limits_{-\infty}^{\infty}\!\!\!\int\!\!\int \Psi^*[\nabla^2(x\Psi) - x\nabla^2\Psi]\, dx\, dy\, dz$$

$$= \frac{\hbar}{Mi} \int\limits_{-\infty}^{\infty}\!\!\!\int\!\!\int \Psi^* \frac{\partial\Psi}{\partial x}\, dx\, dy\, dz. \tag{2–46}$$

Differentiating once more and multiplying by M, we have

$$M \frac{d^2\bar{x}}{dt^2}$$

$$= \frac{\hbar}{i} \int\limits_{-\infty}^{\infty}\!\!\!\int\!\!\int \frac{\partial\Psi^*}{\partial t} \frac{\partial\Psi}{\partial x}\, dx\, dy\, dz + \frac{\hbar}{i} \int\limits_{-\infty}^{\infty}\!\!\!\int\!\!\int \Psi^* \frac{\partial}{\partial x}\left(\frac{\partial\Psi}{\partial t}\right) dx\, dy\, dz$$

$$= \frac{\hbar}{i} \int\limits_{-\infty}^{\infty}\!\!\!\int\!\!\int \frac{\hbar}{2Mi}\left(\nabla^2\Psi^* - \frac{2M}{\hbar^2}U\Psi^*\right) \frac{\partial\Psi}{\partial x}\, dx\, dy\, dz$$

$$- \frac{\hbar}{i} \int\limits_{-\infty}^{\infty}\!\!\!\int\!\!\int \Psi^* \frac{\hbar}{2Mi}\left[\frac{\partial}{\partial x}\nabla^2\Psi - \frac{2M}{\hbar^2}\frac{\partial}{\partial x}(U\Psi)\right] dx\, dy\, dz$$

$$= -\frac{\hbar^2}{2M} \int\limits_{-\infty}^{\infty}\!\!\!\int\!\!\int \left[(\nabla^2\Psi^*)\frac{\partial\Psi}{\partial x} - \Psi^*\nabla^2\left(\frac{\partial\Psi}{\partial x}\right) + \Psi^*\frac{2M}{\hbar^2}\frac{\partial U}{\partial x}\Psi\right] dx\, dy\, dz. \tag{2–47}$$

The first two terms cancel out after we perform the same kind of integration by parts as we have just done. Thus we have

$$M \frac{d^2\bar{x}}{dt^2} = \int\limits_{-\infty}^{\infty}\!\!\!\int\!\!\int \left(-\frac{\partial U}{\partial x}\right) \Psi^*\Psi\, dx\, dy\, dz. \tag{2–48}$$

The right-hand side is just the average value of the force $-\partial U/\partial x$ over the wave packet. When the dimension of the packet is reduced to a point, the above relation states that mass times acceleration is equal to the force acting on the particle, which is Newton's second law of motion.

We note that the classical law is derived on the basis that the wave packet may be regarded as a point. Thus the equivalence is not exact but approximate. As the wave packet is subject to the uncertainty relation, we conclude again that the validity of classical mechanics is limited by the uncertainty relation.

2–7 Quantum mechanics and the Hamiltonian theory.*

While a localized wave function may be easily identified with a classical particle, the meaning of a general wave function is not easily visualized. Still, there is a close relation between quantum mechanics and the Hamiltonian theory of mechanics, and a classical picture of a general wave function may be formed on the basis of this relation.

A general wave function $\Psi(x, y, z, t)$ is a complex quantity and thus may be written in the following form:

$$\Psi(x, y, z, t) = A(x, y, z, t)e^{iS(x,y,z,t)}, \tag{2–49}$$

where $A(x, y, z, t)$ and $S(x, y, z, t)$, representing the modulus and phase, are real. Then,

$$\begin{aligned} \nabla^2\Psi &= \nabla\cdot\nabla(Ae^{iS}) \\ &= \nabla\cdot(e^{iS}\nabla A + Ae^{iS}i\nabla S) \\ &= ie^{iS}\nabla S\cdot\nabla A + e^{iS}\nabla^2 A + i\nabla(Ae^{iS})\cdot\nabla S + iAe^{iS}\nabla^2 S \\ &= 2ie^{iS}\nabla S\cdot\nabla A + e^{iS}\nabla^2 A + iAe^{iS}\nabla^2 S - Ae^{iS}(\nabla S)^2. \end{aligned} \tag{2–50}$$

The time-dependent Schrödinger equation, after substitution of Eqs. (2–49) and (2–50) in Eq. (2–41), becomes

$$2i\nabla S\cdot\nabla A + \nabla^2 A + iA\nabla^2 S - A(\nabla S)^2 - \frac{2M}{\hbar^2}UA + \frac{2Mi}{\hbar}\dot{A} - \frac{2M}{\hbar}A\dot{S} = 0, \tag{2–51}$$

where $\dot{A}$ and $\dot{S}$ denote the partial derivatives of A and S with respect to time. When a complex quantity is equal to zero, both its real and imaginary parts must vanish. Thus we have

$$\nabla^2 A - A(\nabla S)^2 - \frac{2M}{\hbar^2}UA - \frac{2M}{\hbar}A\dot{S} = 0, \tag{2–52}$$

$$2\nabla S\cdot\nabla A + A\nabla^2 S + \frac{2M}{\hbar}\dot{A} = 0. \tag{2–53}$$

---

* For those who are not well prepared in analytical mechanics this section may be omitted without breaking the continuity of this book. However, readers doing so will have to disregard remarks to appear later using the terms "probability element" and "quantum force." These remarks throw side light on the physical meaning of quantum mechanics but they are not essential elements of our presentation and thus may be omitted.

Equation (2–53) may be transformed, after multiplication by A, as follows:

$$\nabla S \cdot \nabla A^2 + A^2 \nabla^2 S + \frac{M}{\hbar} \frac{\partial A^2}{\partial t} = 0, \qquad (2\text{–}54)$$

or

$$\frac{\hbar}{M} \nabla \cdot (A^2 \nabla S) + \frac{\partial A^2}{\partial t} = 0. \qquad (2\text{–}55)$$

We now transform S to a function ϕ as follows:

$$S(x, y, z, t) = \frac{1}{\hbar} \phi(x, y, z, t). \qquad (2\text{–}56)$$

Equations (2–52) and (2–55) then take the following forms:

$$\hbar^2 \nabla^2 A - A(\nabla \phi)^2 - 2MUA - 2MA\dot{\phi} = 0, \qquad (2\text{–}57)$$

$$\nabla \cdot \left(A^2 \frac{\nabla \phi}{M} \right) + \frac{\partial A^2}{\partial t} = 0. \qquad (2\text{–}58)$$

Up to this point the derivation is exact. We now make the approximation of letting $\hbar$ approach zero. Equation (2–57) becomes

$$\frac{1}{2M} (\nabla \phi)^2 + U + \dot{\phi} = 0. \qquad (2\text{–}59)$$

This equation is identical with the Hamilton-Jacobi differential equation in classical mechanics* for a particle of mass M moving in a potential field $U(x, y, z)$:

$$H \left(\frac{\partial \phi}{\partial q}, q \right) + \frac{\partial \phi}{\partial t} = 0. \qquad (2\text{–}60)$$

According to the Hamiltonian theory we have

$$\nabla \phi = \vec{p} = M\vec{v}, \qquad (2\text{–}61)$$

where $\vec{v}$ is the velocity of the particle. Equation (2–58) thus becomes

$$\nabla \cdot (A^2 \vec{v}) + \frac{\partial A^2}{\partial t} = 0, \qquad (2\text{–}62)$$

which takes the form of the equation of continuity for the flow of a compressible fluid with a density function $[A(x, y, z, t)]^2$ and a velocity function $\vec{v}(x, y, z)$. (Note that A^2 equals $\Psi^*\Psi$.)

---

* See, for example, Goldstein, *Classical Mechanics*, Addison-Wesley Publishing Co., Inc.

This interesting result may be interpreted in the following way. Consider the motion of a particle of mass M in a potential field $U(x, y, z)$. To solve this problem according to the method of analytic mechanics, we first set up the Hamilton-Jacobi differential equation which takes the form of Eq. (2–60). The next step is to solve Eq. (2–60) and obtain a complete integral which contains three arbitrary constants c_1, c_2, c_3 (note that we do not want the most general solution containing three arbitrary functions). This solution ϕ gives us the momentum (and thus the velocity) of the particle according to Eq. (2–61). The arbitrary constants c_1, c_2, c_3 contained in ϕ are determined to fit the initial conditions of the particle. The problem is thus completely solved. Now let us vary the three arbitrary constants c_1, c_2, c_3 in ϕ. This results in the generation of a set of trajectories corresponding to a set of initial conditions. Let us artificially assign a large number of particles, all of mass M, each moving in the potential field according to one of the trajectories. It is not necessary that all the trajectories be occupied, and the distribution of the particles may be quite arbitrary. At any time t we find these particles distributed in space in a certain pattern. At a later time, as each particle moves along its trajectory, the spatial distribution of these particles will be different from the previous distribution. Thus we may assign a function ρ to represent the density of these particles which will be a function of position (x, y, z) as well as time t. This group of particles form a compressible fluid moving according to the velocity function $\vec{v}$. The equation of continuity of this fluid takes the form

$$\nabla \cdot (\rho \vec{v}) + \frac{\partial \rho}{\partial t} = 0. \tag{2–63}$$

Comparing Eq. (2–62) with Eq. (2–63), we find that the amplitude A of wave function Ψ may be interpreted as follows: the square of A, which equals $\Psi^*\Psi$, may be considered to represent the density function of a group of classical particles described by one principal function ϕ. At time t_0, by suitable choice of the constants of superposition in Ψ, the spatial distribution of $\Psi^*\Psi$ can be made to take any arbitrary form and thus to take the form of the density distribution of the classical particles. Once thus matched at t_0, they remain the same at any later time by virtue of Eq. (2–62) and Eq. (2–63). The phase S of the wave function Ψ equals the principal function ϕ divided by $\hbar$. Thus the relation between Ψ and the group of classical particles is completely established.

For a wave function Ψ representing a localized wave packet, the amplitude A is a localized function, being zero everywhere except in a small region. As a result, this wave packet may be considered to represent a group of classical particles clustering together in a limited region in space and moving along their classical trajectories. The square of the amplitude

A of the wave packet at any time t represents the density distribution of this cluster of particles. When the size of the packet is reduced to zero, the set of trajectories reduces to one single trajectory; and we arrive once again at the well-known result that a localized wave packet moves along a classical trajectory.

The above interpretation enables us to describe the collective behavior of a group of classical particles by the amplitude function A which is obtained by solving the Schrödinger equation. However, this result has no practical application, as such a classical problem seldom occurs. On the other hand, if we consider this cluster of classical particles not as many real particles, but as representing the possible positions of a single particle at a given time, then quantum mechanics in the limit $\hbar \to 0$ gives us a mathematical description of a classical particle having a number of possible positions at a given time, or having a classical error attached to its position. Proceeding along this line, we may attach to every point in the cluster a probability of finding the real particle at this point. When a point is thus associated with a probability we call it a *probability element.* It is identical with a particle so far as the dynamical properties are concerned but differs in that the appearance of a probability element at a point (x, y, z) means the existence of a certain amount of probability to find the particle at (x, y, z) instead of the actual appearance of the particle at (x, y, z). The future position of a classical particle together with its future error may thus be obtained by following the motion of the probability elements (each moves like a classical particle). Quantum mechanics in the limit $\hbar \to 0$, interpreted in this way, may be used as a calculus of errors within the framework of classical mechanics. The wave function serves as the error distribution function of the position of the classical particle.

The foregoing discussion is based on the assumption that $\hbar \to 0$. When $\hbar$ cannot be regarded as small, we must go back to Eq. (2–57), which may be written in the following form:

$$\frac{1}{2M}\,(\nabla\phi)^2 + U - \frac{\hbar^2}{2M}\,\frac{\nabla^2 A}{A} + \dot{\phi} = 0. \tag{2–64}$$

This equation may be interpreted as the equation of motion of a classical particle in a potential field V given by

$$V = U - \frac{\hbar^2}{2M}\,\frac{\nabla^2 A}{A}. \tag{2–65}$$

In addition to the original potential U there is a second term which is dependent on $\hbar$ and vanishes when $\hbar$ is set to zero. A new term in the potential corresponds to a new force, and we shall loosely call this one the *quantum force.* It is not a real force but is a useful concept for the interpretation of quantum-mechanical results in terms of the familiar concept of force. Thus the exact form of quantum mechanics (not letting $\hbar \to 0$) differs from classical mechanics only in the introduction of the quantum force. Now, the previous statement regarding the meaning of the amplitude A of the wave function Ψ may be changed to the following: If we have a group of particles moving according to classical mechanics in a combined field of a classical potential $U(x, y, z)$ and a *quantum potential* $-(\hbar^2/2M)(\nabla^2 A/A)$, then their density function will be represented by the square of amplitude A of a certain wave function Ψ. This statement is equivalent to saying that the square of the amplitude A of a wave function Ψ describes the probability distribution of the position of a particle whose probability elements move under the influence of both the classical and quantum forces. As the introduction of the quantum force changes the distribution of the probability elements, quantum-mechanical results will be different from those of classical mechanics.

Since quantum mechanics differs from classical mechanics solely in the presence of the quantum force, we should be able to attribute those features of quantum mechanics which are at variance with classical mechanics to the existence of this quantum force. Equation (2–65) shows that the quantum potential depends on A and thus on the distribution of the probability elements. *This may be interpreted to mean that the probability elements exert "forces" on one another or they interfere with one another.* Here lies the basic difference between the classical error and the quantum-mechanical uncertainty. *Classical errors do not interfere with one another. Therefore, quantum mechanics is not just a calculus of errors; it contains something new in which lies the key to the understanding of the quantum phenomena.*

The discussion above is purely mathematical in nature. It shows that quantum mechanics may be worked out within the framework of classical mechanics by the addition of a quantum potential. After all, a system of mechanics is but an empty scheme; the physical law of nature is actually expressed by the potential function. We shall return to this viewpoint in Section 6–3.

2–8 Normalization. The probability current density. According to the statistical interpretation, $|\Psi(x, y, z, t_0)|^2\, dx\, dy\, dz$ represents the probability of finding the particle in the volume $dx\, dy\, dz$ around the point (x, y, z) at the time t_0. Since the total probability equals unity, we require

$$\iiint_{-\infty}^{\infty} |\Psi(x, y, z, t_0)|^2\, dx\, dy\, dz = 1. \tag{2–66}$$

This condition may always be satisfied for the following reason. The Schrödinger equation is a linear equation; any solution of it multiplied by a constant c remains a solution. Let

$$c = \frac{1}{\sqrt{\iiint\limits_{-\infty}^{\infty} |\Psi(x, y, z, t_0)|^2 \, dx \, dy \, dz}} \; ; \qquad (2\text{–}67)$$

the new wave function $c\Psi(x, y, z, t_0)$ will be one satisfying Eq. (2–66). Wave functions satisfying Eq. (2–66) are called *normalized* wave functions; otherwise, they are said to be *unnormalized*. The constant c defined in Eq. (2–67) is called the *normalization constant*. In later discussions we shall always use normalized wave functions, which may be obtained readily from the unnormalized wave functions by multiplying with the proper normalization constants. Such a procedure is called *normalization*.

We note that not all functions of (x, y, z) may be normalized. If the integral in Eq. (2–66) diverges to infinity, then the value of c becomes zero and $c\Psi(x, y, z, t_0)$ becomes a trivial solution. Such a situation may arise when $\Psi(x, y, z, t_0)$ becomes infinite as (x, y, z) approaches infinity.* Wave functions of this kind do not represent a particle with finite probability distribution and therefore do not describe any real physical situation. They will be excluded. Only normalizable wave functions will be considered in quantum mechanics. Thus Eq. (2–66) imposes a restriction on physically acceptable wave functions. In order that the normalization integral in Eq. (2–66) be finite, it is necessary that the wave function $\Psi(x, y, z, t_0)$ approach zero sufficiently rapidly at infinity. Thus the normalization condition imposes a boundary condition at infinity on all physically acceptable wave functions. Later we shall see that this boundary condition leads to quantization of energy. For the same reason it imposes a certain limitation on the singularities that may be allowed to exist in a wave function. For example, if ψ diverges to infinity at the origin according to the relation

$$\psi \sim \frac{1}{r^n}, \qquad (2\text{–}68)$$

where r is the distance from the origin, the normalization condition requires the following integral to be finite in the small region around the origin

$$\iiint |\psi|^2 \, d\tau \sim \int \frac{1}{r^{2n}} \, r^2 \, dr \sim \frac{1}{r^{2n-3}} . \qquad (2\text{–}69)$$

---

* For those wave functions remaining finite but not vanishing at infinity the normalization integral also diverges, but some of them are still physically meaningful, an example of which is given in Section 3–2. Special normalization techniques will have to be introduced for such cases.

It becomes necessary that

$$2n - 3 < 0. \tag{2-70}$$

Therefore a singularity with n less than $\frac{3}{2}$ may be allowed.

The previous normalization procedure applies only at a given time t_0. However, it will be proved presently that the Schrödinger equation has the property that once a wave function is normalized at a time t_0, it remains normalized at any time t; accordingly we need only to normalize the wave function at one given time. The time-dependent Schrödinger equation is a differential equation of the first order in the time variable. Therefore, once Ψ is determined at a time t_0, it will be determined at any later time t. Let us write the time-dependent Schrödinger equation (2–41) and its conjugate equation (2–42):

$$\nabla^2\Psi - \frac{2M}{\hbar^2} U\Psi + \frac{2Mi}{\hbar} \frac{\partial\Psi}{\partial t} = 0, \tag{2-71}$$

$$\nabla^2\Psi^* - \frac{2M}{\hbar^2} U\Psi^* - \frac{2Mi}{\hbar} \frac{\partial\Psi^*}{\partial t} = 0. \tag{2-72}$$

Multiplying Eq. (2–71) by Ψ^* and Eq. (2–72) by Ψ, and then taking the difference, we have

$$\Psi^*\nabla^2\Psi - \Psi\nabla^2\Psi^* + \frac{2Mi}{\hbar} \left(\Psi^* \frac{\partial\Psi}{\partial t} + \Psi \frac{\partial\Psi^*}{\partial t}\right) = 0. \tag{2-73}$$

From a theorem in vector analysis,

$$f\nabla^2 g - g\nabla^2 f = \nabla \cdot (f\nabla g - g\nabla f), \tag{2-74}$$

we can rewrite Eq. (2–73) as follows:

$$\frac{\partial}{\partial t} |\Psi|^2 + \nabla \cdot \left[\frac{\hbar}{2Mi} (\Psi^*\nabla\Psi - \Psi\nabla\Psi^*)\right] = 0. \tag{2-75}$$

Integrating over a volume τ bounded by a surface σ, and applying Green's theorem to change the second volume integral to a surface integral, we obtain

$$\frac{d}{dt} \iiint_\tau |\Psi|^2 \, d\tau + \iint_\sigma \frac{\hbar}{2Mi} (\Psi^*\nabla\Psi - \Psi\nabla\Psi^*) \, d\sigma = 0. \tag{2-76}$$

(After the integration, t is the only variable left; thus we have total differentiation with respect to t in the first term.) When the volume τ increases to infinity, the surface σ also goes to infinity. Since Ψ, and therefore Ψ^*, must vanish at infinity, the surface integral vanishes and we have

$$\frac{d}{dt} \iiint_{\infty} |\Psi|^2 \, d\tau = 0. \tag{2–77}$$

Therefore the normalization integral is a constant in time. If it equals unity at one time t_0, it remains so at any time. Thus once a wave function is normalized, it remains normalized at any time.

The above conclusion may be inferred from the discussion of Section 2–7. Equation (2–62) enables us to interpret $|\Psi|^2$, which equals A^2 in Eq. (2–62), as the density function of a compressible fluid. The normalization integral thus represents the total amount of the fluid, which, according to the equation of continuity, can neither be increased nor decreased and therefore has to remain a constant in time.

The fluid analogy enables us to introduce the concept of current density in quantum mechanics. Equation (2–75) takes the form of the equation of continuity. Since we interpret $|\Psi|^2$ as the probability density, the corresponding current in the equation of continuity, Eq. (2–75), will have to be interpreted as a probability current. Thus we have the expression for the *probability current density*:

$$\vec{j} = \frac{\hbar}{2Mi} \, (\Psi^* \nabla \Psi - \Psi \nabla \Psi^*). \tag{2–78}$$

With this concept we may interpret Eq. (2–76) as meaning that the increase of total probability inside a finite volume τ is due to the influx of the probability current over the surface σ bounding this volume.

In Section 2–7, we interpret the quantity $|\Psi|^2$ in the classical limit as the density function of a group of classical particles moving along their classical trajectories. In the same spirit, we may interpret, in the classical limit, the probability current density given by Eq. (2–78) as the actual current density of this group of classical particles.

2–9 The general solution of the time-dependent Schrödinger equation.

As the time-dependent Schrödinger equation is the equation of motion in quantum mechanics, its general solution is the key to the solutions of all quantum-mechanical problems. Before we discuss the general solution we write down without proof a few important results of the *Fourier analysis* which will be used presently as well as in later discussions.

The fundamental theorem in Fourier analysis states that for a piecewise continuous function $f(x)$, defined in the interval $-\pi \leq x \leq \pi$, having continuous first and second derivatives within each subinterval, the *Fourier series*

$$\frac{a_0}{2} + \sum_{n=1}^{\infty} (a_n \cos nx + b_n \sin nx),$$

where

$$a_n = \frac{1}{\pi} \int_{-\pi}^{\pi} f(x) \cos nx \, dx,$$

$$b_n = \frac{1}{\pi} \int_{-\pi}^{\pi} f(x) \sin nx \, dx,$$

converges to $f(x)$ wherever $f(x)$ is continuous inside the interval, to

$$\tfrac{1}{2}[f(x_{i-}) + f(x_{i+})]$$

at each point of discontinuity x_i inside the interval and to

$$\tfrac{1}{2}[f(\pi_-) + f(\pi_+)]$$

at $x = \pm \pi$, the convergence being uniform in each closed subinterval. This statement is not the most general—for example, the requirement on the continuity of the second derivative may be relaxed—but is sufficient in most applications. The Fourier series may be expressed in the complex form which is more convenient, particularly for our purposes,

$$f(x) = \sum_{-\infty}^{\infty} c_n e^{inx},$$

where the *Fourier coefficient* c_n may be evaluated from $f(x)$ by

$$c_n = \frac{1}{2\pi} \int_{-\pi}^{\pi} f(x) e^{-inx} \, dx.$$

By a suitable limiting process in which the interval extends to $-\infty < x < \infty$, we obtain the *Fourier integral* for a piecewise continuous function $f(x)$, defined in the interval $-\infty < x < \infty$, having continuous first derivative and rendering the integral $\int_{-\infty}^{\infty} |f(x)| \, dx$ convergent,

$$f(x) = \frac{1}{\sqrt{2\pi}} \int_{-\infty}^{\infty} g(t) e^{ixt} \, dt,$$

where the *Fourier transform* $g(t)$ may be calculated from $f(x)$ by

$$g(t) = \frac{1}{\sqrt{2\pi}} \int_{-\infty}^{\infty} f(x) e^{-itx} \, dx.$$

We now consider the general solution Ψ at a fixed point (x_0, y_0, z_0). Its value at this point $\Psi(x_0, y_0, z_0)$ is a function of time and may be expanded in a Fourier series or integral in the time variable,

$$\Psi(x_0, y_0, z_0, t) = \sum_{\omega} a_{\omega} e^{-i\omega t}. \qquad (2\text{-}79)$$

When the point (x_0, y_0, z_0) changes, the Fourier coefficients a_{ω} also change. The dependence of a_{ω} on the point may be expressed as follows:

$$\Psi(x, y, z, t) = \sum_{\omega} a_{\omega}(x, y, z) e^{-i\omega t}. \qquad (2\text{-}80)$$

Substituting Eq. (2–80) in the time-dependent Schrödinger equation, we have

$$\sum_{\omega} e^{-i\omega t} \left[\nabla^2 a_{\omega}(x, y, z) - \frac{2M}{\hbar^2} U a_{\omega}(x, y, z) + \frac{2M\omega}{\hbar} a_{\omega}(x, y, z) \right] = 0. \quad (2\text{-}81)$$

The expression within the brackets is time-independent. In order that Eq. (2–81) be satisfied at all times we must require that all the brackets in the summation vanish:

$$\nabla^2 a_{\omega}(x, y, z) + \frac{2M}{\hbar^2} (\hbar\omega - U) a_{\omega}(x, y, z) = 0. \qquad (2\text{-}82)$$

If we write

$$\hbar\omega = E, \qquad (2\text{-}83)$$

Eq. (2–82) becomes identical with the time-independent Schrödinger equation. The quantity E so introduced may be identified with the energy, and the function $a_{\omega}(x, y, z)$ may be identified with $c_E \psi_E(x, y, z)$, c_E being a constant. Thus the time-independent Schrödinger equation may be considered as the equation which determines the Fourier coefficients of the general solution of the time-dependent Schrödinger equation. Equation (2–83) is consistent with Eq. (2–31), since the angular frequency ω is related to frequency ν by

$$\omega = 2\pi\nu. \qquad (2\text{-}84)$$

The above conclusion means that the general superposition of monochromatic waves considered in Section 2–2 is actually the general solution of the time-dependent Schrödinger equation. Therefore the general solution Eq. (2–80) may be rewritten as follows:

$$\Psi(x, y, z, t) = \sum_{E} c_E \psi_E(x, y, z) e^{-i(E/\hbar)t}. \qquad (2\text{-}85)$$

The normalization condition imposes some restriction on the general solution. Those solutions of the time-independent Schrödinger equation which do not satisfy the normalization condition should be excluded; the general solution is thus a superposition of those solutions which satisfy the normalization condition.

2–10 Summary. We have established the mathematical formalism of quantum mechanics by introducing the time-dependent Schrödinger equation as the general equation of motion. The mathematical solution of this equation on the one hand and the physical state of the material system to be described on the other are related by the two assumptions of Born. Thus we have a complete system of mechanics ready to apply to specific problems. In a specific application the first step is to set up the Schrödinger equation with the given knowledge of the potential function. Next we translate the initial conditions of the physical system into the initial conditions of the Schrödinger equation according to Born's assumptions. Then we are ready for the major mathematical task of finding a solution $\Psi(x, y, z, t)$ of the equation of motion satisfying the initial conditions. Once the solution is obtained, the future information of the physical system may be obtained from it according to Born's assumptions. We have shown that the results so obtained are reducible to those of classical mechanics if we let $\hbar$ approach zero. On the other hand, the difference between quantum mechanics and classical mechanics may be attributed to the incorporation into the former of the uncertainty relation or the quantum force. In the next three chapters we shall discuss the applications of quantum mechanics to three specific problems. We shall verify quantum mechanics in the classical limit by working out some particular solutions. More importantly, we shall find that in these three cases quantum mechanics contains solutions, not contained in classical mechanics, which enable us to explain the quantum phenomena, namely, the wave property of matter, the quantization of energy, and the penetration of potential barrier. The relation between these quantum phenomena and the uncertainty relation or the quantum force will be discussed in Chapter 6.

PROBLEMS

2–1. What is the meaning of the statement that quantum mechanics reduces to classical mechanics when the Planck constant approaches zero, inasmuch as the Planck constant has a definite value 6.625×10^{-27} erg·sec? [*Hint:* Classical mechanics is valid for particles of large masses.]

2–2. By the method of Section 2–6 show that for a wave packet of monochromatic plane waves, $\Psi = \sum_i a_i(1/\sqrt{\Omega}) \exp (i/\hbar)(p_{xi}x + p_{yi}y + p_{zi}z - E_i t)$ where Ω is the volume of the coordinate space (see p. 73 for this method of normalization),

$$M \frac{d\bar{x}}{dt} = \sum_i a_i^* a_i p_{xi}. \tag{2–86}$$

Note that this result is independent of Born's second assumption but corroborates it.

2–3. Discuss the condition under which the concept of group velocity is valid. [*Hint:* Expand the exponent of Eq. (2–9) about (σ_0, ν_0). The zero-order term represents the phase wave; the first-order term gives rise to the group velocity. The concept of group velocity is meaningful when the second-order term may be neglected.]

2–4. Mathematical exercise: Determine the Fourier series of the square wave function,

$$f(x) = \begin{cases} -1, & -\pi \le x < 0, \\ +1, & 0 \le x \le \pi. \end{cases} \tag{2-87}$$

Plot the first three partial sums and compare with the square wave function.

Answer: $f(x) = (4/\pi) \sum_{n=1}^{\infty} [\sin (2n - 1)x]/(2n - 1)$.

2–5. Mathematical exercise: Determine the Fourier integral of

$$f(x) = \begin{cases} 0, & x < 0, \\ 1, & 0 \le x \le 1, \\ 0, & x > 1. \end{cases} \tag{2-88}$$

Answer: $f(x) = \dfrac{1}{\pi} \displaystyle\int_0^{\infty} \dfrac{\sin t \cos xt + (1 - \cos t) \sin xt}{t} \, dt.$

2–6. Determine the Fourier integral of

$$f(x) = e^{-x^2/a^2}, \qquad -\infty < x < \infty, \tag{2-89}$$

and verify the relation

$$\Delta x \, \Delta \sigma \cong 1, \tag{2-90}$$

where Δx and $\Delta \sigma$ are ranges of x and σ (wave number) within which $f(x)$ and its Fourier transform respectively are appreciable. For the present problem, $\Delta x \cong a$. (Compare with Problem 4–2.) What is the relation of this equation to the uncertainty relation?

*2–7. By Eqs. (2–41), (2–42), and (2–78) show that

$$\frac{d}{dt} \iiint_{-\infty}^{\infty} \vec{j} \, dx \, dy \, dz = -\frac{1}{M} \iiint_{-\infty}^{\infty} (\nabla U) \Psi^* \Psi \, dx \, dy \, dz. \tag{2-91}$$

Define

$$u = \frac{\hbar^2}{2M} \nabla \Psi^* \cdot \nabla \Psi. \tag{2-92}$$

Also show that

$$\frac{d}{dt} \iiint\limits_{-\infty}^{\infty} u \, dx \, dy \, dz = -\iiint\limits_{-\infty}^{\infty} U \frac{\partial}{\partial t} (\Psi^*\Psi) \, dx \, dy \, dz. \qquad (2\text{--}93)$$

Note: If Eq. (2–41) is regarded as the equation of motion in a classical wave theory of matter—a point of view we do not consider here but will discuss in Section 12–12, Chapter 12—we may define $M\vec{j}$ and u as the momentum density and energy density respectively for the wave field and consider the above two equations (2–91) and (2–93) as the statements of the laws of conservation of momentum and energy.

---

* Indicates more difficult problems.

CHAPTER 3

THE FREE PARTICLE

As the first example of the applications of quantum mechanics to specific problems, we shall discuss the motion of a free particle. After a review of the classical solution of this problem we shall obtain the general solution in quantum mechanics. From the general solution we can construct a special solution representing a wave packet, the motion of which resembles the motion of a particle obeying classical mechanics. Thus we shall verify the conclusion that quantum mechanics includes classical mechanics as a special case. Also, from the general solution we shall find other solutions which do not correspond to any classical motion, but instead may be used to describe a quantum phenomenon, namely, the wave property of matter.

3–1 The general solution in classical mechanics. For a free particle, the potential function is a constant which may be set equal to zero. Thus the classical equations of motion are

$$M \frac{d^2x}{dt^2} = 0, \qquad M \frac{d^2y}{dt^2} = 0, \qquad M \frac{d^2z}{dt^2} = 0, \qquad (3\text{–}1)$$

where M is the mass of the particle and (x, y, z) are the coordinates of its position. The general solutions of the equations of motion are

$$x = x_0 + v_x t, \qquad y = y_0 + v_y t, \qquad z = z_0 + v_z t, \qquad (3\text{–}2)$$

where $x_0, y_0, z_0, v_x, v_y, v_z$ are integration constants to be determined by the initial conditions [initial position (x_0, y_0, z_0) and initial velocity (v_x, v_y, v_z)]. The momentum components of the particle are

$$p_x = M v_x, \qquad p_y = M v_y, \qquad p_z = M v_z. \qquad (3\text{–}3)$$

The total energy of the particle, which equals the kinetic energy, is

$$E = \frac{M}{2} (v_x^2 + v_y^2 + v_z^2). \qquad (3\text{–}4)$$

3–2 The general solution in quantum mechanics. As the free particle is specified by a potential equal to zero, its quantum-mechanical equation of motion is obtained by setting the potential U equal to zero in the time-dependent Schrödinger equation,

$$\nabla^2\Psi + \frac{2Mi}{\hbar}\frac{\partial\Psi}{\partial t} = 0. \tag{3-5}$$

The general solution of Eq. (3–5), according to Eq. (2–85), is a superposition of monochromatic waves [Eq. (3–6)] the space-dependent parts of which satisfy the time-independent Schrödinger equation [Eq. (3–7)]:

$$\Psi = \sum_E c_E \psi_E(x, y, z)e^{-i(E/\hbar)t}, \qquad E = \hbar\omega, \tag{3-6}$$

$$\nabla^2\psi_E(x, y, z) + \frac{2M}{\hbar^2}E\psi_E(x, y, z) = 0. \tag{3-7}$$

A solution of Eq. (3–7) is readily found by substitution,

$$\psi_E(x, y, z) = e^{(i/\hbar)(p_x x + p_y y + p_z z)}, \tag{3-8}$$

where p_x, p_y, p_z are three arbitrary constants satisfying the following equation:

$$E = \frac{1}{2M}(p_x^2 + p_y^2 + p_z^2). \tag{3-9}$$

Thus we have obtained a set of solutions of the time-independent Schrödinger equation: a set of plane waves; their frequency ν and wave number vector $\vec{\sigma}$ are given by

$$\nu = \frac{E}{h}; \qquad \sigma_x = \frac{p_x}{h}, \qquad \sigma_y = \frac{p_y}{h}, \qquad \sigma_z = \frac{p_z}{h}. \tag{3-10}$$

Comparing Eq. (3–10) with the de Broglie relations, we see that the constants p_x, p_y, and p_z may be identified with the momentum components, and Eq. (3–9) is consistent with it. Substituting Eq. (3–8) in the right-hand side of Eq. (3–6), we have the general solution in the form of a superposition of plane waves, their dispersion equation being given by Eq. (3–10) and Eq. (3–9).

Since the normalization condition determines the acceptable solutions, we now consider its consequence in the present case. At first look, the solution Eq. (3–8) cannot be normalized because

$$\iiint_\tau \psi_E^* \psi_E \, d\tau = \iiint_\tau d\tau. \tag{3-11}$$

When the volume τ increases to infinity, the value of the normalization integral becomes infinite. However, if we are willing to confine the free particle in a large but finite volume Ω, then the normalization integral equals Ω and the normalization constant has a finite value $1/\sqrt{\Omega}$. This unrealistic situation will approach the true situation if we allow Ω to approach infinity. The significant point is that many important results obtained by using such normalized wave functions turn out to be independent of Ω and thus remain valid even when Ω becomes infinite (see applications in Sections 11–3 and 11–4). Thus the procedure of limiting the free particle in a finite volume Ω serves the purpose of completing many important calculations and therefore is perfectly legitimate. One important observation is that although the normalization integral of a plane wave diverges, a wave packet formed by many plane waves is normalizable, because it vanishes at infinity. Thus the general solution may well be regarded as a superposition of many wave packets. Along this line of thought, in fact, a different method of normalization for the free-particle wave functions has been devised, which will not be discussed here.*

The normalization of plane wave functions is justified from a physical consideration. A particle the probability distribution of which is everywhere the same is perfectly possible from the physical point of view. The wave function corresponding to this distribution, i.e., the plane wave, therefore should be considered acceptable.

According to the present normalization procedure, all plane wave solutions of Eq. (3–7) are acceptable, there being no restriction placed on the value of E so long as it is positive. Thus the energy E may take any numerical value. In other words, the energy of a free particle is not quantized.

3–3 Classical mechanics as a special case of quantum mechanics.
For simplicity we limit ourselves to one-dimensional motion. Our purpose is to find from the general solution a special solution which reproduces the results of classical mechanics and thereby verifies the general theorem of Chapter 2 for the special case of a free particle.

The general solution for one-dimensional motion is a general superposition of one-dimensional plane waves,

$$\Psi(x, t) = \int_{-\infty}^{\infty} A(\sigma_x)e^{2\pi i(\sigma_x x - \nu t)}\, d\sigma_x. \tag{3–12}$$

The summation is replaced by an integral, since the wave number σ can take any value from $-\infty$ to ∞. This results from the fact that the energy value E is not quantized. Equation (3–9), taking the following form in

---

* An exhaustive discussion on the normalization of plane wave functions is not imperative or pedagogically profitable at this stage.

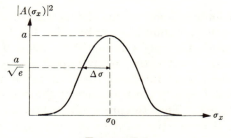

FIGURE 3–1

the one-dimensional case,

$$E = \frac{1}{2M} p_x^2,$$ (3–13)

shows that the value of p_x can be any from $-\infty$ to ∞. The same is true for the value of σ_x, since

$$\sigma_x = \frac{p_x}{h}.$$ (3–14)

We consider a particular solution for which $A(\sigma_x)$, the *spectral distribution function*, is a Gaussian function,

$$A(\sigma_x) = ae^{-\alpha(2\pi\sigma_x - 2\pi\sigma_0)^2},$$ (3–15)

α being a constant. Figure 3–1 shows the probability distribution $|A(\sigma_x)|^2$ corresponding to Eq. (3–15). This particular solution represents a wave packet composed of waves with wave number σ restricted in the immediate neighborhood of a fixed value σ_0. The range of distribution $\Delta\sigma$, defined by the standard deviation $\Delta\sigma$ of the Gaussian distribution $[(1/\sqrt{2\pi}\,\Delta\sigma)e^{-(\sigma-\sigma_0)^2/2(\Delta\sigma)^2}]$, is given by

$$\Delta\sigma = \frac{1}{4\pi\sqrt{\alpha}}.$$ (3–16)

In carrying out the integration in Eq. (3–12) we remember that frequency ν is related to energy E and thus to p_x and σ_x through Eq. (3–13) and Eq. (3–14). Therefore,

$$2\pi i\nu t = it\frac{\hbar}{2M}(2\pi\sigma_x)^2.$$ (3–17)

A new variable y is introduced for convenience:

$$y = 2\pi\sigma_x - 2\pi\sigma_0.$$ (3–18)

The integration is performed by making a perfect square in the exponent

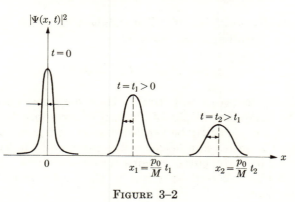

FIGURE 3-2

after which we may make use of the known result of the Gaussian integral:

$$\Psi(x, t) = \int_{-\infty}^{\infty} a e^{-\alpha(2\pi\sigma_x - 2\pi\sigma_0)^2 + 2\pi i(\sigma_x x - \nu t)} \, d\sigma_x$$

$$= \frac{a}{2\pi} e^{2\pi i \sigma_0 x} \int_{-\infty}^{\infty} e^{-\alpha y^2 + ixy - (i\hbar t/2M)(y + 2\pi\sigma_0)^2} \, dy$$

$$= \frac{a}{2\pi} e^{2\pi i(\sigma_0 x - \nu_0 t)} \int_{-\infty}^{\infty} e^{-[\alpha + (i\hbar t/2M)]y^2 + i[x - (2\pi\hbar t/M)\sigma_0]y} \, dy$$

$$= \frac{a}{2\pi} e^{2\pi i(\sigma_0 x - \nu_0 t)} e^{-[x - (\hbar\sigma_0/M)t]^2/4[\alpha + (i\hbar/2M)t]}$$

$$\times \int_{-\infty}^{\infty} e^{-\{\sqrt{\alpha + (i\hbar/2M)t}\, y + i[x - (\hbar\sigma_0/M)t]/2\sqrt{\alpha + (i\hbar/2M)t}\}^2} \, dy$$

$$= \frac{a}{2\pi} e^{2\pi i(\sigma_0 x - \nu_0 t)} \sqrt{\frac{\pi}{\alpha + (i\hbar t/2M)}}\, e^{-[x - (\hbar\sigma_0/M)t]^2/4[\alpha + (i\hbar t/2M)]}.$$

$$(3\text{-}19)$$

Define p_0 by

$$p_0 = \hbar\sigma_0. \qquad (3\text{-}20)$$

The probability density function is then

$$|\Psi(x, t)|^2 = \frac{a^2}{4\pi} \frac{1}{\sqrt{\alpha^2 + (\hbar^2 t^2/4M^2)}}\, e^{-[x - (p_0/M)t]^2/2[\alpha + (\hbar^2 t^2/4\alpha M^2)]}. \quad (3\text{-}21)$$

This function is plotted in Fig. 3-2 for three values of the time t. At a given time, $|\Psi(x, t)|^2$ gives a Gaussian curve centered at a point x equal to $(p_0/M)t$. As time goes on, this center moves along the x-axis with a velocity p_0/M. Thus the wave packet moves like a classical particle moving with a constant velocity p_0/M. The standard deviation of the position distribution is, according to Eq. (3–21),

$$\Delta x = \sqrt{[\alpha + (\hbar^2 t^2/4\alpha M^2)]}, \qquad (3\text{–}22)$$

which gradually increases with time. Thus the wave packet tends to spread. The maximum height of $|\Psi(x, t)|^2$, according to Eq. (3–21), decreases with time. It may be verified that as time goes on the increase in width and the decrease in height of the curve occur in such a manner that the area under the curve remains constant in time (Problem 3–2). This result is also expected by virtue of the normalization property of the wave function.

After having established the kinematical equivalence of the results of quantum mechanics and classical mechanics by making use of Born's first assumption, we want to establish the dynamical equivalence by Born's second assumption. According to the second assumption the wave packet represents a particle having probable momentum values equal to the values of $h\sigma_x$ of the component waves with a probability distribution given by the squares of the amplitudes of the component waves. Therefore the spectral distribution in Fig. 3–1 represents the momentum distribution which is centered around the value $h\sigma_0$, or p_0. If the distribution width $\Delta\sigma_x$ is small enough (the value of α is large enough), we may roughly say that the particle has a definite momentum p_0. This momentum corresponds to a velocity p_0/M, and we expect the particle to move with such a velocity. That the wave packet actually moves with a velocity p_0/M establishes the equivalence of quantum mechanics and classical mechanics.

The momentum spread of this particle is

$$\Delta p_x = h\,\Delta\sigma_x = h\,\frac{1}{4\pi\sqrt{\alpha}}. \qquad (3\text{–}23)$$

At the time $t = 0$, the position spread is smallest. Its value, according to Eq. (3–22), is

$$\Delta x = \sqrt{\alpha}. \qquad (3\text{–}24)$$

Therefore we have

$$\Delta p_x\,\Delta x \geqq \frac{\hbar}{2}, \quad \text{for} \quad t \geqq 0, \qquad (3\text{–}25)$$

and the uncertainty relation is verified (see also Problem 2–6). The free particle described by a wave packet in quantum mechanics thus cannot have exact values of both position and momentum, and a complementary relation exists between the uncertainties of these two quantities. The spread in momentum results in a spread in velocity,

$$\Delta v_x = \frac{\Delta p_x}{M} = \frac{\hbar}{2\sqrt{\alpha}\,M}.$$ (3–26)

As time goes on, this spread in velocity causes a spread in position by the amount $t\,\Delta v_x$ which equals $\hbar t/2\sqrt{\alpha}M$. On the other hand, when t is large, Eq. (3–22) gives a spread in position also equal to $\hbar t/2\sqrt{\alpha}M$, consistent with the above result.

The momentum distribution determines the energy distribution, since the two are related by Eq. (3–13). Thus the spectral distribution function determines the energy distribution through the relation $E = h\nu$.

3–4 Explanation of a quantum phenomenon: the wave property of matter. We return to the general solution of Section 3–2, i.e.,

$$\Psi(x, y, z, t) = \sum_{p_x, p_y, p_z, E} a_{p_x, p_y, p_z, E}\, e^{i/\hbar(p_x x + p_y y + p_z z - E t)}.$$ (3–27)

The quantities E, p_x, p_y, p_z in this expression are integration constants. They determine the frequencies and wave numbers of the monochromatic plane waves. By the de Broglie relations we identify the parameter E as the energy and p_x, p_y, p_z as the momentum components. The relative probability of finding the particle to have energy E and momentum components p_x, p_y, p_z is $|a_{p_x, p_y, p_z, E}|^2$.

The particular solution discussed in Section 3–3 leads to results identical with those of classical mechanics. In this section we shall discuss another particular solution which describes a quantum phenomenon. This particular solution is a monochromatic wave, obtained by letting $a_{p_x, p_y, p_z, E}$ equal unity for a particular set of indices p_{x0}, p_{y0}, p_{z0}, E_0, and equal zero for all others.

The physical meaning of this solution may be obtained by Born's two assumptions. According to the second assumption, this wave function represents a particle with definite values of energy and momentum, i.e., E_0 and p_{x0}, p_{y0}, p_{z0}. According to the first assumption, it represents a particle, the probability of finding which at a given point in space is everywhere and always the same, since $|\Psi(x, y, z, t)|^2$ is a constant. In classical physics, we have never encountered a particle identified by such a description. However, in Chapter 6, we shall discuss why such a description is allowed and even required in quantum mechanics. (In Chapter 6 we shall point out that once the momentum value of a particle is *exactly* known, the position can no longer be exactly located. This result has a negligible effect on particles of macroscopic size, hence we were unaware of its existence. However, it becomes important in microscopic physics and it is consistent with the description given by a mono-chromatic wave.)

The fact that the probability density $|\Psi(x, y, z, t)|^2$ is constant in space and time gives rise to an illusion that the particle is not moving at all. This is not so, because the particle has momentum and energy. The motion of such a particle may be made evident if we consider the proba-bility current density,

$$\vec{j} = \frac{\hbar}{2Mi} (\Psi^* \nabla \Psi - \Psi \nabla \Psi^*) = \frac{\vec{p}}{M} \Psi^* \Psi. \tag{3-28}$$

Equation (3–28) may be interpreted to mean that each probability density element $\Psi^* \Psi \, d\tau$ is moving with a velocity $\vec{p}/M$, i.e., the actual velocity of the particle calculated from its momentum. Therefore, instead of the classical picture of a single point moving with a velocity v, the mono-chromatic wave depicts a motion of the whole probability density distri-bution with a uniform velocity v. Since the initial density distribution is uniform, such a uniform motion leads to a new density distribution iden-tical to the previous one. The situation resembles the flow of a stationary hydrodynamical current; the density remains constant in time while the current flows continuously.

The above discussion makes it evident that a monochromatic plane wave is a natural representation of a uniform beam of particles all having the same energy. In many modern experiments, electrons or other particles are generated from a source and then accelerated by a potential to produce a uniform beam of a given energy. Each of the electrons, being a particle having definite energy and momentum, has its probability distribution of position described by a monochromatic wave. For a large number of electrons in a beam, the probability distribution of one electron gives the actual density distribution of all electrons, provided the beam is uniform. Therefore quantum-mechanical results may be interpreted in terms of the actual density of electrons instead of the probability density of a single electron, when a beam is represented by a monochromatic plane wave. Such results may be compared directly with the experimental results to verify the theory.

We may imagine an experiment in which a beam of electrons is imping-ing on a single slit, with a screen placed behind to record the arrival of the electrons. If such an experiment were ever possible (if we could find a slit narrow enough to make it feasible), we would observe the electrons falling on the screen with a density distribution exactly the same as the single-slit diffraction pattern of light with a wavelength given by the de Broglie relation. Since the behavior of the electron beam in this case is the same as that of a light wave, and diffraction phenomena were thought to be explainable only in wave theory, a wave property is ascribed to the electron beam. (Actually the wave property of electrons is established experimentally by passing electrons through crystals, which may be regarded as gratings of extremely narrow spacings.) The discovery of the wave property of electrons poses a fundamental problem: Are electrons particles or waves? The answer to this problem by quantum mechanics is as follows.

The motion of an electron as a particle is described by a wave function. Before reaching the slit, the wave function is a monochromatic plane wave; after passing through the slit, the wave function has to satisfy the time-dependent Schrödinger equation as well as the boundary conditions on the diaphragm and over the slit. Let us find such a solution having a frequency equal to the original one. The space-dependent part of this wave function must satisfy

$$\nabla^2 \psi(x, y, z) = -(2\pi\sigma)^2 \psi(x, y, z). \tag{3-29}$$

This is the same as the differential equation for the diffraction of light. The boundary conditions are also the same as in the diffraction of light by a single slit. Therefore the wave function $\psi(x, y, z)$ behind the slit will be the same as the light wave amplitude behind the slit and $|\psi(x, y, z)|^2$ the same as the light intensity. Since $|\psi(x, y, z)|^2$ represents the electron density on the screen, the above argument shows that the electron density is the same as the diffraction pattern of light of wave number σ. Thus the experimental result of electron diffraction (electrons behave like a wave of wavelength $1/\sigma$ specified by the de Broglie relation) is theoretically derived from the equation of motion of the electron in quantum mechanics.

The purpose of a physical theory is to predict correctly experimental results by the equation of motion which represents the fundamental physical law. To this end quantum mechanics has succeeded in "explaining" the wave property of matter, for the latter follows the Schrödinger equation as a necessary consequence. To the question whether the electron is a particle or a wave, we may say that the electron is a particle whose motion is governed by the Schrödinger equation. The wave property of the electron is just one manifestation of the Schrödinger equation. It is important to note that no *physical wave* of any sort is introduced here to explain a *wave phenomenon*. The emphasis is placed on an explanation of the distribution pattern instead of on the wave which causes it. The monochromatic wave $\Psi(x, y, z, t)$ is not a real wave, not a physical entity. At best we may call it a probability wave which has only an abstract meaning.

From a purely experimental point of view the particle and the wave aspects of matter (and also radiation) seem to be on an equal footing. *Quantum mechanics of a particle* discussed here takes an attitude stressing the particle aspect of matter; the symmetry of the wave-particle duality is somewhat distorted. This attitude is reversed in the *quantum theory of wave fields*, which is outside the scope of this book (see Section 12–12 for a very brief description).

The electron diffraction theory discussed in this section is useful for pedagogical purposes only. To obtain a theory able to produce experimentally verifiable results, we have to start from the scattering theory (see perturbation theory in Chapter 11). The experimentally observed electron diffraction pattern is a result of superposition of all scattered waves by all atoms in a crystal through which the electron beam passes.

Many conclusions obtained in this chapter are expected from the general considerations of the last chapter (for example, the motion of the wave packet and the uncertainty relation). We elaborate them in this special case to illustrate the general principles of Chapter 2 and to familiarize the reader with some of the typical mathematical operations.

PROBLEMS

3–1. Consider the planet Mars as a particle, the motion of which is described by a wave packet. If Mars were free from the gravitational force of the sun, how much would the width of its wave packet increase in the period from its birth about 5 billion years ago to the present time? Assume that the original width at its birth is 1 cm (a fantastically small quantity in astronomy). The mass of Mars is 6.58×10^{26} gm.

3–2. Verify that the areas under the curves in Fig. 3–2 are the same.

3–3. What are the quantum-mechanical descriptions of the following phenomena which, in classical theory, must be described in terms of conflicting theories? (1), An α-particle produces a straight track in a Wilson cloud chamber. (2), A beam of α-particles passing through a crystal produces a diffraction pattern on a screen. *Note:* A detailed discussion of the cloud chamber tracks cannot be made here; compare with Problem 6–6 and Problem 12–11 later.

CHAPTER 4

THE LINEAR HARMONIC OSCILLATOR

As the second example we consider the one-dimensional harmonic oscillator. After reviewing the results of classical mechanics we proceed to find the general solution in quantum mechanics. Once again a special solution representing a wave packet may be found which exhibits the classical properties of the oscillator. On the other hand, this example shows that quantum mechanics is able to account for another quantum phenomenon, the quantization of energy.

4–1 The general solution in classical mechanics. The linear harmonic oscillator is specified by a potential proportional to the square of the displacement,

$$U = \tfrac{1}{2}kx^2, \qquad (4\text{–}1)$$

where k is the force constant. The classical equation of motion is thus

$$M \frac{d^2x}{dt^2} = -kx, \qquad (4\text{–}2)$$

the general solution of which may be written, where the arbitrary phase constant is set equal to zero for simplicity,

$$x = A \cos\left(\sqrt{(k/M)}\, t\right). \qquad (4\text{–}3)$$

The angular frequency of oscillation, independent of the amplitude A, is

$$\omega = \sqrt{k/M}. \qquad (4\text{–}4)$$

The kinetic energy at any time t is

$$K = \frac{1}{2} M \dot{x}^2 = \frac{k}{2} A^2 \sin^2\left(\sqrt{\frac{k}{M}}\, t\right). \qquad (4\text{–}5)$$

The total energy is a constant independent of time,

$$E = K + U = \tfrac{1}{2}kA^2. \qquad (4\text{–}6)$$

4–2 The general solution in quantum mechanics. The equation of motion for the linear harmonic oscillator is the time-dependent Schrödinger equation with the potential function specified by Eq. (4–1). Its general solution, according to Eq. (2–85), may be written as follows:

$$\Psi(x, y, z, t) = \sum_E c_E \, \psi_E(x, y, z) e^{-i(E/\hbar)t} \qquad (4\text{–}7)$$

where c_E is a constant coefficient and $\psi_E(x, y, z)$ satisfies the time-inde-pendent Schrödinger equation

$$\frac{d^2}{dx^2} \, \psi_E(x, y, z) + \frac{2M}{\hbar^2} \left(E - \frac{1}{2} \, kx^2 \right) \psi_E(x, y, z) = 0. \qquad (4\text{–}8)$$

The Laplacian ∇^2 reduces to one term, as this is a one-dimensional prob-lem; ψ_E is thus a function of x only. The task before us is to find solutions of Eq. (4–8) which satisfy the normalization condition.

To simplify the mathematical form of Eq. (4–8) we introduce a new variable ξ and a new constant ϵ defined as follows:

$$\xi = \sqrt{(M\omega/\hbar)} \; x, \qquad (4\text{–}9)$$

$$\epsilon = \frac{2E}{\hbar\omega}. \qquad (4\text{–}10)$$

Equation (4–8) thus takes the form:

$$\frac{d^2\psi}{d\xi^2} + (\epsilon - \xi^2)\psi = 0. \qquad (4\text{–}11)$$

In a region far away from the origin, the constant ϵ is much smaller than ξ^2 (which increases as x^2) and may be neglected. The resulting equa-tion,

$$\frac{d^2\psi_\infty}{d\xi^2} - \xi^2\psi_\infty = 0,$$

thus determines the asymptotic behavior of ψ at infinity. An approximate solution of this equation for large values of ξ is found by substitution:

$$\psi_\infty = e^{-\xi^2/2}. \qquad (4\text{–}12)$$

When substituting and carrying out the differentiation, a term $-e^{-\xi^2/2}$ appears which may be neglected in comparison with $\xi^2 e^{-\xi^2/2}$. Another solution, $\psi_\infty = e^{+\xi^2/2}$, is excluded because it diverges at infinity. Equa-tion (4–12) gives the limit to which ψ approaches as x goes to infinity. It is then reasonable to assume that the complete solution ψ takes the following form:

$$\psi = v(\xi)e^{-\xi^2/2}, \qquad (4\text{–}13)$$

where $v(\xi)$ is a function of ξ, presumably of a simpler structure than $\psi(\xi)$.

Substituting Eq. (4–13) in Eq. (4–11), we obtain the following differential equation for $v(\xi)$:

$$\frac{d^2v}{d\xi^2} - 2\xi \frac{dv}{d\xi} + (\epsilon - 1)v = 0. \tag{4–14}$$

This equation may be solved by the power series method. Let the solution $v(\xi)$ be written in a power series

$$v(\xi) = a_0 + a_1\xi + a_2\xi^2 + \cdots + a_n\xi^n + \cdots. \tag{4–15}$$

We have

$$\frac{d^2v}{d\xi^2} = 2 \cdot 1 \cdot a_2 + 3 \cdot 2 \cdot a_3\xi + \cdots + (n+2)(n+1)a_{n+2}\xi^n + \cdots, \tag{4–16}$$

$$-2\xi \frac{dv}{d\xi} = -2a_1\xi - \cdots - 2na_n\xi^n - \cdots, \tag{4–17}$$

$$(\epsilon - 1)v = (\epsilon - 1)a_0 + (\epsilon - 1)a_1\xi + \cdots + (\epsilon - 1)a_n\xi^n + \cdots. \tag{4–18}$$

In order to satisfy Eq. (4–14) the sum of the above three series must be zero, which is possible only when the coefficients of all powers of ξ in the sum vanish. Thus

$$\left.\begin{aligned} 2a_2 + (\epsilon - 1)a_0 &= 0 \\ 3 \cdot 2a_3 + (\epsilon - 3)a_1 &= 0 \\ \cdots \cdots \cdots \cdots \cdots \cdots \\ (n+2)(n+1)a_{n+2} + (\epsilon - 2n - 1)a_n &= 0 \end{aligned}\right\}. \tag{4–19}$$

The last equation is called the *recursion formula*, from which a_n may be calculated from a_{n-2}. By successive application of the recursion formula we may obtain all coefficients a_n, once a_0 and a_1 are known. The results are

$$a_n = \begin{cases} \dfrac{(1 - \epsilon)(5 - \epsilon) \cdots (2n - 3 - \epsilon)}{n!} \, a_0, & \text{for } n \text{ even,} \\[3mm] \dfrac{(3 - \epsilon)(7 - \epsilon) \cdots (2n - 3 - \epsilon)}{n!} \, a_1, & \text{for } n \text{ odd.} \end{cases} \tag{4–20}$$

The solution $v(\xi)$ thus may be written

$$v(\xi) = a_0 \left(\sum_{n \text{ even}} \frac{(1 - \epsilon)(5 - \epsilon) \cdots (2n - 3 - \epsilon)}{n!} \xi^n \right)$$

$$+ a_1 \left(\sum_{n \text{ odd}} \frac{(3 - \epsilon)(7 - \epsilon) \cdots (2n - 3 - \epsilon)}{n!} \xi^n \right). \tag{4–21}$$

Since a solution containing two arbitrary constants is the general solution of a differential equation of the second order, Eq. (4–21) is the general solution of Eq. (4–14). This solution contains two series, each containing terms increasing by ξ^2. The ratio of the coefficients of two successive terms of either series, according to the recursion formula, takes the following limiting form when n is large:

$$\frac{a_{n+2}}{a_n} \to \frac{2}{n}. \qquad (4\text{–}22)$$

Consider the well-known series

$$e^{\xi^2} = 1 + \xi^2 + \frac{\xi^4}{2!} + \cdots + \frac{\xi^n}{(n/2)!} + \frac{\xi^{n+2}}{[(n/2)+1]!} + \cdots,$$

which also contains terms increasing by ξ^2. The ratio of the coefficients of successive terms approaches the following limit when n is large:

$$\frac{a_{n+2}}{a_n} \to \frac{2}{n}. \qquad (4\text{–}23)$$

As Eq. (4–23) is identical with Eq. (4–22), we conclude that both series of Eq. (4–21) behave like e^{ξ^2} when ξ approaches infinity. Therefore the general solution of ψ, according to Eq. (4–13), behaves at infinity like the function $e^{+\xi^2/2}$, and thus diverges. As a result, the general solution cannot be normalized and is not acceptable as a solution of the Schrödinger equation.

However, the general solution, Eq. (4–21), includes some special solutions, for which the infinite series happens to terminate at a certain term and becomes a polynomial. The corresponding ψ, being the product of a polynomial and the factor $e^{-\xi^2/2}$, approaches zero rapidly enough at infinity and thus can be normalized. The termination of one of the two series happens when one of the coefficients a_n equals zero, which makes all the following coefficients, a_{n+2}, a_{n+4}, ..., vanish in accordance with the recursion formula. Such a situation may arise when ϵ assumes a certain particular value, since the value of the coefficient a_n is dependent on ϵ. Actually, from the recursion formula we learn that if

$$\epsilon = 2n + 1, \qquad n = 0, 1, 2, \ldots, \qquad (4\text{–}24)$$

the coefficient a_{n+2} will be zero and one of the series will terminate at the term a_n. Since ϵ is related to the energy E, Eq. (4–24) imposes a condition on E. The condition is that the energy E must take on one of the following values:

$$E_n = \hbar\omega(n + \tfrac{1}{2}), \qquad n = 0, 1, 2, \ldots. \qquad (4\text{–}25)$$

When E equals one of the above values E_n, one of the two series of Eq. (4–21) becomes a polynomial. By setting the coefficient (a_0 or a_1) of the other series equal to zero we obtain a particular solution of Eq. (4–21) in the form of a polynomial. These polynomials are denoted by the notation $H_n(\xi)$. A few of them are listed below:

$$H_0(\xi) = 1$$
$$H_1(\xi) = 2\xi$$
$$H_2(\xi) = 4\xi^2 - 2 \qquad (4\text{–}26)$$
$$H_3(\xi) = 8\xi^3 - 12\xi$$
$$H_4(\xi) = 16\xi^4 - 48\xi^2 + 12$$

$$\cdot \quad \cdot \quad \cdot \quad \cdot \quad \cdot \quad \cdot \quad \cdot \quad \cdot \quad \cdot \quad \cdot$$

For each polynomial $H_n(\xi)$, Eq. (4–13) gives one solution of Eq. (4–8), $\psi_n(x)$, labeled by the index n corresponding to the energy value E_n. The acceptable particular solutions are to be found only among the polynomials $H_n(\xi)$. The general solution Eq. (4–7), in order to be normalizable, must therefore contain only terms corresponding to these polynomial solutions, i.e.,

$$\Psi(\xi, t) = \sum_{n=0}^{\infty} c_n H_n(\xi) e^{-\xi^2/2} e^{-i(E_n/\hbar)t}. \qquad (4\text{–}27)$$

Transforming the variable ξ back to the variable x, we express the general solution thus:

$$\Psi(x, t) = \sum_{n=0}^{\infty} c_n H_n(\sqrt{(M\omega/\hbar)}\, x) e^{-(M\omega/2\hbar)x^2} e^{-i\omega[n+(1/2)]t}. \qquad (4\text{–}28)$$

The polynomials $H_n(\xi)$ are known as the *Hermite polynomials*, the mathematical properties of which will not be discussed in detail here. A few important results will be stated without proof as follows.

1. The explicit form of the Hermite polynomials may be obtained from either

$$H_n(\xi) = (-1)^n e^{\xi^2} \frac{d^n}{d\xi^n} e^{-\xi^2} \qquad (4\text{–}29)$$

or

$$e^{\xi^2 - (s-\xi)^2} = \sum_{n=0}^{\infty} \frac{H_n(\xi)}{n!} s^n. \qquad (4\text{–}30)$$

The left-hand side of Eq. (4–30) is called the *generating function* of the Hermite polynomials.

2. The Hermite polynomials satisfy the following recursion formulas:

$$\frac{dH_n(\xi)}{d\xi} = 2nH_{n-1}(\xi), \tag{4–31}$$

and

$$H_{n+1}(\xi) = 2\xi H_n(\xi) - 2nH_{n-1}. \tag{4–32}$$

3. They give rise to the following result:

$$\int_{-\infty}^{\infty} [H_n(\xi)e^{-\xi^2/2}]^2 \, d\xi = \sqrt{\pi}\, 2^n n!. \tag{4–33}$$

This equation helps us evaluate the normalization integral,

$$\int_{-\infty}^{\infty} \psi_n^*(x)\psi_n(x) \, dx = \sqrt{(\hbar/M\omega)}\, \sqrt{\pi}\, 2^n n!.$$

The normalized time-independent wave function $\psi_n(x)$ is thus

$$\psi_n(x) = \left(\frac{M\omega}{\hbar}\right)^{1/4} \frac{1}{(\sqrt{\pi}\, 2^n n!)^{1/2}} \, H_n(\sqrt{(M\omega/\hbar)}\, x)e^{-(M\omega/2\hbar)x^2}. \tag{4–34}$$

A few of the wave functions are represented graphically in Fig. 4–1.

We now consider the *orthogonal property* of the wave functions. The wave functions will be shown presently to satisfy the following equation:

$$\int_{-\infty}^{\infty} \psi_n\psi_m \, dx = 0, \qquad n \neq m. \tag{4–35}$$

Functions satisfying Eq. (4–35) are said to be *orthogonal* to each other. If also normalized, a set of functions, such as that represented by Eq. (4–34), is called an *orthonormal set*. Orthogonality is a general property of the solutions of the wave equation. Equation (4–35) is a special case. To prove it we first write the differential equations for two solutions ψ_n and ψ_m:

$$\frac{d^2\psi_n}{d\xi^2} + (\alpha_n - \xi^2)\psi_n = 0, \tag{4–36}$$

$$\frac{d^2\psi_m}{d\xi^2} + (\alpha_m - \xi^2)\psi_m = 0. \tag{4–37}$$

Multiply Eq. (4–36) by ψ_m and Eq. (4–37) by ψ_n and then take the difference. The result is

$$\psi_m \frac{d^2\psi_n}{d\xi^2} - \psi_n \frac{d^2\psi_m}{d\xi^2} + (\alpha_n - \alpha_m)\psi_n\psi_m = 0. \tag{4–38}$$

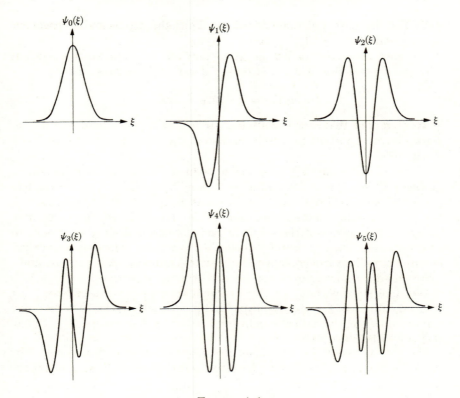

FIGURE 4–1

Integration over ξ from $-\infty$ to ∞ gives the following result:

$$\int_{-\infty}^{\infty} \frac{d}{d\xi} \left(\psi_m \frac{d\psi_n}{d\xi} - \psi_n \frac{d\psi_m}{d\xi} \right) d\xi + (\alpha_n - \alpha_m) \int_{-\infty}^{\infty} \psi_n \psi_m \, d\xi = 0, \qquad (4\text{–}39)$$

or

$$\left[\psi_m \frac{d\psi_n}{d\xi} - \psi_n \frac{d\psi_m}{d\xi} \right]_{-\infty}^{\infty} + (\alpha_n - \alpha_m) \int_{-\infty}^{\infty} \psi_n \psi_m \, d\xi = 0. \qquad (4\text{–}40)$$

Since ψ_n and ψ_m vanish at infinity, the first term reduces to zero. Thus,

$$\int_{-\infty}^{\infty} \psi_n \psi_m \, d\xi = 0, \qquad \text{if} \quad \alpha_n \neq \alpha_m. \qquad (4\text{–}41)$$

When $\alpha_n = \alpha_m$, the integral may not be zero; in fact when $\psi_n = \psi_m$ the integral is always greater than zero. We note that the proof remains valid for any potential function $U(x)$. Thus the orthogonality property is a general property of the wave functions.

We note that $\psi_n(x)$ of Eq. (4–34) is an even function when n is even, and an odd function when n is odd. This may be expressed by

$$\psi_n(x) = (-1)^n\psi_n(-x).$$

Wave functions that are even are said to have *even parity;* odd, *odd parity. Parity* is an important property characterizing wave functions (see Problem 4–10).

A few terms commonly used are introduced here. The mathematical problem of solving a differential equation, like Eq. (4–11), which has permitted solutions only when a constant contained in it, such as ϵ in Eq. (4–11), assumes certain particular values, is called a *characteristic value problem.* The values of the constant and the corresponding solutions are called *characteristic values* and *characteristic functions.* They are also called *eigenvalue problems, eigenvalues,* and *eigenfunctions.* A quantum state represented by an eigenfunction is said to be an *eigenstate.*

We shall state without proof a theorem which will be used later. An arbitrary function $f(\xi)$ may be expanded in a series of $\psi_n(\xi)$ (we shall not elaborate the mathematical conditions restricting $f(\xi)$ except to say that $f(\xi)$ is well-behaved):

$$f(\xi) = \sum_{n=0}^{\infty} c_n\psi_n(\xi), \tag{4–42}$$

where the coefficients of expansion may be evaluated as follows:

$$c_n = \int_{-\infty}^{\infty} f(\xi)\psi_n(\xi)\, d\xi. \tag{4–43}$$

Equation (4–43) is easily verified by multiplying Eq. (4–42) by $\psi_n(\xi)$ and then performing integration. The above theorem is a special case of a more general theorem* stating that an arbitrary function (well-behaved) may be expanded in a series of a *complete set of orthogonal functions.* It may be noted that the Fourier series expansion is a special case of this general theorem.

4–3 Classical mechanics as a special case of quantum mechanics. We shall verify the general theorem of Chapter 2 again for the special case of the oscillator not only for illustration but also for later application (see Section 12–11). Let us write the general solution of the Schrödinger equa-

---

* See, for example, Kaplan's *Advanced Calculus,* pp. 414–423. Addison-Wesley Publishing Company, Inc.

tion for the linear harmonic oscillator in terms of the normalized wave functions ψ_n,

$$\Psi(x, t)$$

$$= \sum_{n=0}^{\infty} c_n \psi_n(x) e^{-i(E_n/\hbar)t}$$

$$= \sum_{n=0}^{\infty} c_n \left\{ \left(\frac{M\omega}{\hbar} \right)^{1/4} \frac{1}{(\sqrt{\pi}\, 2^n n!)^{1/2}} H_n \left(\sqrt{\frac{M\omega}{\hbar}}\, x \right) e^{-M\omega x^2/2\hbar} \right\} e^{-i\omega[n+(1/2)]t},$$

$$(4\text{–}44)$$

where c_n is the coefficient of superposition. From the orthonormal property of ψ_n we find

$$\int_{-\infty}^{\infty} |\Psi(x, t)|^2 \, dx = \sum_{n=0}^{\infty} |c_n|^2. \qquad (4\text{–}45)$$

The normalization condition thus imposes a restriction on the values of c_n:

$$\sum_{n=0}^{\infty} |c_n|^2 = 1. \qquad (4\text{–}46)$$

The physical meaning of the general solution is specified by Born's two assumptions. According to the first assumption, $|\Psi(x, t)|^2$ represents the probability distribution of the position of the particle. The second assumption, as stated in Chapter 2, applies only to a superposition of plane waves; it has to be generalized for the interpretation of the other wave functions including the present one. As assumptions are justified by their consequences, we are free to generalize so long as the generalized assumption includes the previous one as a special case and produces results in agreement with experiments. The generalized version* may be stated as follows for the present purpose:

"In a superposition of wave functions such as in Eq. (4–44), the value of $|c_n|^2$ represents the probability of finding the particle, upon an act of observation, to have energy value equal to E_n."

When applied to plane waves this statement is equivalent to the previous one. Furthermore, Eq. (4–46) is consistent with the fact that the total probability is 1.

---

* The general statement of the Born assumptions is to be found in Section 12–4, which includes all previous statements as special cases. We begin with special cases and then generalize it step by step in the hope that the physical meaning may easily be brought out.

We now proceed to construct from Eq. (4–44) a special solution which, according to the interpretation by Born's two assumptions, reproduces all essential results of classical mechanics set forth in Section 4–1.

This particular solution is specified by the following coefficients of superposition:

$$c_n = c \frac{\alpha^n}{\sqrt{2^n n!}}, \qquad \alpha \gg 1, \qquad (4\text{–}47)$$

where c is a constant adjusted to satisfy the normalization condition Eq. (4–46), and α is a fixed number much greater than unity. Substituting Eq. (4–47) in Eq. (4–44) we simplify the expression by making use of the properties of the generating function expressed in Eq. (4–30):

$$\Psi(x, t) = \sum_{n=0}^{\infty} c \frac{\alpha^n}{\sqrt{2^n n!}} \left(\frac{M\omega}{\hbar}\right)^{1/4} \frac{1}{(\sqrt{\pi}\, 2^n n!)^{1/2}} H_n(\xi) e^{-\xi^2/2} e^{-i\omega[n+(1/2)]t}$$

$$= \frac{c}{\pi^{1/4}} \left(\frac{M\omega}{\hbar}\right)^{1/4} e^{-(\xi^2/2)-(i/2)\omega t} \sum_{n=0}^{\infty} \frac{H_n(\xi)}{n!} \left(\frac{\alpha}{2} e^{-i\omega t}\right)^n$$

$$= \frac{c}{\pi^{1/4}} \left(\frac{M\omega}{\hbar}\right)^{1/4} e^{-(\xi^2/2)-(i/2)\omega t} e^{\xi^2 - [(\alpha/2) e^{-i\omega t}-\xi]^2}$$

$$= \frac{c}{\pi^{1/4}} \left(\frac{M\omega}{\hbar}\right)^{1/4} e^{-(\xi^2/2)-(i/2)\omega t+\xi^2 - [(\alpha/2\cos\omega t-\xi)-i(\alpha/2)\sin\omega t]^2}$$

$$= \frac{c}{\pi^{1/4}} \left(\frac{M\omega}{\hbar}\right)^{1/4}$$
$$\times e^{-(\xi^2/2)+\xi\alpha\cos\omega t-(\alpha^2/4)\cos^2\omega t+(\alpha^2/4)\sin^2\omega t-i\{(\omega t/2)-\alpha\sin\omega t[(\alpha/2)\cos\omega t-\xi]\}}$$

$$= \frac{c}{\pi^{1/4}} \left(\frac{M\omega}{\hbar}\right)^{1/4} e^{-1/2(\xi-\alpha\cos\omega t)^2+(\alpha^2/4)} e^{-i\{(\omega t/2)-\alpha\sin\omega t[(\alpha/2)\cos\omega t-\xi]\}}.$$

$$(4\text{–}48)$$

The probability distribution of position is thus

$$|\Psi(x, t)|^2 = \text{constant } e^{-(\xi-\alpha\cos\omega t)^2}.$$

Introducing a number A by

$$\alpha = \sqrt{(M\omega/\hbar)}\; A,$$

we have

$$|\Psi(x, t)|^2 = \text{constant } e^{-(M\omega/\hbar)(x-A\cos\omega t)^2}, \qquad (4\text{–}49)$$

which is a Gaussian distribution function, centered upon a point the position of which is given by

$$x = A \cos \omega t. \tag{4–50}$$

Once again, we have a wave packet in the form of a Gaussian function, the center of which changes in time in exactly the same way as a classical linear harmonic oscillator, as Eq. (4–50) and Eq. (4–3) are identical. Equation (4–50) enables us to interpret the constant A introduced above as the amplitude of oscillation. The *Gaussian width* of the wave packet is easily found to be

$$\Delta x = \sqrt{\hbar/M\omega}. \tag{4–51}$$

For macroscopic systems, for example, a mass of one gram and a frequency of one cycle per second, Δx is of the order of magnitude of 10^{-14} cm, an extremely small quantity. The wave packet is therefore practically a point. Thus the kinematics of the classical theory is reproduced. However, for subatomic systems Δx may not be a negligible quantity.

While Born's first assumption enables us to obtain the *kinematical* information of a particle from its wave function, Born's second assumption enables us to obtain the *dynamical* information. Accordingly, the probability distribution of the energy value of the particle is given by

$$|c_n|^2 = c^2 \frac{[(M\omega/2\hbar) A^2]^n}{n!} \equiv c^2 \frac{(\alpha^2/2)^n}{n!}. \tag{4–52}$$

Since α is a number much greater than 1, $|c_n|^2$ increases with n if $n \ll \alpha^2/2$ but decreases if $n \gg \alpha^2/2$. Thus Eq. (4–52) represents a peaked distribution, the maximum of which may be found by differentiating $|c_n|^2$ with respect to n and setting the result equal to zero. In differentiating the factorial $n!$, we make use of the well-known Stirling formula

$$\ln n! = n \ln n - n, \qquad n \gg 1, \tag{4–53}$$

which is valid for large values of n.

An approximate derivation of this formula will be given here.

$$n! = n(n-1)!, \tag{4–54}$$

$$\ln n! - \ln (n-1)! = \ln n, \tag{4–55}$$

$$\frac{d}{dn} (\ln n!) = \ln n, \tag{4–56}$$

$$\ln n! = \int \ln n \, dn \tag{4–57}$$

$$= n \ln n - n, \qquad \text{Q.E.D.}$$

From Eq. (4–53) we find

$$\frac{d}{dn} n! = (\ln n)(n!), \qquad n \gg 1. \tag{4–58}$$

The maximum of $|c_n|^2$ is found, with the help of Eq. (4–58), to be specified by

$$n' = \frac{\alpha^2}{2} \equiv \frac{M\omega}{2\hbar} A^2. \tag{4–59}$$

The corresponding energy value may be obtained from Eq. (4–25). Since α is much greater than unity, n' must also be much greater than unity. Therefore the term $\frac{1}{2}$ in Eq. (4–25) may be neglected:

$$E_{n'} = \hbar\omega n'$$

$$= \hbar\omega \frac{M\omega}{2\hbar} A^2$$

$$= \tfrac{1}{2}M\omega^2 A^2$$

$$= \tfrac{1}{2}kA^2. \tag{4–60}$$

Born's second assumption thus leads to the conclusion that the most probable value of energy is the same as the classical value, as Eq. (4–60) and Eq. (4–6) are identical. The probability distribution of energy, like that of position, exhibits a spread around the most probable value. The width of the spread may be estimated as follows. We first determine the number n'' such that $|c_{n''}|^2$ is less than $|c_{n'}|^2$ by a factor of e, the base of natural logarithms. The difference $E_{n''} - E_{n'}$ is then the Gaussian width of the energy spread. We note that

$$|c_{n'+s}|^2$$

$$= |c_{n'}|^2 \frac{(n')^s}{(n'+1)(n'+2)\cdots(n'+s)}$$

$$= |c_{n'}|^2 \frac{1}{[1+(1/n')][1+(2/n')]\cdots[1+(s/n')]} \quad \text{for } s \ll n',\ n' \gg 1$$

$$\cong |c_{n'}|^2 \frac{1}{[1+(1/n')][1+(1/n')]^2\cdots[1+(1/n')]^s}$$

$$= |c_{n'}|^2 \frac{1}{[1+(1/n')]^{s(s+1)/2}}. \tag{4–61}$$

Use was made of the fact that $1/n'$ is much smaller than unity, so that the binomial expansion of $[1+(1/n')]^s$ may be approximated by $[1+(s/n')]$. When

$$\frac{s(s+1)}{2} = n', \tag{4–62}$$

the denominator in Eq. (4–61) approaches e. Thus Eq. (4–62) determines the value of s for which $|c_n|^2$ decreases by a factor of e. As $n' \gg 1$, the solution of Eq. (4–62) is

$$s \cong \sqrt{2n'}. \tag{4–63}$$

Defining α by

$$\alpha \equiv \sqrt{2n'},$$

we have

$$n'' = n' + \alpha. \tag{4–64}$$

The energy spread is thus

$$\Delta E = \hbar \omega \alpha$$

$$= \frac{2}{\alpha} E. \tag{4–65}$$

Since α is much larger than 1, the energy spread is a minute fraction of the total energy. The percentage error is thus small, and for practical purposes we may consider the energy as having one definite value given by Eq. (4–60).

The solution specified by Eq. (4–47) thus represents a particle having "well-defined" position and energy values, both of which obey the classical law. The spread in both position and energy, for macroscopic objects, is so small that it is beyond detection. This solution thus reproduces all the classical results, verifying that classical mechanics is a special case of quantum mechanics.

We emphasize that the constant α in this solution must be a number much greater than unity. Following this, n' must be a large number. Thus only in the region of high quantum numbers may quantum mechanics be reduced to classical mechanics.

Beginning students may be amazed at this moment by the beautiful display of mathematical operations leading to such elegant results. Inquisitive ones may ask, could it not be just a happy mathematical accident? What was going on backstage to make such a performance possible? To answer these questions, it may be pointed out that the results of this section may be obtained without any detailed knowledge of the mathematical properties of the Hermite polynomials. From the considerations of Section 2–7, we know that a localized wave function remains localized for some time and moves as a whole without spreading; the square of the amplitude of a wave function may be considered to represent the density distribution of a group of classical particles which move in a pack. According to the expansion theorem, an arbitrary function may be expanded in a series of wave functions. Thus a localized function around the point $x = A$ may be written as a series of ψ_n's. If in this series we attach to each ψ_n its time factor $e^{-i(E_n/\hbar)t}$, the resultant series will be a particular solution of the time-dependent Schrödinger equation. It represents a localized function around the point $x = A$ at time $t = 0$; furthermore as time goes on the function it represents remains localized but the point of localization moves back and forth like a classical particle. Thus Eq. (4–49) is expected from this qualitative consideration. It may be added parenthetically that the dominant wave functions in the expansion of the localized function at $x = A$ are those whose values near $x = A$ are large (see Fig. 4–3, below). By the so-called WKB method to be discussed in Chapter 7 it can be shown that in order that the maximum of $\psi_n(x)$ may be at $x = A$, n must be such that $E_n \cong \frac{1}{2}kA^2$. The quantum-mechanical energy information thus agrees with the classical energy value.

The mathematical complexity of the Hermite polynomials gives the impression that the solutions of the Schrödinger equation for the linear harmonic oscillator are drastically different from those of the free particle. In fact, a close kinship between them is discernible. A glance at Fig. 4–1 and Fig. 4–3 later shows the wave nature of the functions $\psi_n(\xi)$ when n is large. Actually, for large values of n the function $\psi_n(\xi)$ is appreciably different from zero in a large range of ξ. We may divide this range into small sections; within each of them the function ξ^2 varies only a little and $(E - \xi^2)$ may be regarded as constant. The solution of Eq. (4–11) within a small section near $x = x_0$ is thus a sinusoidal function $e^{i\sqrt{(2m/\hbar^2)[E_n - (k/2)x_0^2]}\,x}$. The fact that the function ξ^2 actually varies from section to section leads merely to the result that the sinusoidal wave is frequency-modulated as well as amplitude-modulated. Apart from such modulations, the behavior of ψ_n when n is large is essentially sinusoidal. Thus the ψ_n exhibits many of the properties of the plane waves discussed in Chapter 3. It is only natural that a superposition of a group of ψ_n, the n values of which are large and are limited within a narrow band, may result in a localized wave packet just as a narrow band of plane waves may. The point of phase agreement will move according to a velocity given by

$$
\begin{aligned}
v_g &= \frac{d\nu}{d\sigma} \\
&= \frac{d(E/\hbar)}{d\sqrt{(2M/\hbar^2)[E - (k/2)x_0^2]}} \\
&= \left(\frac{dE}{dp}\right)_{x=x_0} \\
&= v(x_0),
\end{aligned}
\tag{4–66}
$$

where p is the classical momentum and $v(x_0)$ is the classical velocity at x_0. Equation (4–66) demonstrates once again that the wave packet will move at x_0 with a velocity equal to the classical velocity at x_0. (From this consideration the theorem that an arbitrary function may be expanded in a series of wave functions may be regarded as a natural generalization of the Fourier theorem.) These general considerations apply not only to the linear harmonic oscillator but also to other problems specified by other kinds of potential function $U(x)$. No matter how complicated the solutions may be in a particular problem, it is generally possible to form a particular solution in the form of a wave packet which reproduces the classical results and thus describes the classical motion.

4–4 Explanation of a quantum phenomenon: the quantization of energy. The interpretation of the general solution, Eq. (4–44), based on Born's second assumption, implies that the energy value of an oscillator, when determined by an act of observation, cannot be arbitrary but must be one of the following values:

$$E_n = \hbar\omega(n + \tfrac{1}{2}), \qquad n = 0, 1, 2, \ldots . \tag{4–67}$$

The set of constants c_n in Eq. (4–44) specifies a *state of motion* (just as a set of initial conditions specifies a state of motion in classical mechanics). For different states of motion the probability distribution of energy is different, but the permitted energy values are limited to the above set. Thus the energy is quantized. The energy level scheme is shown in Fig. 4–2.

On the experimental side we consider one example: the vibrational motion of a diatomic molecule about its equilibrium position. In this motion the interatomic distance of the diatomic molecule increases and decreases periodically. For small vibrations the diatomic molecule behaves like a linear harmonic oscillator. The energies of the vibrations, or rather the differences among them, may be obtained by analyzing molecular spectra. It was found experimentally that the energy is quantized to equal intervals. The Wilson-Sommerfeld quantum condition leads to energy values $E_n = n\hbar\omega$. It gives the correct energy spacings between successive levels just as they are given in Eq. (4–67). However, it differs from Eq. (4–67) by leaving out the half integer $1/2$. It was found experimentally that in order to account for the observed isotope displacements

FIGURE 4–2

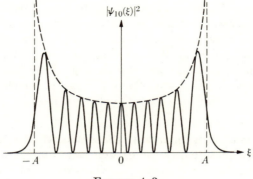

FIGURE 4–3

in the spectra of the diatomic molecules, the half integer is necessary.*
Therefore quantum mechanics leads to the correct rule of quantization
for the linear harmonic oscillator.

The rule of quantization is a derived result in quantum mechanics
instead of an arbitrary restriction imposed on an established theory as in
the old quantum theory. The origin of energy quantization may be traced
back to the normalization condition which requires the wave function
not to diverge at infinity. This condition is natural to assume, and it leads
to energy quantization in many other mechanical systems. Thus quantum
mechanics offers a systematic method for study of the general problem of
energy quantization.

We now consider a special case of the general solution Eq. (4–44) in
which the summation consists of only one term, ψ_n. According to Born's
second assumption, the particle described by this solution has unity
probability of having an energy value E_n, and zero probability for others;
therefore the particle has a definite energy E_n. According to Born's first
assumption, the probability distribution in position is given by $|\psi_n(x)|^2$.
For $n = 10$ this distribution is shown in Fig. 4–3. It is seen that the
probability distribution is not localized and has an exponential tail extending to infinity. This also brings out the fact that a certainty in the
energy information is accompanied with a large uncertainty in the position information. This is similar to the fact that a certainty in momentum
is accompanied with a large uncertainty in position according to the
uncertainty relation. The classical amplitude A of an oscillator whose
energy is E_{10} is calculated and marked out in Fig. 4–3 by dotted lines.
According to classical mechanics, the oscillator should move only in the
region between the two dotted lines. The quantum-mechanical probability
distribution is largely confined in this region, but there is a small proba-

* R. S. Mulliken, *Phys. Rev.* **25**, 259 (1925).

bility outside of it. At the two endpoints of the classically permitted region the kinetic energy of the classical oscillator reduces to zero and the potential energy equals total energy. Outside this region the potential energy is higher than the total energy, and the particle would have a "negative kinetic energy" which is never possible in classical mechanics. The "explanation" in quantum mechanics for the existence of a small probability of finding a particle in a classically forbidden region will be postponed to Chapter 6.

There is a close relationship between the probability distribution of position and the classical motion. Consider the motion of a classical oscillator. The relative probability P_{cl} of finding it in a range dx in the neighborhood of x is proportional to the time it spends in dx,

$$P_{cl} \sim \frac{dx}{v} = \frac{dx}{\sqrt{2/M[E - (1/2)kx_0^2]}}. \tag{4–68}$$

P_{cl} is plotted as the dotted curve in Fig. 4–3. This curve is like an *envelope* of $|\psi_{10}(\xi)|^2$ except near the ends. This result holds true for wave functions ψ_n with large values of n and may be proved mathematically (see the WKB method in Chapter 7). Furthermore, for large n, the number of peaks in the corrugated structure of $|\psi_n(\xi)|^2$ (which is $n + 1$) is also large. The corrugated structure thus loses its practical significance and the relative probability of finding a particle in a region from x to $x + dx$, which may include a large number of corrugation peaks, is practically determined by the envelope curve at x. Thus for large values of n, the quantum-mechanical density distribution is the same as the classical distribution (except at the endpoints and the fine structure of the corrugation).

Is this merely a happy mathematical accident? No. From the discussion of Section 2–7, we know that $|\psi_n(x)e^{-i(E_n/\hbar)t}|^2$ may be considered to represent the density distribution of a group of particles having the same energy E_n and moving according to classical mechanics. Since $|\psi_n(x)e^{-i(E_n/\hbar)t}|^2$ turns out to be time independent, the corresponding group of particles must arrange themselves in such a way that the density distribution is stationary (time independent). Such a situation may be realized if we imagine that a series of particles are being ejected at the origin with the same initial velocity pointing along the positive x-axis, one after another at an even pace for a time interval equal to the period of oscillation. The density distribution of this group of classical particles will be time independent, and its spatial dependence will be just like P_{cl}. Therefore, when n is much greater than unity, the quantum-mechanical probability distribution $|\psi_n(x)|^2$ agrees with the classical distribution P_{cl}.

The inquisitive student will hasten to ask why the two distributions differ at the endpoints and what is the origin of the corrugation of the quantum-mechanical probability distribution. It may be remembered that the analogy with a group of independent classical particles is valid only in the classical limit when the Planck constant h may be regarded as small (Section 2–7). If h is not negligible, the analogy will have to be modified in such a way that the group of particles or the probability elements are subject to a quantum force derived from the quantum potential of Eq. (2–65), in addition to the classical force derived from the potential $U(x)$. The difference between the classical and quantum-mechanical results may thus be attributed to this quantum force. We may therefore satisfy ourselves by saying that the quantum force is acting on the group of hypothetical particles as a result of which the particles are re-grouped in a corrugated distribution, and at the endpoints some particles are sent by this force into the classically forbidden region, making the density at the endpoints finite instead of infinite as would be the case if the classical force alone were acting.

Problems

4–1. A simple pendulum may be regarded as a linear harmonic oscillator to the first approximation. Discuss the motion of a simple pendulum consisting of a mass of 1 gm attached to a string of length 980 cm, according to quantum mechanics. Deduce the quantum-mechanical results. Find the quantum-mechanical solution corresponding to a classical motion of amplitude 1 cm.

4–2. Expand $f(\xi) = e^{-\xi^2}$ in series of ψ_n of Section 4–2. Calculate the first five coefficients. Compare the series and the function graphically by taking the first 1, 3, 5 terms of the series.

4–3. Mathematical exercise: Verify the generating function of the Hermite polynomials.

4–4. Mathematical exercise: Verify the recursion formulas of the Hermite polynomials.

*4–5. Find the momentum distribution of the energy eigenstate ψ_n of a linear harmonic oscillator by determining the Fourier coefficients (see also Problem 12–9).

4–6. Find the algebraic equation determining the positions ξ_i of the $n + 1$ maxima of $\psi_n(\xi)$.

4–7. Let $\xi_1, \xi_2, \ldots, \xi_n$, be the zeros of $\psi_n(\xi)$. Evaluate $\sum_{i=1}^{n} \xi_i^2$. Answer: $\frac{1}{2}n(n-1)$.

4–8. Can a particle pass through the zeros $\xi_1, \xi_2, \ldots, \xi_n$?

4–9. Determine the probability current density of an energy eigenstate.

*4–10. Determine the probability current density of the solution discussed in Section 4–3.

4–11. Under what conditions does a wave function exhibit a definite parity?

---

* Indicates more difficult problems.

CHAPTER 5

ONE-DIMENSIONAL POTENTIAL BARRIER PROBLEMS

So far we have considered problems in which the potential is a continuous function of the coordinates. In this chapter we shall discuss another kind of one-dimensional problem in which the potential is discontinuous and is such that between two points of discontinuity it is a constant. Two special cases, the potential well and the potential barrier of finite thickness, illustrating the essential features of this kind of problem, will be discussed. As these problems differ from those discussed previously in the existence of discontinuities in the potential function, the continuity property of the wave function at these points of discontinuity is an important point of consideration. We shall not repeat the effort of deriving classical results from quantum mechanics; we shall discuss some special cases of practical importance and show that quantum mechanics accounts for another quantum phenomenon: the penetration of a particle through a potential barrier.

5–1 The potential well problem. A potential well is a potential function which is zero in a certain region of the coordinate but assumes a positive constant value V_0 elsewhere (Fig. 5–1),

$$U = \begin{cases} 0, & \text{for } -a < x < a, \\ V_0, & \text{for } -a > x > a, \end{cases} \tag{5–1}$$

$$V_0 > 0.$$

The classical solution of the motion of a particle in a potential well is easily obtained. If the particle has an energy E less than V_0, it will move with a constant velocity $\sqrt{2E/M}$ between $x = -a$ and $x = a$,

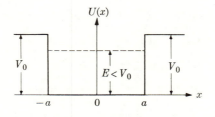

FIGURE 5–1

and will reverse its direction of motion whenever it collides with the potential barrier at $x = \pm a$. It will never go into the regions $x > a$ and $x < -a$. If the energy E is greater than V_0, then the particle will move to $+\infty$ or $-\infty$ depending on its initial direction of motion. The velocity will be $\sqrt{2E/M}$ for $-a < x < a$ and $\sqrt{2(E - V_0)/M}$ for the regions $x < -a$ and $x > a$.

The quantum-mechanical general solution may be obtained from the time-dependent Schrödinger equation and may be written in a series:

$$\Psi(x, t) = \sum_{n=0}^{\infty} c_n \psi_n(x) e^{-(i/\hbar) E_n t}, \tag{5-2}$$

where $\psi_n(x)$ is a solution of the time-independent Schrödinger equation,

$$\frac{d^2 \psi_n}{dx^2} + \frac{2M}{\hbar^2} [E_n - U(x)] \psi_n = 0. \tag{5-3}$$

Inserting the potential function of Eq. (5–1) and dropping the index n for generality, we may replace Eq. (5–3) by three equations in the three regions:

$$\frac{d^2 \psi}{dx^2} + \frac{2M}{\hbar^2} (E - V_0) \psi = 0, \qquad \text{for } x < -a,$$

$$\frac{d^2 \psi}{dx^2} + \frac{2M}{\hbar^2} E \psi = 0, \qquad \text{for } -a < x < a, \tag{5-4}$$

$$\frac{d^2 \psi}{dx^2} + \frac{2M}{\hbar^2} (E - V_0) \psi = 0, \qquad \text{for } x > a.$$

As in the classical solution we have to distinguish two cases: $E < V_0$ and $E > V_0$. For the first case, $E < V_0$, we define two constants α and β as follows:

$$\alpha^2 \equiv \frac{2M}{\hbar^2} E,$$

$$\beta^2 \equiv \frac{2M}{\hbar^2} (V_0 - E). \tag{5-5}$$

The general solution of Eq. (5–4) may then be expressed as follows:

$$\psi = A e^{\beta x} + A' e^{-\beta x}, \qquad \text{for } x < -a,$$

$$\psi = B e^{i\alpha x} + C e^{-i\alpha x}, \qquad \text{for } -a < x < a, \tag{5-6}$$

$$\psi = D e^{-\beta x} + D' e^{\beta x}, \qquad \text{for } x > a.$$

The condition that ψ does not diverge at $\pm\infty$ requires that the constants

A' and D' be zero. Thus we have four arbitrary constants, A, B, C and D, left. These constants will have to be related in some way so that the three functions above may join together to form an acceptable solution. Before we can determine these relations we have to investigate the continuity property of the wave function at the points where the potential $U(x)$ is discontinuous. When the potential is represented by an analytic function, the wave function is also analytic,* i.e., $\psi(x)$ and all its derivatives, $d\psi/dx$, $d^2\psi/dx^2, \ldots$, are continuous. If the potential $U(x)$ has a discontinuity at x_0, then $d^2\psi/dx^2$, being equal to $-(2M/\hbar^2)(E - U)\psi$ in accordance with the Schrödinger equation, is also discontinuous at x_0, and so are all the higher derivatives. Therefore the best we can ask for is that $\psi(x)$ and $d\psi(x)/dx$ are continuous at x_0. We shall show that this is actually the continuity condition ψ must satisfy at x_0. If ψ or $d\psi/dx$ is not continuous at x_0, the probability current density, given by Eq. (2–78), will be discontinuous at x_0. As a result, x_0 will be a source or a sink of the probability current; this means that matter will be created or destroyed with a given rate at x_0. Since matter cannot be created or destroyed (at least in low energy physics) we must require the continuity of ψ and $d\psi/dx$ at the points where the potential is discontinuous. The above consideration may be restated as follows: a discontinuous potential function does not prevent us from obtaining physically meaningful solutions of the wave equation if ψ and $d\psi/dx$ are required to be continuous at the points of discontinuity.

Returning to Eq. (5–6) we note that the continuity of $\psi(x)$ at $x = -a$ and $x = +a$ requires that

$$
\begin{aligned}
Ae^{-\beta a} &= Be^{-i\alpha a} + Ce^{+i\alpha a}, \\
De^{-\beta a} &= Be^{+i\alpha a} + Ce^{-i\alpha a}.
\end{aligned} \tag{5–7}
$$

The continuity of $d\psi/dx$ at $x = -a$ and $x = +a$ requires that

$$
\begin{aligned}
\beta Ae^{-\beta a} &= i\alpha Be^{-i\alpha a} - i\alpha Ce^{+i\alpha a}, \\
-\beta De^{-\beta a} &= i\alpha Be^{+i\alpha a} - i\alpha Ce^{-i\alpha a}.
\end{aligned} \tag{5–8}
$$

Thus we have four linear homogeneous equations for determining the four constants A, B, C, and D. In order that a set of nonvanishing solutions

---

* The differential equation (5–3) defines an analytic function $\psi(Z)$ over a region in the complex plane where $U(Z)$ is analytic. By the theory of functions of a complex variable we learn the continuity property of ψ. However, only on the real axis where $U(Z)$ and $\psi(Z)$ reduced to $U(x)$ and $\psi(x)$ do these functions have physical significance.

may exist the determinant of these simultaneous equations must vanish, which leads to the following equation:

$$\left(\tan \alpha a - \frac{\beta}{\alpha}\right)\left(\tan \alpha a + \frac{\alpha}{\beta}\right) = 0. \tag{5-9}$$

Since α and β contain the energy E, Eq. (5–9) imposes a restriction on the energy value. Only when the energy equals one of the values satisfying Eq. (5–9), can we obtain a set of solutions A, B, C, D, from Eq. (5–7) and Eq. (5–8), and so an acceptable solution of Eq. (5–4). The energy values satisfying Eq. (5–9) are thus the allowed energy values and therefore the energy is quantized. The solutions of Eq. (5–9) consist of the solutions of the following equations:

$$\tan \alpha a = \frac{\beta}{\alpha}, \tag{5-10}$$

$$\tan \alpha a = -\frac{\alpha}{\beta}. \tag{5-11}$$

These equations may be solved for the energy eigenvalues by graphical or numerical methods (Problem 5–6). For the solutions of Eq. (5–10), the corresponding solutions of A, B, C, and D are

$$A = D, \quad B = C, \quad A = 2B \cos (\alpha a)e^{\beta a}. \tag{5-12}$$

Note that for a set of homogeneous equations the solution gives only the ratios of the variables. One of the four constants A, B, C, D is left arbitrary. The corresponding wave function is

$$\psi = \begin{cases} 2B \cos (\alpha a)e^{\beta(x+a)}, & \text{for } x < -a, \\ 2B \cos (\alpha x), & \text{for } -a < x < a, \\ 2B \cos (\alpha a)e^{-\beta(x-a)}, & \text{for } x > a. \end{cases} \tag{5-13}$$

The constant B will be determined eventually by normalization. The wave function of the lowest energy in this group is graphically represented in Fig. 5–2. As the energy eigenvalue increases, the cosine function in the middle region makes more crossings over the x-axis. The points where the curve crosses the x-axis are called *nodal points*. From Eq. (5–13) we note that all wave functions in this group are *symmetric*, i.e., $\psi(x) = \psi(-x)$, and thus have even parity.

For the solutions of Eq. (5–11) the corresponding solutions of A, B, C, and D are

$$A = -D, \quad B = -C, \quad A = -2iB \sin (\alpha a)e^{\beta a}. \tag{5-14}$$

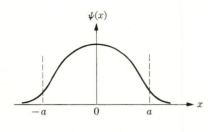

FIGURE 5–2

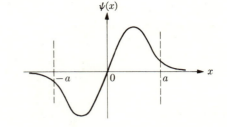

FIGURE 5–3

The corresponding wave function is

$$
\psi = \begin{cases}
-2iB \sin (\alpha a)e^{\beta(x+a)}, & \text{for } x < -a, \\
2iB \sin (\alpha x), & \text{for } -a < x < a, \\
2iB \sin (\alpha a)e^{-\beta(x-a)}, & \text{for } x > a.
\end{cases} \tag{5–15}
$$

The wave function of the lowest energy in this group is graphically represented in Fig. 5–3. As the energy eigenvalue increases, the sine function in the middle region makes more crossings over the x-axis. All wave functions in this group are *antisymmetric*, i.e., $\psi(x) = -\psi(-x)$, and thus have odd parity.

To summarize for the case $E < V_0$: We have four constants A, B, C, and D, and four continuity conditions. But only three of the four constants are adjustable, the other being determined by normalization. Hence we have more conditions than necessary for determining the three constants and in general no solution exists. By allowing the energy value to vary we introduce an additional variable and the conditions may be satisfied. Thus acceptable solutions may be obtained only for a selected set of energy values and the energy is quantized.

We consider a special case in which V_0 approaches infinity, i.e., the potential barrier is infinitely high. For the first group of solutions,

$$
\tan \alpha a = \frac{\beta}{\alpha} \to \infty, \tag{5–16}
$$

$$
\therefore \alpha a = (2n + 1) \frac{\pi}{2}. \tag{5–17}
$$

By Eq. (5–17) the characteristic energy values are

$$
E_s = \frac{(2n + 1)^2 \pi^2 \hbar^2}{8Ma^2}. \tag{5–18}
$$

The corresponding wave functions, the symmetric ones, reduce to pure cosine functions between the barriers and vanish at the boundary and beyond ($x = \pm a$ are the nodal points of the cosine function). For the second group of solutions,

$$\tan \alpha a = -\frac{\alpha}{\beta} \to 0, \tag{5-19}$$

$$\therefore \alpha a = n\pi. \tag{5-20}$$

The characteristic energy values are

$$E_a = \frac{(2n)^2 \pi^2 \hbar^2}{8Ma^2}. \tag{5-21}$$

The corresponding wave functions, the antisymmetric ones, reduce to pure sine functions between the barriers and vanish at the boundary and beyond ($x = \pm a$ are the nodal points of the sine function).

Having discussed the $E < V_0$ case, we now turn our attention to the $E > V_0$ case. The solutions may be obtained by a similar procedure. They differ from those of the previous case in that the wave functions in all three regions are oscillatory, there being no exponential functions. Thus the boundary condition at infinity does not rule out any solution as unacceptable and we have a total of six arbitrary constants. This is more than necessary to satisfy the four continuity conditions in addition to that of normalization, and we can find acceptable solutions for any value of energy. Therefore the energy is not quantized. The details of this problem are left to the reader (Problem 5–1).

5–2 The potential barrier problem. We now consider a problem in which the potential is zero everywhere except in a finite region $0 < x < a$, where the potential is a positive constant V_0. This potential is represented graphically in Fig. 5–4. In classical mechanics a particle having energy E less than V_0, in region I ($x < 0$), if it moves in the positive x-axis direction, will proceed with a constant velocity until it reaches the origin, after

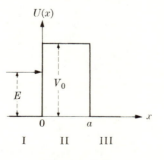

FIGURE 5–4

which the direction of motion will be reversed while the speed will remain the same. The potential thus acts like a barrier which a particle with energy E less than V_0 cannot penetrate. On the other hand, a particle with energy E greater than V_0 will move right into region II ($0 < x < a$) with a reduction of its kinetic energy. The kinetic energy will be restored to its original value when the particle moves into region III ($x > a$).

In quantum mechanics, the motion of a particle in this potential field is described by a wave function determined by the Schrödinger equation. The time-independent Schrödinger equation for this potential is

$$\frac{d^2\psi}{dx^2} + \frac{2M}{\hbar^2} E\psi = 0 \qquad \text{for region I,}$$

$$\frac{d^2\psi}{dx^2} + \frac{2M}{\hbar^2} (E - V_0)\psi = 0 \qquad \text{for region II,} \qquad (5\text{--}22)$$

$$\frac{d^2\psi}{dx^2} + \frac{2M}{\hbar^2} E\psi = 0 \qquad \text{for region III.}$$

Consider first the case $E < V_0$. We define two constants α and β as follows:

$$\alpha^2 \equiv \frac{2ME}{\hbar^2}, \qquad (5\text{--}23)$$

$$\beta^2 \equiv \frac{2M(V_0 - E)}{\hbar^2}. \qquad (5\text{--}24)$$

The solutions of Eqs. (5–22) in the three regions are

$$\psi = Ae^{i\alpha x} + Be^{-i\alpha x} \qquad \text{in region I,}$$

$$\psi = Ce^{\beta x} + De^{-\beta x} \qquad \text{in region II,} \qquad (5\text{--}25)$$

$$\psi = Ee^{i\alpha x} + Fe^{-i\alpha x} \qquad \text{in region III.}$$

Since the solutions are oscillatory at infinity instead of exponential, the boundary condition does not rule out any solution as unacceptable. Thus we have six arbitrary constants. As before, the continuity conditions for ψ and $d\psi/dx$ at $x = 0$ and $x = a$ (where the potential is discontinuous) must be satisfied. This imposes four conditions among the six constants. Again the requirement of normalization imposes one additional condition. With only five conditions to satisfy, one of the six constants may be left arbitrary. We will consider one special solution in which the constant F is arbitrarily set to zero. This special solution has an application which interests us, while the general solution ($F \neq 0$) does not correspond to any practical physical situation (explained later). When $F = 0$, the other five constants are uniquely determined by the five conditions mentioned before. The four continuity conditions are as follows:

for continuity of ψ at $x = 0$, $A + B = C + D$,   (5–26)

for continuity of $\dfrac{d\psi}{dx}$ at $x = 0$, $i\alpha(A - B) = \beta(C - D)$,   (5–27)

for continuity of ψ at $x = a$, $Ce^{\beta a} + De^{-\beta a} = Ee^{i\alpha a}$,   (5–28)

for continuity of $\dfrac{d\psi}{dx}$ at $x = a$, $\beta(Ce^{\beta a} - De^{-\beta a}) = i\alpha Ee^{i\alpha a}$.   (5–29)

From these four equations we solve for B, C, D, E, in terms of A. The results are:

$$B = A\, \frac{[1 + (\beta^2/\alpha^2)](e^{-\beta a} - e^{\beta a})}{\Delta},$$

$$C = A\, \frac{2[1 + (\beta/i\alpha)]e^{-\beta a}}{\Delta},$$

$$D = A\, \frac{-2[1 - (\beta/i\alpha)]e^{\beta a}}{\Delta},$$

$$E = A\, \frac{4(\beta/i\alpha)e^{-i\alpha a}}{\Delta},$$

(5–30)

where

$$\Delta = \begin{vmatrix} 1 + (\beta/i\alpha) & 1 - (\beta/i\alpha) \\ [1 - (\beta/i\alpha)]e^{\beta a} & [1 + (\beta/i\alpha)]e^{-\beta a} \end{vmatrix}.$$ (5–31)

What kind of physical situation does this special solution represent? In region III, the wave function is $Ee^{i\alpha x}$. Applying Eq. (2–78), we find the current density to be

$$j_t = \frac{\hbar}{M}\, \alpha E^* E.$$ (5–32)

Its direction points along the positive x-axis. Analogously, the terms $Ae^{i\alpha x}$ and $Be^{-i\alpha x}$ represent two currents:

$$j_i = \frac{\hbar}{M}\, \alpha A^* A, \qquad j_r = -\frac{\hbar}{M}\, \alpha B^* B,$$ (5–33)

in region I, with j_i in the $+x$-direction and j_r in the opposite direction. Furthermore, it may be shown by straightforward calculation that

$$A^* A = B^* B + E^* E,$$ (5–34)

so that

$$|j_i| = |j_r| + |j_t|.$$ (5–35)

From these results we may interpret this solution as follows. A uniform beam of particles (j_i) traveling along the $+x$-axis in region I is split into two beams when it reaches the potential barrier, one (j_r) traveling in the backward direction in region I corresponding to a reflected beam, the other (j_t) traveling in the forward direction in region III corresponding to a transmitted beam. Quantum mechanics thus implies that when a beam of particles with energy E less than V_0 impinges upon a potential barrier of finite thickness, a part of the beam may go through the barrier and reappear on the other side of it. If there is only one particle in the beam, then the intensity of the transmitted beam represents the probability of finding this particle passing through the potential barrier. According to classical mechanics, the whole beam will be reflected and none of it will go through. This quantum-mechanical result, contrary to classical mechanics, is the basis of the explanation of a quantum phenomenon—the penetration of potential barrier in radioactive decay, which will be discussed in the next section.

We see from Eqs. (5–26) to (5–29) that if $E = 0, C$ and D must be zero and consequently A and B must be zero. Therefore it is impossible to construct a nontrivial wave function without a transmitted beam. In other words, any beam falling upon a potential barrier of finite thickness will have to be partially transmitted. In the general solution in which the constant F is not zero, the additional term simply represents another incident beam impinging upon the potential barrier from the right in addition to the incident beam from the left. This beam, in turn, will be partially reflected and partially transmitted for the same reason. Thus the general solution does not introduce anything physically new. It merely represents a physical situation which rarely occurs in practice.

Let us define a *transmission coefficient* T by the ratio of the current densities of the transmitted and incident beams:

$$T \equiv \frac{j_t}{j_i}.$$

T may be obtained by straightforward calculation (Problem 5–8), the result being

$$T = \frac{16\beta^2}{\alpha^2} \frac{1}{|\Delta|^2}. \tag{5–36}$$

The rate of transmission depends on the energy E, the barrier height V_0, and the barrier width a, these quantities entering the transmission coefficient through α, β, and Δ.

Since we have enough constants to satisfy the continuity and normalization conditions, an acceptable solution may be obtained for any energy value E. Therefore, the energy is not quantized.

It remains to discuss the solution for the case $E > V_0$. The solution is oscillatory in all three regions. The energy is not quantized. The fact that the solution must satisfy continuity conditions leads to the result that an incident beam will be partially reflected at the boundary $x = 0$. This result is contrary to the classical result that the whole beam will enter region II without reflection. The details of this problem will be left to the reader (Problem 5–2).

5–3 Explanation of a quantum phenomenon: the penetration of potential barrier in radioactive decay. In radioactive decay, some materials such as radium spontaneously emit α-particles which are doubly-charged particles identified as the nuclei of helium atoms. This phenomenon poses great difficulty to classical mechanics.

The first difficulty concerns the energy of the α-particle in the field of the nucleus. From Rutherford's scattering experiments in which α-particles were used as a tool to probe the interior of a heavy atom, it was found that the force experienced by the α-particles is a Coulomb force down to a very small distance ($\sim 10^{-12}$ cm) from the center of the nucleus of the atom. The point at which Coulomb's law breaks down may be considered as the boundary of the nucleus. Thus the potential experienced by an α-particle outside the nucleus is a Coulomb potential as shown in Fig. 5–5. When an α-particle enters the nucleus ($0 < r < a$), it will experience very strong attractive nuclear forces from the nuclear constituents. These forces nearly balance out in the interior of the nucleus but show up strongly at the surface. Hence they may be approximately represented by a potential well as shown in Fig. 5–5. Taking the radius of the radium nucleus to be 9.1×10^{-13} cm we calculate the Coulomb potential energy of an α-particle just outside the nucleus to be 4.4×10^{-5} erg or 27.8 Mev. However, in the radioactive decay of radium, the

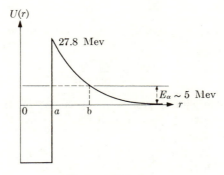

FIGURE 5–5

α-particle emitted from the nucleus has an energy E_α equal to 4.88 Mev, which is much lower than the Coulomb energy barrier. The following question thus arises: How can an α-particle of 4.88 Mev energy go through a potential barrier of 27.8 Mev?

The second difficulty concerns the law of causality. The rate of emission of α-particles from a radioactive material is governed by the law of radioactive decay according to which the number of α-particles emitted per unit time is proportional to the number of radioactive atoms present. For radium the decay rate is such that the amount of radium will be reduced by half after 1620 years. After several periods of 1620 years there will still be a small fraction of radium left. Since all radium atoms are supposed to be exactly the same, the following question arises: Why do some radium atoms decay in the first few years while others survive for thousands of years? In other words, why do the same initial conditions lead to different results, contrary to the law of causality? It may be noted that this problem arises because of the *indivisibility* of the atom. One atom must act as a whole; it cannot be partly radium and partly radon. If matter (radium) were continuous, instead of being quantized in the unit of atom, this problem would not arise. Once again we see that indeterminism is intimately related to quantization.

These difficulties of classical mechanics disappear in quantum mechanics. In the last section we have seen that a particle impinging upon a potential barrier, according to quantum mechanics, has a certain probability to go through it even though the energy of the particle is less than the barrier height. Thus the first difficulty ceases to exist. Moreover, the result of the last section implies that the number of α-particles emitted per unit time is equal to the total number of radioactive atoms multiplied by the probability of penetration through the potential barrier per unit time. The latter quantity equals the transmission coefficient multiplied by the frequency of bombardment; it is a constant in time. Thus the law of radioactive decay is derived and we are able to derive the radioactive constant on a theoretical basis. The question of causality does not arise, for the behavior of a particle in quantum mechanics is probabilistic. What can be predicted in quantum mechanics about a particle is its probability of doing something; this probability may be realized in a few years or after many thousand years. These points will be discussed further in Chapter 6.

The success of the quantum-mechanical treatment of α-decay soon after the establishment of quantum mechanics was considered a major triumph of the new theory. We shall return to this topic in Chapter 8 because the emission of an α-particle is actually a three-dimensional problem. However, some qualitative features of the theory may be inferred from the expression of the transmission coefficient, Eq. (5–36). If α and β

are of the same order of magnitude, we have very roughly

$$T \sim e^{-2\beta a} = e^{-2\sqrt{(2M/\hbar^2)(V_0-E)}a},\qquad(5\text{-}37)$$

provided $2\beta a$ is much larger than 1. Equation (5-37) shows that the probability of transmission, being an exponential function, is strongly dependent on the thickness of the barrier a as well as on $(V_0 - E)$, the barrier height above the energy of the α-particle. Therefore, in α-decay, we expect a strong dependence of the radioactive constant on the energy E_α of the α-particle. In fact, the variation of E_α among the α-emitters is not more than a few Mev, while the variation of half-life ranges from a fraction of a second to several billion years.

Problems

5-1. What is a fast-moving automobile's chance of being rebounded at a river bank instead of plunging into the river? [*Hint:* Approximate this problem by a potential well problem in which $E > V_0$.]

5-2. Discuss the potential barrier problem for the case $E > V_0$.

*5-3. Two identical potential wells are separated by a distance b comparable to the well width a (Fig. 5-6). Discuss the motion of a particle in this potential field. When $b \to 0$, the solution of this problem reduces to that of Section 5-1. When $a \to \infty$, the solution reduces to that of Section 5-2.

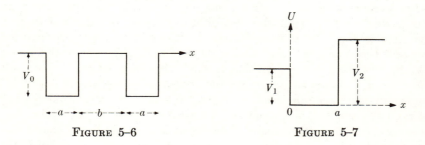

FIGURE 5-6   FIGURE 5-7

*5-4. Determine the energy eigenvalues of a particle in the potential of Fig. 5-7.

*5-5. Discuss the problem in which the potential function is as shown in Fig. 5-8. Show that in the energy range $0 < E < V_0$ there exist a number of energy values E_i in the neighborhood of which the corresponding wave functions have much larger amplitude in the region between the two barriers than elsewhere. A proper superposition of such wave functions within a range ΔE_i results in a wave packet confined between the two barriers. However, it gradually "leaks" out of the confinement. The particle described by the wave packet

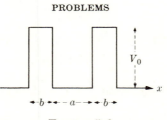

FIGURE 5–8

is said to be in a *virtual state*, the energy being E_i with an uncertainty ΔE_i. Let Δt_i be the time the wave packet takes to "leak" out. Show that for a virtual state of energy E_i, the following equation holds:

$$\Delta E_i \, \Delta t_i \sim h. \tag{5–38}$$

This is another uncertainty relation which will be discussed further in Chapter 6. When a is held constant and b is made to approach infinity, show that the unbound virtual states E_i approach the bound energy eigenstates of the potential well problem discussed in Section 5–1.

5–6. Solve Eq. (5–10) by the graphical method. Assume M to be the neutron mass, a to be the radius of the uranium nucleus, and V_0 to be 30 Mev. Find the first three energy eigenvalues. How many bound states are there in this problem?

5–7. In the last problem let M be the electron mass. How many bound states are there? What conclusion may one draw concerning the possibility of the existence of an electron inside the nucleus as a nuclear constituent? *Note:* The actual problem is three-dimensional.

5–8. Derive the expression for the transmission coefficient, Eq. (5–36).

*5–9.[#] *Nuclear physics. Spontaneous fission* is definitely a potential barrier penetration problem with a potential barrier established by *neutron induced fission*. But in the spontaneous fission the two fragments both assume the roles of evictor and evictee simultaneously and the simple theory developed here cannot be applied. Find a way out and then prove the following known facts: The double-humped mass distribution curve of fission products in neutron fission is carried over in spontaneous fission intact except the peak-to-valley yield ratio is greatly increased. Assume Fong's theory for neutron fission.

5–10.[#] *Electronics.* Study the tunnel diode (Esaki diode, 1957) as experimental evidence of electron tunneling in solids.

5–11.[#] *Electronics.* Explain the high field electron emission from cold metal by the tunneling effect and its use in tunnel electron microscope.

5–12.[#] *Electronics.* A semiconductor provides a potential well, which can be changed by etching, doping and so on in many ways and be used as the basis to develop an integrated circuit, which revolutionizes electronics. Study the outline of this development.

---

* Indicates more difficult problems.

[#] Indicates new problems added to the expanded edition.

CHAPTER 6

THE PHYSICAL MEANING OF QUANTUM MECHANICS

We have introduced the mathematical formalism of quantum mechanics and discussed some of its applications. In this chapter we turn our attention to its physical meaning.

As mentioned in the historical introduction (Chapter 1), the mathematical formalism of quantum mechanics was developed rapidly around 1925. Its many successful applications left no doubt of its correctness, yet its abstractness had created a feeling of uneasiness among physicists. In 1927 Heisenberg* first propounded the *uncertainty principle;* then Bohr† advocated a point of view termed *complementarity.* Their efforts not only gave rise to a rational interpretation of the mathematical formalism of quantum mechanics but also aroused a widespread and deeply felt discussion on the philosophy of science. Their study, despite the dissident opinions expressed by a few physicists,‡ has been regarded by many as conclusive and their views are adopted conventionally. The basic elements of their work will be discussed in Sections 6–1 and 6–2.

Section 6–1 is concerned with a critical analysis of some of the basic concepts used in classical mechanics, which leads to the uncertainty principle. The nature of the quantum-mechanical description of physical processes is expounded in Section 6–2. As mentioned in the summary of Chapter 2, the relation between the quantum phenomena and the uncertainty relation or the quantum force will be discussed here in Section 6–3, which is intended to explain the paradoxes often associated with the quantum phenomena. The chapter ends with a few concluding remarks in Section 6–4.

6–1 Critiques of the classical concepts in mechanics. Heisenberg's uncertainty principle. The validity of classical mechanics would not be doubted were it not for the discovery of the quantum phenomena, which are incomprehensible in terms of classical concepts and laws. It appears that if these quantum phenomena are to be accounted for in a coherent physical theory, some elements of classical mechanics will have to be renounced. In the last few chapters we have seen in a few examples that quantum mechanics accounts for the quantum phenomena successfully.

---

* W. Heisenberg, *Z. Phys.* **43,** 172 (1927); *The Physical Principles of the Quantum Theory,* University of Chicago Press, 1930.

† N. Bohr, *Naturwiss.* **16,** 245 (1928); **17,** 483 (1929); **18,** 73 (1930); *Atomic Theory and the Description of Nature,* Cambridge University Press, 1934.

‡ Including A. Einstein.

On the other hand, it introduces some concepts alien to the classical theory, notably the adoption of a probabilistic point of view. We recognize that the use of probability in describing and predicting physical processes is the basic tenet of quantum mechanics. This attitude is contrary to the deterministic attitude in classical theory according to which all dynamical quantities have definite values at any time and they change in time according to deterministic laws—in short, the motion of a particle may be described by a trajectory. Our argument thus unavoidably leads to the following question: Are we prepared to renounce the determinism of classical mechanics?

In considering this basic problem it may well be kept in mind that the use of probability is not completely new in physics. Statistical mechanics is based on the probability theory; however the probability theory is used as a mathematical expedient for treating very complicated mechanical systems, the fundamental physical laws remaining deterministic. This leads some physicists to suspect that the use of probability in quantum theory might also be an expedient and that determinism might prevail in fundamental laws on a deeper level yet to be discovered.

The conventional attitude following Heisenberg and Bohr, on the contrary, is to renounce determinism (as defined above), to accept the probabilistic point of view as fundamental, and to proceed to correlate and coordinate the vast amount of experimental knowledge in atomic physics within the formalism of quantum mechanics. This view is strengthened by an analysis of some of the classical concepts from a point of view of the actual possibility of verifying them by experimental procedure. After all, experimental verification is the last judgment of any physical theory. The results of the analysis, to be described below, seem to show that there is no reason why some of the classical concepts should not be renounced and that the formalism of quantum mechanics provides exactly what we may rightly hope for in a physically meaningful description of natural processes.

Emphasis on the possibility of experimental verification has played an important role in the development of the theory of relativity. Although the Lorentz transformation follows inescapably any sensible discussion of the Maxwell equations, its implication for the space-time transformation was greatly clarified when Einstein pointed out the relativity of simultaneity by considering the experimental procedures of synchronizing clocks at different places. The concept of absolute simultaneity, not being definable by experimental procedures, is said to have no *operational meaning*, and is considered useless in physics. This line of thought leads to the development of *an operational point of view** according to which

---

* Percy Bridgman, *The Logic of Modern Physics.*

any quantity appearing in a physical theory must be definable in terms of the result of measurement in an actual experiment. It has become fashionable to re-examine the many physical concepts which we use by tradition, without assurance of their being capable of experimental verification.

In classical mechanics, the position of a particle in one-dimensional motion is described by a function $x(t)$ and the momentum is described by a function $p(t)$. These functions obey the classical equation of motion and thus are obtainable by solving the equation of motion with given initial conditions. At any time t these two functions give definite values of position and momentum of the particle. The development of the system in time is thus completely determined by the initial conditions (determinism). We consider the classical theory to be verifiable by experiments in the following sense. We are able to perform a series of observations on the position and momentum at a number of times $t_1, t_2, \ldots$, and to compare the results obtained with those calculated from $x(t)$ and $p(t)$. If they agree, the theory is verified. Actually classical mechanics has been verified this way, for example, in its applications in astronomy. The position of a planet may be observed night after night for many years, and the observed data agree remarkably well with the calculations of classical mechanics.

In the above argument we have rightly ignored one complication: the disturbance of the object being observed by the process of observation. For example, in order to observe the position of an object, we have to direct a beam of light on it; the energy and momentum carried by the light beam may change the motion of the object so that after the act of observation it may not follow what was predicted before. However, in astronomical observations, a planet, because of its massiveness, is hardly disturbed by acts of observation no matter how numerous they may be.

In terrestrial observations involving objects much less massive than a planet, the disturbance due to an act of observation may be noticeable. In measuring the temperature of an object the thermometer exchanges heat with the measured system and thus disturbs it. However, this disturbance is controllable. The heat loss or gain may be corrected for, and the original temperature may be deduced by calculation. Throughout classical physics, disturbance due to observation is ignored; it becomes possible to make the tacit assumption that physical properties of a material object have an objective existence and are accessible upon observation.

This concept, unwittingly formed in macroscopic physics, was carried over into microscopic physics in its early period. An atom was pictured as a planetary system modeled on the solar system. However, if we try to ascertain the electronic orbit in the atom, following the method we used to find a planetary orbit in the solar system, we meet a number of diffi-

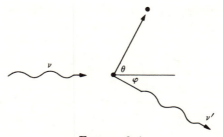

FIGURE 6–1

culties. In order to determine the position of an electron we may direct a beam of light on it. The energy and momentum of the light disturbs the electron and changes its position and momentum. If light were continuous, as the wave theory had supposed, we could reduce the disturbance to as little as we desired by reducing indefinitely the light intensity. But this is not feasible because of the fact that light has a discontinuous structure: it can transfer energy and momentum only in whole units of a quantum, not fractions thereof. (We are not considering the corpuscular *theory* of light but simply stating an experimental fact.) Since a photon* cannot be divided further, the disturbance produced by the minimum available amount of light, i.e., a photon, is *a finite amount* which cannot be reduced further.

Another difficulty will be brought out in connection with the Compton scattering. In this process the electron may receive an amount of energy E from the light and thus move in a direction making an angle θ with the incident direction (Fig. 6–1). The light may change its frequency from ν to ν' and then propagate along a direction making an angle φ with the incident direction. There are four unknown quantities involved, E, ν', θ, and φ, but we have only three equations, obtained by considering conservation of energy and momentum, to solve them (Problem 1–1). Thus the outcome of the interaction is not uniquely determined and the electron may move in any direction θ (once θ is specified, the other quantities, E, ν', φ, are determined). Actually electrons are found, in Compton scattering experiments, in all directions with definite proportions. Therefore, *we cannot predict deterministically the final result of the interaction;* the best we can say is that the electron has a certain probability to be scattered in a certain direction θ (and thus to receive a certain amount of energy E, etc.). In other words, the disturbance imparted to the electron by the photon is *uncontrollable*. Unlike the disturbance produced by a thermometer, it

---

* The photon, or the light quantum, is the fundamental unit of energy and momentum of light; it need not be associated with the kinematical attributes of a particle.

cannot be calculated exactly. *Therefore we have no way of eliminating it from the observed results.* The reader may wonder why a similar problem in classical mechanics, i.e., the collision of two billiard balls, has a deterministic solution. In that case, we have to know the line of collision, i.e., the line joining the centers of the two balls at the moment of collision. If the photon and the electron were tiny billiard balls and we had ways of ascertaining along which line they collide, then the outcome of the collision would be uniquely determined and the disturbance would be controllable. If, in addition to the conservation laws of energy and momentum, there were another physical law we could apply to determine the outcome of the collision, then the disturbance would also be controllable. As such conditions have not been realized, we must conclude that the disturbance is uncontrollable. In the final analysis this indeterminacy may again be traced to the *indivisibility of the quantum.* If the photon could be subdivided without limit, the recoil of the electron might be reduced to zero and there would be no uncontrollable disturbance. If the light energy were not quantized, as the classical theory asserted, then the energy of the light would be spread over the beam continuously and the light beam would be scattered continuously in all directions. The outcome of the process would be deterministically ascertainable: the electron would be accelerated in the $\theta = 0$ direction due to the pressure of light and even the angular dependence of the wavelength change would be accountable by the Doppler effect of the forward motion of the electron (Problem 6–2).

As the disturbance incurred by an act of observation is *finite* and *uncontrollable, cannot be eliminated* and *separated* from the intrinsic property of the system being observed, it seems meaningless to speak of the intrinsic property independent of any act of observation, for such a property is not verifiable by experimental procedure. Moreover, it becomes perfectly legitimate to consider the "observer" and the "observed" as forming one inseparable system and to consider as the only physically meaningful information the probabilistic description of the possible outcomes of an act of observation. Successive measurements will not serve the purpose of verifying a theory predicting the change of the intrinsic properties (position $x(t)$ and momentum $p(t)$) in the course of time because each measurement disturbs the system and changes it into a new state of motion with which the previous theoretical prediction has no connection at all. Such a theory thus seems of little use. Its correctness does not help us and its falseness does not hurt us. The more important concern seems to be: *What are the possible results that may happen after an act of observation is carried out and what are the relative probabilities for them to happen?* Since a system is disturbed to an uncontrollable amount after an act of observation, the prediction based on the original system no longer applies afterward, and we must treat the future development as a new system.

Whereas in classical mechanics we could predict the future behavior of a system at all times, we now have to satisfy ourselves in making predictions for the outcome of the next act of observation. These changes of viewpoint, in the final analysis, all originate from the indivisibility of the quantum.

Returning to the mathematical formalism of quantum mechanics, we note that the viewpoints made mandatory in the above analysis are already embodied in it. The preceding discussion thus implies that the quantum-mechanical description is actually the physically meaningful description.

In classical mechanics it is tacitly assumed that we can exactly specify several quantities simultaneously. Thus both the position and momentum of a particle at a given time may be precisely specified simultaneously. In atomic physics, because of the uncontrollable disturbance introduced by an act of observation, this concept must come under careful scrutiny. When a measurement is made to ascertain the *position* of a particle, the consequent uncontrollable disturbance changes the physical system being observed to a new state uncorrelated to the original state, so that the original *momentum* information sought is lost before a measurement may be made to ascertain it. The accessibility of simultaneous information may be realized in exceptional cases (when the quantities are "compatible"; see commutative operators in Section 12–2) but cannot be taken for granted. To illustrate how a measurement of position may affect the information on momentum, let us consider the following hypothetical experiment.

A beam of electrons is directed along the x-axis (see Fig. 6–2). A single slit is placed at the point $(0, y)$ and a fluorescent screen is placed behind it. If a scintillation is observed on the screen we know that an electron has passed through the slit and that its position along the y-direction when it passes the $x = 0$ line is y. However, its position is not exactly y because

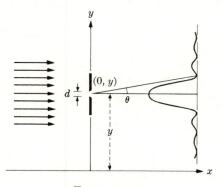

FIGURE 6–2

the slit has a width of d. We know only that the electron has passed through the slit, but we do not know exactly where within the slit it passed. Therefore the position y is accompanied with an uncertainty

$$\Delta y = d. \tag{6-1}$$

If electrons were miniature billiard balls, the beam would cast a sharp image of the slit on the screen. However, it is an experimental *fact* that a beam of electrons passing through a slit does not cast a sharp image of the slit, but instead forms a pattern on the screen similar to the diffraction pattern of light of wavelength equal to h/p where h is the Planck constant and p is the momentum of the electron. (We are not considering the wave *theory* of the electron but simply stating a *fact*. We regard the electrons as *particles* and are interested in the necessary modifications of the concept of particle which may enable us to explain all experimental facts.) The central maximum of this pattern subtends an angle 2θ at the center of the slit with

$$\sin \theta = \frac{\lambda}{d} = \frac{h}{pd}, \tag{6-2}$$

a well-known result in the diffraction of light. The intensity distribution of this pattern represents the density distribution of the electrons arriving at the screen. If the beam is made so weak that only one electron arrives at a time, we shall not see the diffraction pattern but shall see only one scintillation at a time appearing in the area which used to be covered by the diffraction pattern. If a camera with its shutter fixed open is used to record the scintillations for a long time, the diffraction pattern will be reproduced. We must conclude that an individual electron may go anywhere within the diffraction pattern and that there is a certain probability of finding the electron at a given spot, its value being proportional to the intensity of the diffraction pattern at this spot. *The behavior of an individual electron is thus probabilistic*, and the best we can predict is the probabilities of its possible positions. (This is merely a logically consistent way to modify the concept of a particle to accommodate the diffraction phenomena without having to assume a physical wave.) Since we cannot predict definitely where the individual electron lands on the screen or where it originates in the slit, we do not know along which line the electron travels from the slit to the screen. Since most of the probability is associated with the central maximum, we may say roughly that the electron travels in a direction within the angle 2θ. The uncertainty of the direction due to the width of the slit is negligible, this being usually the case in diffraction experiments. For a given direction, the momentum of the electron has a component along the y-axis. The uncertainty of the landing place of the electron within the central maximum thus gives rise to an uncertainty of the y-component of the momentum, the order of magnitude of which is $p \sin \theta$. Thus

$$\Delta p_y \sim p \sin \theta. \tag{6-3}$$

To summarize: The act of observation (using a single slit) enables one to locate the y-position of an electron, at the time t_0 when it passes the slit, within an uncertainty Δy. At the same time, the y-component of momentum of the electron after and including time t_0 will have an uncertainty Δp_y. The product of these two uncertainties (at time t_0) is

$$\Delta p_y \, \Delta y \sim pd \sin \theta = pd \, \frac{h}{pd} . \tag{6-4}$$

Therefore

$$\Delta p_y \, \Delta y \gtrsim h. \tag{6-5}$$

The $>$ sign is added because actual uncertainties may be greater than the theoretical minimum values discussed above. Equation (6-5) means that the product of the uncertainty of position along one direction and the uncertainty of momentum *along the same direction* at time t_0 is always greater than a finite quantity h. If one is made smaller, the other will become larger. By using a narrower slit, the position information may be known more accurately but the diffraction pattern will spread out more, causing a larger uncertainty in momentum information. In order to reduce the spread of the diffraction pattern, the width of the slit must be made large; but this worsens the position information. It is not impossible to make one piece of information known exactly (Δy or Δp_y equal to zero) but this will be accompanied by a complete ignorance of the other information (Δp_y or Δy equal to infinity). The simultaneous exact determination of position and momentum is thus impossible, the errors being subject to the restriction of Eq. (6-5).

Heisenberg and Bohr conceived a number of hypothetical experiments of a similar nature; some will be discussed later in this section. In all cases the same relation of Eq. (6-5) is obtained. These results mean that the simultaneous specification of position and momentum (x and p_x, y and p_y, z and p_z) by actual experimental procedures (not by procedures impossible to carry out or involving entities nonexistent in nature) cannot be more accurate than that specified by Eq. (6-5) and that the lowest limit of accuracy is the same in all such procedures. In classical mechanics we tacitly assume that both x and p_x may be precisely specified simultaneously (or both Δx and Δp_x may be reduced to zero). In making such an assertion we have not considered if it can be verified by actual experimental procedures; we merely assumed it, though its plausibility in macroscopic physics is obvious. In view of the above discussion, such a specification cannot be realized in practice (is void of operational meaning) and the theory based on it runs the risk of leading to results contrary to the actual situation. Although the classical theory is logically self-consistent, eventually a time must come when the theory is to be compared with results obtained by concrete experimental procedures. Any restriction on the experimental possibilities may not be ignored in the theory or contradictions may arise. The restriction of Eq. (6–5) may be ignored in considering objects of macroscopic size; it is precisely for this reason that the classical theory may be successfully applied in macroscopic physics. Usually concepts and theories established in a limited range of validity are unthinkingly assumed to be valid beyond their proper areas of application, the uncritical generalization eventually leading to contradictions. The uncritical use of the classical concepts, particularly the precise specification of x and p_x, in atomic physics, is a good example. Ignoring the restriction of Eq. (6–5) gives rise to a number of paradoxes associated with the quantum phenomena. In investigating the operational meaning of the classical concepts we try to find just how far such concepts may be applied; the result is that the classical concepts must be restricted by Eq. (6–5) in order not to lead to erroneous conclusions.

We note that *the lowest limit of accuracy specified by Eq. (6–5) turns out to be identical with Eq. (2–38), the uncertainty relation derived from the solutions of the Schrödinger equation.* This equivalence means that the physical principles underlining the results of the hypothetical experiments are already contained in the mathematical formalism of quantum mechanics and that the hypothetical experiments furnish concrete illustrations of the uncertainty relation. The discussions of the hypothetical experiments thus provide us with the physical interpretation of the mathematical result of quantum mechanics. As mentioned earlier, the uncertainty relation thus serves as a point of departure for exploring the physical meaning of the new mechanics. In addition, it sets a limit within which the classical concepts may be valid. It is therefore convenient as well as legitimate to postulate the validity of Eq. (6–5) as a general principle. When we do so, the principle embodied in Eq. (6–5) is called the *uncertainty principle*.

Classical mechanics makes deterministic predictions of the future of a system once the initial conditions are given. To specify the initial conditions we require the simultaneous information of position and momentum. Equation (6–5) shows that the simultaneous, exact specification of position and momentum at a given time t_0 is impossible and therefore we have no way of making deterministic predictions.

Another point may be made here in connection with the single-slit experiment. Before entering the slit, the electron has no velocity component along the y-direction. After passing through the slit, it has one. Thus the momentum of the electron is changed by the interposition of the slit. This is an illustration of the assertion that an act of observation, in this case the measurement of position by means of a slit, disturbs the system being observed and changes it from one state of motion to another.

It may be emphasized that Eq. (6–5) places a restriction on the uncertainties of position and momentum along the same direction. It does not imply the same for position and momentum along different directions, e.g., Δx and Δp_y (Problem 6–3).

In the single-slit experiment, the registration of the electron on the screen may be regarded as a second measurement of position. After this is done, we may draw a line from the slit to the spot on the screen and consider this line as the direction of motion of the electron.* Thus the momentum of the electron in the time period between the two successive position measurements at t_0 and t_1 may be known exactly. Consequently the position and momentum of this particular electron immediately after the first position measurement, $t = t_0 + 0$, may be known exactly by retroactive calculation. This result seems to contradict the uncertainty principle. However, it may be remarked that such simultaneous information does not help the cause of determinism for predicting the future from specified initial conditions. The simultaneous information of position and momentum obtained immediately after t_0 is not useful for predicting the system from t_0 to t_1, because the information is not yet available before the second measurement is performed at t_1. Also it is not useful for predicting future behavior after t_1, because the act of observation at t_1 disturbs the state of motion so that the information at t_0 is no longer the correct initial condition for the system after t_1. Accordingly, the uncertainty principle should be understood to mean that *it is impossible to specify both position and momentum exactly for the purpose of predicting the future behavior of a physical system.*

We shall now discuss another hypothetical experiment to illustrate the Heisenberg uncertainty principle. A microscope is used as the apparatus to determine the position of an electron† (see Fig. 6–3). Light must be used to illuminate the object; let the wavelength be λ, the direction being along the y-axis. Assume that the objective lens subtends an angle of 2θ at the point where the electron is situated. The theory of microscope tells us that the resolving power of the apparatus is such that a distance

$$\Delta x \sim \frac{\lambda}{\sin \theta} \tag{6–6}$$

---

* The uncertainty of this direction, due to the width of the slit which is assumed to be much narrower than the central band of the diffraction pattern as is usually the case in diffraction experiments, may be lessened by increasing the distance between the slit and the screen.

† N. Bohr, *Nature*, **121**, 580 (1928).

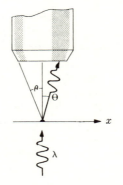

FIGURE 6–3

may be resolved. In other words, the uncertainty in the position determination in the present experimental arrangement is $\lambda/\sin\theta$. On the other hand, light may transfer momentum to the electron. As noted in the discussion of the Compton effect, the disturbance is finite and uncontrollable. After the interaction the photon may propogate in any direction Θ within the angle 2θ to enter the microscope. By conservation of momentum the electron is to receive a recoil momentum along the x-axis the amount of which, $(h/\lambda)\sin\Theta$, depends on the direction of the scattered photon Θ and therefore is also uncertain. The range of this recoil momentum is from 0 to $(h/\lambda)\sin\theta$. Thus

$$\Delta p_x \sim \frac{h}{\lambda}\sin\theta. \tag{6–7}$$

It follows that

$$\Delta p_x\,\Delta x \sim h, \tag{6–8}$$

and Eq. (6–5) is again reproduced.

When more than one photon is used, the uncertainty of position will be reduced but that of momentum will be enlarged. Though the recoil momenta of the photons add at random they do not cancel exactly. It may be shown that Eq. (6–8) holds also for the multiphoton experiment (Problem 6–4).

From the uncertainty relation between position and momentum, another relation may be derived. Let v and E be the velocity and energy corresponding to momentum p_x. Then,

$$v\,\Delta p_x\,\frac{\Delta x}{v} \gtrsim h, \tag{6–9}$$

or

$$\Delta E\,\Delta t \gtrsim h. \tag{6–10}$$

where ΔE is the uncertainty of energy corresponding to the uncertainty of momentum Δp_x, and Δt is the uncertainty in time within which the particle (or the wave packet) passes over a fixed point on the x-axis. Equation (6–10) expresses the uncertainty relation between energy and time.

We shall not describe the many other hypothetical experiments for the measurement of position (by scintillations, cloud chamber tracks, etc.), of momentum (by the Doppler effect, magnetic deflections, etc.), of time and energy. In all cases the validity of the uncertainty relation, Eq. (6–5) or (6–10), may be established.* Although no general proof is given, the following argument will help to show the plausibility of the general validity of the uncertainty relation.

A physical measurement in atomic physics is based on an interaction, or an exchange of energy and momentum, between the measuring apparatus and the atomic system being measured. By the change of the energy and momentum of the apparatus, we may deduce some information about the system by invoking conservation of energy and momentum. If a system does not interact with any of our apparatus and thus leaves no trace of any kind, we will never know of its existence and it will not make its appearance in physics. We may use as probing tools either matter or radiation, both carriers of energy and momentum. Let us first consider radiation. The propagation of radiation may be described by a wave, and a wavelength may be assigned. The energy and momentum of the wave may be localized in space-time (as in a wave packet) but the space extension is at least one wavelength and the time extension is at least one period of vibration. When interaction takes place we know only that the energy and momentum exchange is taking place within this space-time extension, but we have no way of knowing exactly at which point and at what time. Therefore the event of interaction is to be registered in space-time within the following uncertainties:

$$\Delta x \gtrsim \lambda, \qquad \Delta t \gtrsim \frac{1}{\nu}. \tag{6–11}$$

These are the uncertainties of the space-time information of the system being observed. On the other hand, the minimum amount of energy and momentum that may participate in an interaction are, according to the Einstein relations,

$$p_0 = \frac{h}{\lambda}, \qquad E_0 = h\nu. \tag{6–12}$$

After the interaction the uncertain amounts of momentum and energy transferred may be anything from 0 to p_0 and 0 to E_0 respectively. Thus

---

* See Heisenberg, *op. cit.*, and Bohr, *op. cit.*

the uncertainties of momentum and energy of the system being observed are

$$\Delta p \gtrsim \frac{h}{\lambda}, \qquad \Delta E \gtrsim h\nu. \tag{6-13}$$

From Eqs. (6–11) and (6–13) it follows that

$$\Delta p \, \Delta x \gtrsim h, \qquad \Delta E \, \Delta t \gtrsim h; \tag{6-14}$$

these are the Heisenberg uncertainty relations. Next let us consider matter as an agent of observation. Since the dual behavior of matter expressed in the de Broglie relations is exactly the same as that of radiation expressed in the Einstein relations, the same results follow. Thus, we have the Heisenberg uncertainty relations again. As matter and radiation seem to be the only agents available for making measurements, the conclusion seems unavoidable that in all measurements the uncertainty relations hold. This argument makes it clear that the uncertainty relations are necessary consequences of the Einstein relations and the de Broglie relations, and are a manifestation of the wave-particle duality of all physical entities. In fact, in all of the hypothetical experiments ever considered, both the wave and the particle properties of either radiation or matter are invoked to arrive at the uncertainty relation. This free use of two conflicting views of a physical entity does not involve any logical contradiction, since we simply take the dual property of both radiation and matter as *facts* and in no case have we used two conflicting *theories*.

That our knowledge concerning an atomic system is derived from the interaction between the system and an apparatus was emphasized by Bohr who advocated a point of view termed *complementarity*. If the disturbance incurred in a process of measurement cannot be eliminated and separated from the intrinsic behavior of the system, then different sets of information obtained under different experimental arrangements may not be correlated and understood within the framework of a single picture, either that of a wave or of a particle. They bring out various aspects of the system and are complementary to one another. It is characteristic that when one piece of information becomes more accurate, another becomes more obscure. Thus in the microscope experiment as well as in the single-slit experiment, the more exact the position information is, the more uncertain the momentum information, and vice versa. The position and momentum information thus are complementary. Following this we conclude that the space-time description and the causal description (energy and momentum conservation) are complementary; kinematics and dynamics are complementary; the wave picture and the particle picture are complementary. In classical physics where the interaction between the observed and the observer is ignored, all physical quantities, kinematical and dynamical, such as position and momentum, may be specified precisely. A complete description of a physical system is characterized by a complete, precise specification of all physical quantities involved. Such a specification gives rise to a clear-cut physical picture, either that of a wave or of a particle. On the other hand, in atomic physics where the interaction between the observed and the observer is considered to be important, only an incomplete specification of pairs of complementary quantities is possible. The complete description of a physical system can be formed only by putting together all such complementary information. This inexactness also allows us to make a complementary use of two conflicting pictures, i.e., the wave picture and the particle picture, for describing a given physical system. Having emphasized the complementary nature of the physical description, Bohr then pointed out that such a description is provided precisely by the mathematical formalism of quantum mechanics.

6–2 Quantum-mechanical description of physical processes. The law of causality. Since the physical quantities may not be exactly specified, a measurement of a quantity may result in a number of possible values, each having a definite probability of occurring. Therefore, the main concern of a physical theory is the prediction of the possible outcomes of an observation. Also, an act of observation disturbs the system being observed to a finite, uncontrollable amount, so that the state of motion of the system after the measurement becomes completely unrelated to that before. Thus the most a theory can predict seems to be the outcome of a particular measurement at a particular time instead of a complete picture of the system's future. These conclusions, derived from an analysis of the elementary physical concepts, are actually embodied in the formalism of quantum mechanics. There, the important quantity is the wave function $\Psi(x, t)$ which, according to Born's assumptions, tells us which positions may be observed with what probabilities upon a measurement of position at time t, and which momentum or energy values may be observed with what probabilities upon a measurement of momentum or energy at time t. *We shall assume further in the formalism of quantum mechanics that the prediction given by a wave function $\Psi(x, t)$ applies only for the time before an actual measurement is carried out.* After a measurement has been performed, the physical system is changed to a new state of motion, and we have to find a new wave function, unrelated to the one before the measurement, to represent the new state of motion of the system.

The assumption that a sudden change of the wave function takes place after a measurement is an integral part of the formalism of quantum mechanics. Since quantum mechanics is committed to describe *discontinuous* processes (quantum jumps) by a *continuous* differential equation (the Schrödinger equation), an element of discontinuity is unavoidable in the interpretation of the solution of the differential equation. From the operational point of view, a quantum jump is ascertained only by an act of observation. It is thus most natural to introduce the discontinuity by assuming a sudden change of the wave function to be associated with an act of observation. Einstein once expressed his concern over the statistical interpretation of the de Broglie wave: if an electron is observed at point A, this immediately wipes out any probability of its being observed at point B. Such a correlation is not provided for in the ordinary wave theory. The assumption of a sudden change of the wave function after a measurement provides precisely the mathematical instrument for describing such a correlation. It also made it possible to describe discontinuous processes by a continuous differential equation.

Quantum mechanics is a complete system in the sense that it has answers for all questions that may legitimately be asked within the system. (In this sense the old quantum theory is not complete.) At a given initial time t_0 the particle may be known with some position and momentum information described by a wave function $\Psi(x, y, z, t_0)$. The time-dependent Schrödinger equation, which is the equation of motion in quantum mechanics, may be written as follows:

$$\frac{\partial \Psi}{\partial t} = - \frac{\hbar}{2Mi} \left(\nabla^2 \Psi - \frac{2M}{\hbar^2} U\Psi \right). \tag{6–15}$$

Since the solution of a first-order differential equation in t is completely determined by the initial condition at t_0, the specification of $\Psi(x, y, z, t_0)$ completely determines $\Psi(x, y, z, t)$ by virtue of the equation of motion. Therefore the specification of initial position and momentum information by a wave function $\Psi(x, y, z, t_0)$ completely determines the possible outcomes of any measurement at any time t by the wave function $\Psi(x, y, z, t)$. The problem is thus completely solved. After a measurement has been performed, we are not able to predict anything from $\Psi(x, y, z, t_0)$, because of the uncontrollable disturbance incurred.

It seems appropriate at this moment to discuss the *law of causality* in quantum mechanics. In classical mechanics, the specification of initial conditions $(x_0, y_0, z_0;\ p_{x0}, p_{y0}, p_{z0})$ completely determines the future of a dynamical system $[x(t), y(t), z(t);\ p_x(t), p_y(t), p_z(t)]$. This fact may be regarded as the mathematical expression of the *causality law in classical mechanics*, which is deterministic. A similar situation exists in quantum mechanics. The specification of initial wave function $\Psi(x, y, z, t_0)$ completely determines the wave function $\Psi(x, y, z, t)$ which describes the future of the dynamical system. This fact may thus be regarded as the mathematical expression of the *causality law in quantum mechanics,* which is probabilistic. The physical meaning of the two causality laws is different. In classical mechanics, it means that for identically prepared systems, the results of similar observations are the same. This follows directly from the mathematical expression of the causality law and the assumption that the disturbance due to an act of observation may be ignored. In quantum mechanics, for identically prepared systems, the results of similar observations may not be the same. Thus the causality law of classical mechanics does not hold in quantum mechanics. On the other hand, the fact that a definite probability distribution of dynamical information at an initial time t_0 specified by $\Psi(x, y, z, t_0)$ leads to a definite probability distribution of the future dynamical information specified by $\Psi(x, y, z, t)$, seems to indicate the existence of a certain sense of causality. It is on this basis that the quantum-mechanical causality law is established. The following discussion should make clear the operational meaning of causality in quantum mechanics.

Let us consider the scattering of α-particles by a heavy atom. We know how to prepare a beam of α-particles by using a source and several collimating slits. The force field of the nucleus of the heavy atom is also known. Our problem is to find what will happen when a beam so prepared passes through this force field. The experimental results are such that the particles are scattered in all directions with definite relative intensities. Thus starting from a well-defined initial situation (a specified beam), we end up with a definite final result (differential scattering cross section). This is a good illustration of the operational meaning of the causality law. If we can determine $\Psi(x, y, z, t_0)$ from the initial information describing the beam (namely, the source of the α-particles and the arrangement of the collimating slits, etc.—enough information to specify the beam uniquely), then we are able to make definite predictions of the differential scattering cross section, for the mathematical machinery of quantum mechanics completely determines $\Psi(x, y, z, t)$ from $\Psi(x, y, z, t_0)$, and the differential cross section may be derived from $\Psi(x, y, z, t)$ according to the physical interpretation of the wave function. It is thus clear that the causality law in quantum mechanics has an operational meaning if we know how to determine the wave function $\Psi(x, y, z, t_0)$ from the initial physical information.

The beam of α-particles described above has a definite density distribution which is proportioned to $|\Psi(x, y, z, t_0)|^2$. If we know the density distribution experimentally, the amplitude of the wave function $|\Psi(x, y, z, t_0)|$ may be determined. To ascertain the density distribution, we may prepare a large number of identical beams according to the same specifications, perform accurate position observations on each of them, and obtain the statistical distribution of the positions. The result approaches the density distribution function when the statistical accuracy is indefinitely increased. However, the amplitude

$|\Psi(x, y, z, t_0)|$ alone is not enough to determine the wave function because we have not determined the phase $\varphi(x, y, z, t_0)$ of the wave function:

$$\Psi(x, y, z, t_0) = |\Psi(x, y, z, t_0)|e^{i\varphi(x, y, z, t_0)}. \tag{6-16}$$

(The phase indeterminacy representing incomplete initial conditions has an important role to play in quantum statistical mechanics.) In order to determine $\varphi(x, y, z, t_0)$ we have to get additional information about the beam, such as momentum. It is thus necessary to prepare another set of identical beams, perform accurate momentum measurements on each of them, and obtain the statistical distribution of the momentum values. The result approaches the probability distribution function of momentum. This procedure enables us to determine experimentally the squares of the Fourier coefficients of $\Psi(x, y, z, t_0)$; we thus have obtained additional information for the purpose of determining $\varphi(x, y, z, t_0)$. Mathematically, given two *arbitrary* distribution functions of position and momentum for the purpose of determining $\Psi(x, y, z, t_0)$, we may have too many conditions to satisfy. For example, when the two arbitrary distributions did not conform to the uncertainty relation, the corresponding $\Psi(x, y, z, t_0)$ simply did not exist. Thus the two distributions cannot be independent of each other; they have to be compatible (at least, the uncertainty relation must be satisfied). Therefore the experimental information obtained in the two sets of operations is more than enough to determine $\Psi(x, y, z, t_0)$. Once $\Psi(x, y, z, t_0)$ has been determined, the causality law in quantum mechanics has an operational meaning, as we have shown above. Incidentally, if there existed a system, the position and momentum information of which, obtained according to the above procedure, could not be expressed in terms of a wave function, then quantum mechanics would not be applicable to this system and would have to be revised. That both position and momentum information are needed to specify the initial condition $\Psi(x, y, z, t_0)$ is evident in the classical limit, for classical mechanics needs both position and momentum to specify the initial condition.

We shall illustrate in the following and in Fig. 6–4 the way by which physical processes are described according to quantum mechanics. Consider a particle of mass M moving in a potential field. We first write down the time-dependent Schrödinger equation. At the initial time $t = 0$, if we know exactly the wave function $\Psi(x, y, z, t_0)$ describing the motion of the particle (obtained by the series of operations described above), then the time-dependent Schrödinger equation, with $\Psi(x, y, z, t_0)$ as initial condition, gives rise to a unique solution $\Psi(x, y, z, t)$ from which we may predict the position and momentum information of the particle at any later time. At a later time t_1 the wave function is $\Psi(x, y, z, t_1)$ which means: (1), if a position measurement is made at t_1, the probability of finding the particle at (x, y, z) within $dx\, dy\, dz$ is $|\Psi(x, y, z, t_1)|^2\, dx\, dy\, dz$; (2), if a momentum measurement is made at t_1, the probability of finding the particle having momentum p is the square of the absolute value of the Fourier

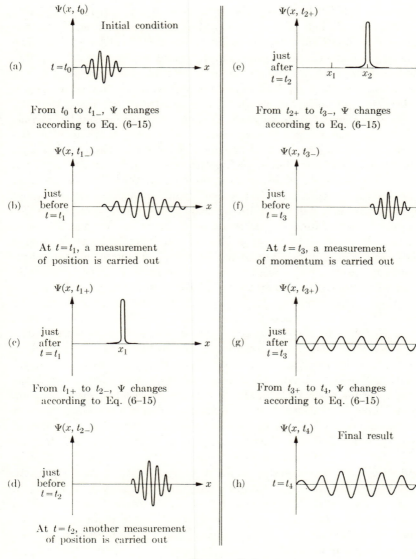

FIGURE 6–4

coefficient a_p of $\Psi(x, y, z, t_1)$. This is the legitimate information that we may hope to obtain and it is provided by quantum mechanics. If a position measurement is carried out at the time t_1, as a result of which the particle is found to be at x_1 (consider the x-coordinate only for simplicity), then the position and momentum information of the particle just after the measurement will be drastically different from that just before. Prior to time t_1 the particle has various probabilities of being observed at various positions [Figs. 6–4(a) and 6–4(b)]; after t_1, the particle is definitely known to be at x_1. Because of this change in position and momentum information, we need a new wave function to describe the particle just after t_1. A wave packet centered at x_1 with a very narrow width, denoted by $\Psi(x, t_{1+})$, may serve the purpose [Fig. 6–4(c)]. This illustrates the point that the wave function describing a physical system undergoes a sudden change right after a measurement is carried out. The wave function after the measurement is completely unrelated to that before, as the position x_1 is not dictated by the previous wave function and is quite arbitrary. Thus the prediction of the Schrödinger equation is valid only up to the moment just before a measurement, after which a new wave function unrelated to the previous one has to be introduced. This sudden, unpredictable change of the wave function, as mentioned earlier, is the mathematical expression of the fact that an act of observation disturbs the system being observed and changes its state of motion by an uncontrollable amount. The sudden change of the wave function represents a sudden change in position and momentum information of the system due to the act of observation. After the time t_1, the system will evolve according to the dictation of the time-dependent Schrödinger equation with the wave function $\Psi(x, t_{1+})$ as the initial condition until a second measurement is made. At a later time, t_2, the solution of the wave equation, denoted by $\Psi(x, t_{2-})$, specifies the probability distribution of position and momentum values [Fig. 6–4(d)]. If another measurement of position is carried out at t_2, as a result of which the particle is found to be at position x_2, then the wave function $\Psi(x, t_{2-})$ will undergo a sudden change and become a new wave function $\Psi(x, t_{2+})$, which will be a narrow wave packet centered around x_2, completely unrelated to $\Psi(x, t_{2-})$ [Fig. 6–4(e)]. After the time t_2, the system evolves again according to the dictation of the time-dependent Schrödinger equation with $\Psi(x, t_{2+})$ as the initial condition. The solution of the wave equation at a later time t_3, denoted by $\Psi(x, t_{3-})$, determines the probability distribution of the position and momentum information of the system at t_3 [Fig. 6–4(f)]. If now a momentum measurement is carried out and the momentum value is found to be p_1, then the wave function right after the measurement, denoted by $\Psi(x, t_{3+})$, will be a plane wave $e^{ip_1 x}$ [Fig. 6–4(g)]; once again the wave function has suffered a sudden change due to an act of observation. Since the result of the momentum measurement may be $p_2, p_3, \ldots$ as well as p_1, the wave function after t_3 may be plane waves of other wavelengths $\lambda_2, \lambda_3, \ldots$ as well as λ_1. The wave function $\Psi(x, t_{3+})$ is thus unrelated to the previous wave function $\Psi(x, t_{3-})$. After the time t_3, the system again evolves according to the dictation of the time-dependent Schrödinger equation with $\Psi(x, t_{3+})$

as the initial condition [Fig. 6–4(h)]. The same situation will arise over and over again whenever measurements are made.

It may be remarked that either an exact specification of the position or an exact specification of the momentum constitutes a complete specification of the initial condition, and the question of phase indeterminacy does not arise. When the position is exactly x_1, the wave function $\psi(x)e^{i\varphi(x)}$ is zero everywhere except at x_1, and the phase factor $e^{i\varphi(x)}$ reduces to a trivial constant $e^{i\varphi(x_1)}$. When the momentum is exactly p_1, the wave function is $e^{ip_1x+i\varphi_0}$ (usually φ_0 is set equal to zero) and the phase factor $e^{i\varphi(x)}$ is again a trivial constant $e^{i\varphi_0}$. Consequently, in each of the above-discussed measurements the position or momentum determination gives us enough information to determine the wave function to be used as the initial condition for the prediction of the future of the system.

6–3 Explanation of the quantum phenomena. Let it be stated categorically at the very outset that quantum phenomena owe no explanation for their appearance. These phenomena, for example, the wave property of matter, the quantization of energy, and the penetration of potential barrier, were regarded as paradoxical because they were not comprehensible in terms of the concepts and laws of classical mechanics. However, they are satisfactorily accounted for in quantum mechanics and in the new theory they are no longer associated with conceptual difficulties. Still, it is instructive and illuminating to see just how the difficulties of classical mechanics disappear in quantum mechanics. As noted earlier, quantum mechanics implies the uncertainty relation while classical mechanics does not. It is thus quite possible that the quantum phenomena are related to the uncertainty relation. Furthermore, we have mentioned that the classical concepts must be restricted by the uncertainty relation. In the following we shall show that when the classical concepts, particularly that of the trajectory of a particle, are limited by the uncertainty relation, the paradoxes associated with the quantum phenomena disappear. Loosely speaking, the quantum phenomena may thus be explained by the uncertainty principle. However, the uncertainty relation is not an exact equation (it is not stated by an equality sign, but by "$\gtrsim$"). It is not the expression of a fundamental law; it is merely a loosely stated relation marking out one outstanding feature of the new mechanics. Therefore we cannot hope to deduce the quantitative details of all quantum phenomena from the uncertainty relation. (For the same reason we cannot hope to *derive* a system of mechanics starting with the uncertainty principle as the fundamental postulate.) In addition to this approach based on the uncertainty principle, we may take an alternative approach to explain the quantum phenomena. The discussion of Section 2–7 shows that quantum mechanics may be analyzed within the framework of the Hamiltonian theory of classical mechanics and that quantum mechanics differs from classical mechanics by the introduction of a quantum potential operating among the probability elements. The quantum potential, according to Eq. (2–65), is expressed by

$$U_q = - \frac{\hbar^2}{2M} \frac{\nabla^2 A(x, y, z)}{A(x, y, z)}, \qquad (6\text{–}17)$$

where A is the amplitude of the wave function (or the square root of the density distribution function of the probability elements). From this mathematical point of view, anything found in quantum mechanics but not found in classical mechanics may have its origin rooted in the quantum potential. As a result, the quantum phenomena may be related to the quantum potential and in the following discussion this relation will be spelled out in detail. The two approaches probe two facets of the mathematical structure of the new theory, and we hope they will shed light on the physical meaning of quantum mechanics.

(A) *The wave property of matter.* To show that the wave property of matter may be inferred from the uncertainty principle, we may reverse the argument of the single-slit experiment discussed in Section 6–1. Since the slit limits the position of the electron to within Δy, there is, according to the uncertainty principle, a corresponding Δp_y equal to about $h/\Delta y$; the electron thus emerges from the slit within an angle of about $h/p \, \Delta y$. The electron distribution is thus similar to the central maximum of a diffraction pattern of light with wavelength h/p. Thus the wave property may be inferred and the de Broglie relation obtained. However, the uncertainty principle does not enable us to predict the intensity distribution quantitatively, or to predict the existence of the secondary maxima beyond the central maximum. As mentioned before, the reason is that the uncertainty relation is not the expression of an exact law.

On the other hand, from the viewpoint of the Hamiltonian theory, the wave property of matter may be shown to be a result of the quantum potential of Eq. (6–17). A system of particles are in equilibrium when the resultant forces on them are zero. For a free particle, the distribution of probability elements of an energy eigenfunction is stationary and therefore is in equilibrium. Hence, the quantum force must be zero. This requires the quantum potential to be a constant independent of position. For U_q equal to a constant, A must be an oscillatory or exponential function. When A is an oscillatory function, the distribution of probability elements exhibits a wave character. Thus the quantum potential leads to the appearance of what was called the wave property of matter.

This argument also helps us to understand the appearance of oscillatory and exponential functions in the wave functions discussed in the last three chapters. It also justifies our previous conjecture regarding the corrugated structure of the wave functions of the linear harmonic oscillator. The corrugated structure (representing a modulated oscillatory function) is necessary to make the quantum force vanish.

(B) *The quantization of energy.* We consider the linear harmonic oscillator as an example. According to classical mechanics, the oscillator may have any amount of energy from 0 to ∞. The state of motion corresponding to the smallest possible energy $E = 0$ is no motion at all. In this special case we know exactly the position of the oscillator; it stays always at the origin. We also know exactly the momentum of the oscillator, which is zero. Therefore, both Δx and Δp_x are zero and their product is zero. According to the uncertainty principle, this product must not be less than h, a finite quantity. Therefore, the state of motion specified by $E = 0$ is impossible. If E is not zero but barely above zero, the oscillator, according to classical mechanics, will be confined in a very small region in the neighborhood of the origin and the momentum will be barely above zero. Thus Δx and Δp_x are both very small. If the product of Δx and Δp_x is smaller than h, such a motion still will be excluded by the uncertainty principle. Hence no motion will be possible for energy values from zero up to a certain value E_0 large enough that the uncertainty relation may be satisfied. This energy E_0 is the energy of the first quantum state. Thus the energy is quantized, i.e., it cannot take on any arbitrary value, at least in the immediate neighborhood of $E = 0$ on the energy scale. Actually, for the first quantum state of the linear harmonic oscillator, the Δx may be estimated to be about $\sqrt{\hbar/M\omega}$. The momentum corresponding to the energy $\frac{1}{2}\hbar\omega$ is $\sqrt{2M\frac{1}{2}\hbar\omega}$. Therefore the momentum ranges from 0 to $\sqrt{M\hbar\omega}$. Consequently,

$$\Delta x \, \Delta p_x \sim \sqrt{(\hbar/M\omega)} \, \sqrt{M\hbar\omega} = \hbar, \qquad (6\text{–}18)$$

showing that the appearance of the first quantum state actually is dictated by the uncertainty principle. However, by the uncertainty principle alone, we cannot show that the energy after the first quantum state is also quantized. Again the reason is that the uncertainty principle is not an exact law.

We cannot present very illuminating arguments for the quantization of energy from the viewpoint of the quantum potential, yet it is not impossible to argue in this direction. For an energy value not on the list of energy eigenvalues, the solution of the time-independent Schrödinger equation diverges at infinity. We may argue that the quantum force among the probability elements acts in such a way that all probability elements are repelled to infinity. When the energy equals one of the eigenvalues, the quantum force balances itself out with all probability elements staying in a finite region. Therefore, all oscillators observed in a finite region must have energy values on the list of eigenvalues, i.e., the energy is quantized.

(C) *The penetration of potential barrier.* We consider the emission of an α-particle by a radioactive nucleus as an example. The α-particle is confined in a potential well surrounded by a Coulomb potential barrier (Fig. 5–5). If the α-particle stays in the well all the time, then we know the position of this α-particle, i.e., inside the nucleus, to an accuracy Δx equal to the width D of the potential well. According to the uncertainty principle, $\Delta x = D$ is accompanied by $\Delta p \cong h/D$. The spread in momentum value causes a spread in energy value. But experimental results are that the α-particles do not have such a spread in energy. The premise $\Delta x = D$ thus cannot be true. Therefore the α-particle cannot stay always inside the nucleus, but instead may be found outside the potential barrier. Another argument based on the uncertainty principle may be made concerning the relation between energy and time. Since the α-particles are observed to have the same energy, we have $\Delta E = 0$. This leads to the result $\Delta t \to \infty$, where Δt is the time interval within which the particle goes through a certain point in space (say, the edge of the nucleus); this time interval is thus infinitely long. This means that we have no way of predicting when the α-particle may come out of the nucleus, whether in the next hour or a thousand years hence. The bearing of this point on causality has been discussed in Chapter 5.

There is one question left unanswered: How can an α-particle go through a classically forbidden region where the potential energy is higher than the total energy of the particle? (In other words, how can an α-particle have a negative kinetic energy in this region?) Actually this situation is found not only in α-radioactivity. In the linear harmonic oscillator, all eigenfunctions have a tail extending to infinity in the classically forbidden region. In the potential well problem we also know that the wave function is not vanishing in the classically forbidden region. The existence of a nonvanishing wave function in a forbidden region means there is a nonvanishing probability to have the particle observed there; the question might be asked how can a particle be found in such a region.

The origin of this question is the classical concept that motion is continuous. However, we can best answer it from the point of view of the quantum force. Within the framework of the Hamiltonian theory, the wave function may be regarded as representing the distribution of the probability elements, each moving under the influence of the classical potential plus the quantum potential. The above question may thus be reformulated as follows: How can some probability elements move into the classically forbidden region? The answer is that *the quantum force is not a conservative force.* Therefore conservation of energy in the classical sense is not valid. Those probability elements moving into the classically forbidden region receive the necessary energy to do so from the nonconservative quantum force. We hasten to add that this has no physical significance. The quantum force is but a mathematical fiction. This explanation merely brings out the formal mathematical relations and points out the origin of the paradox.

One other point remains to be mentioned. Suppose that an observation on position is made in the classically forbidden region and according to the probability specified by the wave function, a particle is actually observed there. What is the energy value of the particle? Take the linear harmonic oscillator as an example. Suppose the wave function concerned is an eigenfunction of energy value E_n. Before the observation, the energy value of the particle is definitely E_n and its position distribution is $|\psi_n(x)|^2$, according to Born's assumptions. After the observation, if the particle is found at a point in the forbidden region where the potential energy is greater than E_n, the particle is to be represented by a new wave function which is a narrow wave packet. The wave packet expresses the momentum information from which we derive a probability distribution of the kinetic energy. The total energy, being the sum of the kinetic energy, which is always positive, and the potential energy, undoubtedly will be greater than E_n, for the potential energy alone is greater. We may ask how can the energy be increased and where has the energy come from. The answer is that an act of observation disturbs the system being observed and imparts to it an uncontrollable amount of energy and momentum. The particle observed in the classically forbidden region makes its observation possible at the expense of the energy of the measuring apparatus.

Combining the mathematical and physical aspects, we may say that quantum mechanics generalizes classical mechanics by (1), introducing the probability elements in kinematics; and (2), introducing the quantum force in dynamics; the latter is responsible for the amount of energy or momentum imparted to a system by an act of observation and the former makes the amount uncertain.

What about the law of conservation of energy? Does it hold in quantum mechanics? First of all, the energy of a system after a measurement is made, in general, is not the same as before because of the transfer of an uncontrollable amount of energy from the measuring apparatus. However, between two measurements, when no energy transfer takes place, the energy information, specified by the probabilities $|a_n e^{-i(E/\hbar)t}|^2$ of finding the energy value equal E_n, is independent of time. That the distribution $|a_n|^2$ is time independent is considered as the quantum-mechanical law of energy conservation. (See Chapter 12 for further discussion.) We note parenthetically that in the classical limit this law reduces to the classical law of energy conservation.

6–4 Concluding remarks.* The development of quantum mechanics has elicited drastic changes in our physical concepts, mathematical methods, and philosophical thought. Many issues have fascinated, if not confused, its students: duality (wave-particle) versus unity, abstract algebra† versus mathematical analysis, indeterminism versus determinism, etc.

However, the revolutionary tide does not completely sweep away classical mechanics; instead, quantum mechanics finds for classical mechanics its rightful place in physical theory and clarifies the conditions under which it is valid. Quantum mechanics, successful in the microscopic realm, is actually a rational generalization of classical mechanics, incorporating in it new elements which are forced on us in view of the existence of·the quantum phenomena, though these elements may be rightfully ignored in the macroscopic realm.

A basic element thus introduced is the concept of *discontinuity* or more specifically the concept of *quantization (the indivisibility of the quantum)*. It is a fundamental concept because the Planck constant h is introduced, not derived. It cannot be subject to further analysis unless the Planck constant can be expressed in terms of other universal constants. It does not originate from any epistemological considerations; it is forced upon us by the experimental evidence which we described in some detail in Chapter 1. We have seen case after case where physical quantities are quantized. Once quantization is established, it becomes inevitable to regard a physical process as the progression of a series of individual, discontinuous, elementary processes (quantum jumps), each of which cannot be analyzed further. This then leads to the conclusion that a completely specified initial condition does not necessarily lead to a definite result at a later time. These concepts and results are incompatible with classical mechanics, in which a physical system evolves in a deterministic, continuous way. In order to incorporate these concepts in the framework of classical mechanics the concept of *probability* has to be introduced, perhaps less as a physical assumption but more as a logical instrument. In Chapters 1, 5, and 6, we have expounded time and again the relation between quantization and probability. Once we elect a probabilistic point of view, we are still left with the freedom of choice between at least two alternatives. Either we regard the concept of probability as an elementary concept which cannot be analyzed further or we consider it as a concept built upon another, more fundamental concept—*the probability amplitude*. The use of probability in classical statistical mechanics follows the first alternative. The choice we have to make in quantum mechanics is again based not on some epistemological considerations, but on the experimental

---

* Students are advised to review this section after having finished Chapter 12.
† See Chapter 12.

evidence. The concept of probability amplitude is forced upon us because it manifests itself in a variety of *interference phenomena:* in the diffraction of electrons, atoms, and molecules (discussed in Chapter 3), in the properties of the helium atom, in the homopolar bond of chemical binding, and in the phenomena of scattering of identical particles (these three we do not discuss in this volume). In the further development of quantum mechanics additional new elements have to be introduced, such as the *spin* and the *statistics*, which are not included in this volume. For the present purpose we may conclude from the above discussion that the concepts of *quantization* and *interference* are two independent, basic concepts that distinguish quantum mechanics from classical mechanics.

The development of the quantum theory thus revolves around the theme: how to generalize classical mechanics to accommodate the concepts of quantization and interference. The old quantum theory embraces the concept of quantization by incorporating in it the Wilson-Sommerfeld quantum condition which introduces Planck's constant h. But the concept of interference is not included. Though the concept of probability is introduced by the correspondence principle, the probability amplitude is not introduced. Thus it is only natural that the helium atom should present the first major failure of the old quantum theory. De Broglie's important contribution lies in the introduction of the concept of interference by means of the matter wave. In our formulation of quantum mechanics as presented in Chapter 2, Eq. (2–30), introducing h into the theory, serves the purpose of the quantum condition and gives rise to quantization. The concept of interference is incorporated into the theory through Born's two assumptions in which the probability is related to the *square* of the wave function itself or its Fourier coefficients. In matrix mechanics and in the general formulation of quantum mechanics, briefly described in Chapter 12, quantization is achieved by introducing the so-called *commutation relations* [see Eqs. (12–92, 93, 94)]. These relations are the quantum conditions because not only the Planck constant is thereby introduced but also they reduce to the Wilson-Sommerfeld quantum condition in the high quantum number region. On the other hand, the concept of interference finds expression in the *principle of superposition* (Section 12–4), of which Born's two assumptions are special cases. Throughout the history of the quantum theory, classical mechanics is always in the background and is generalized in each step of the development.

A word concerning the role of the uncertainty principle will be in order here. This principle is often considered as the fundamental principle, if not the fundamental law, of quantum mechanics and is usually accorded a place in the first page of a treatise. In this connection it may be mentioned that this principle as conventionally stated does not imply the concept of interference. In fact, the demand of the uncertainty principle may be satisfied by *a* classical, statistical mechanics by virtue of the Liouville theorem. This principle is a result of the quantum condition (Theorem 8, Section 12–2). Therefore it speaks part of the truth contained in the quantum condition (quantization). Its essence seems to be, as illustrated on previous occasions, that there exists a first quantum state, for atomic systems as well as for radiation fields, characterized by one unit of the universal quantum of action *h*.

That the quantum condition in its various forms is the mathematical expression of the physical fact of quantization, and the principle of superposition or its equivalent is the mathematical expression of the physical fact of interference, is the essence of the physical meaning of quantum mechanics.

Problems

6–1. The wave packet of a free particle tends to spread, whereas that of a linear harmonic oscillator does not. Will the wave packet of a particle moving in the potential $U = Kx^4$ spread? What is the general condition governing the spread of the wave packet?

6–2. Discuss the classical theory of the Compton effect. The angular dependence of the wavelength change may be attributable to the Doppler effect of the recoil of the electron.

6–3. Discuss the possibility of measuring x and p_y simultaneously.

*6–4. Derive the uncertainty relation in the multiphoton microscope experiment. [*Hint:* The recoil momenta due to the photons add themselves at random but do not cancel themselves out exactly. Make use of the result of the random walk problem in statistics.]

*6–5. A superposition of energy eigenfunctions of the linear harmonic oscillator, with their time factors attached, in the high quantum number region but limited to the even quantum numbers n, results in a wave packet at the origin at $t = 0$. Predict the future of this quantum-mechanical state, assuming that position, momentum, and energy measurements are being made at different times in the future in various order of succession. (The arbitrary phases of the eigenfunctions are assumed to be zero.)

6–6. Discuss the determination of the "trajectory" of an α-particle by its cloud chamber track from the point of view discussed in Section 6–2, each ionization of the water molecule being considered as a position measurement. Discuss the straggling of the track. (See also Problem 12–11.)

6–7. In determining the position of an electron by a microscope, the amount of momentum transferred to the electron is related to the momentum of the photon. If a measurement of the recoil of the microscope (now considered movable) is carried out to ascertain the momentum of the photon, are we able to escape the restriction of the uncertainty relation?

6–8. Discuss the result of Problem 5–5, i.e., Eq. (5–38), from the point of view of the uncertainty principle.

6–9. A beam of electrons passing through a single crystal displays a diffraction pattern of Laue spots. Considering the interaction between an electron and the atoms in the crystal, those atoms along the path of the electron affect the motion of the electron most. On the other hand, the sharpness of the diffraction spots, according to the theory of the diffraction grating, is essentially due to the existence of the atoms far away from the path of the electron. How is this paradox explained?

6–10. Show by an example that a wave function localized in a time interval Δt may be analyzed into monochromatic waves with a frequency range $\Delta \nu$ such that

$$\Delta t \, \Delta \nu \cong 1.$$

What is the relation between this equation and the uncertainty relation between energy and time? [*Hint:* Make use of the result of Problem 2–6 with the substitution of the variable x by t.]

*6–11. A stationary, monochromatic light source is placed in front of a slit which is opened and closed by a shutter. If the slit is left open for a time interval Δt, the photon appearing behind the slit will have an energy uncertainty ΔE, or equivalently the light will have a frequency uncertainty $\Delta \nu$; governed by an uncertainty relation. Derive this uncertainty relation by the wave theory point of view which considers space-time propagation, and then by the corpuscular theory which considers energy-momentum conservation. Show that the two viewpoints are equivalent. This equivalence means that the wave and the particle pictures are compatible if uncertainty relations are taken into consideration.

*6–12. Repeat the above problem with the following changes: no movable shutter is used and the light source is made to move with a velocity v parallel to the diaphragm so that only when the source moves over the opening of the slit does an appreciable amount of light come through the slit (the Einstein-Rupp experiment). The moving light source produces the same effect caused by the moving shutter in the preceding problem.

*6–13.# *Quantum theory.* Study the Einstein, Podolsky and Rosen issue on the relation between quantum theory and relativity theory.

*6–14.# *Quantum theory.* Einstein raised a question on a possible violation of the uncertainty relation between energy (mass of a clock) and time (indicated by the moving hand of the clock), on which Niels Bohr came up with an answer. Einstein was convinced that quantum theory is at least *self consistent*.

*6–15.# *Quantum theory.* Review the issue of Schrödinger's cat.

*6–16.# *Quantum theory.* Review the issue of Fermi's earth. Fermi used to tell students at Chicago that he did not understand why the wave packet representing the earth did not disperse to make the earth disappear. The problem has been studied by scientists including Murry Gell-Mann. The idea is that every of the infinitely many photons of the background microwave radiation hitting the earth forms a position measurement to restore the wave packet.

*6–17.# *Quantum theory.* Review the *Bell Inequality* that rules out any hidden variable in the interpretation of the quantum theory.

6–18.# Astrophysics. Does the uncertainty principle of position and momentum as related to the initial condition of the Big Bang have any relevance to the history of the subsequent expansion?

---

* Indicates more difficult problems.

\# Indicates new problems added to the expanded edition.

CHAPTER 7

GENERAL METHODS FOR ONE-DIMENSIONAL PROBLEMS

In Chapters 3, 4, and 5 we discussed three one-dimensional problems. The wave functions in these three problems are found to have many properties in common. We shall see that these are general properties of the one-dimensional wave function and that they may be derived qualitatively from the Schrödinger equation. The quantitative solution of a general one-dimensional problem is complicated. However, an approximate method, known as the WKB method, usually enables us to obtain the essential information on wave functions and energy eigenvalues. This method will be discussed.

7–1 Qualitative properties of the one-dimensional wave function. In Section 5–1 we saw that the wave function in the potential well problem consists of a section of a sine (or cosine) curve joined with two exponential curves which extend to infinity. The sine curve may cross the x-axis a number of times. We find that a similar situation exists in the wave function of the linear harmonic oscillator. In the classically permitted region the eigenfunction has a wavelike structure, crossing the x-axis a number of times; it extends with exponential tails into the classically forbidden region. These features will be shown to be general properties of all one-dimensional wave functions.

The Schrödinger equation for a one-dimensional problem specified by a potential function $U(x)$ is

$$\frac{d^2\psi}{dx^2} + \frac{2M}{\hbar^2} (E - U)\psi = 0. \tag{7-1}$$

Without solving the differential equation, many properties of the solution ψ may be obtained qualitatively. For convenience we define

$$g(x) = \frac{2M}{\hbar^2} [E - U(x)]. \tag{7-2}$$

Equation (7–1) may thus be written as follows:

$$\frac{d^2\psi}{dx^2} = -g\psi. \tag{7-3}$$

Equation (7–3) means that the curvature of $\psi(x)$ is determined by the

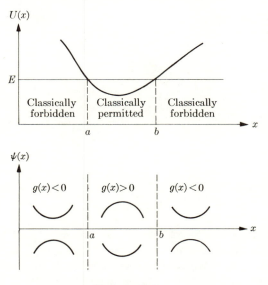

FIGURE 7–1

product of $g(x)$ and $\psi(x)$. In the classically permitted region where $g(x) > 0$, positive $\psi(x)$ corresponds to negative $d^2\psi/dx^2$, and consequently the curve of $\psi(x)$ above the x-axis is concave downward. Likewise, negative ψ corresponds to positive $d^2\psi/dx^2$, and the curve below the x-axis is concave upward. These facts may be summarized by saying that in the classically permitted region the curve of $\psi(x)$ is concave toward the x-axis. On the other hand, in the classically forbidden region where $g(x) < 0$, positive $\psi(x)$ corresponds to positive $d^2\psi/dx^2$ and the curve above the x-axis is concave upward. Likewise, negative ψ corresponds to negative $d^2\psi/dx^2$ and the curve of $\psi(x)$ below the x-axis is concave downward. These facts may be summarized by saying that in the classically forbidden region the curve $\psi(x)$ is concave away from the x-axis. The conclusions are represented graphically in Fig. 7–1.

Consider the solution of Eq. (7–1) for a given energy value E. First of all, the value of E cannot be less than the minimum of $U(x)$ if $U(x)$ has a minimum. Otherwise $g(x)$ would be negative for all values of x and ψ would diverge to infinity as $x \to \infty$. For E greater than the minimum of $U(x)$, the horizontal line representing energy E intersects the potential curve $U(x)$ at least twice. Let the two points of intersection be a and b. The region between a and b is classically permitted; the rest is classically forbidden. In the region $x > b$, if the curve is above the x-axis, its slope cannot be positive; otherwise, ψ diverges when $x \to \infty$. Similarly, if the curve is below the x-axis, its slope cannot be negative. Furthermore, the

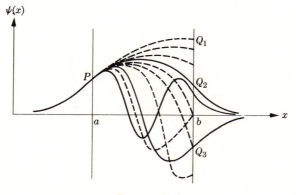

FIGURE 7–2

curve cannot cross the x-axis, since the curvature and slope after the crossing are such that ψ increases to infinity when $x \to \infty$. Consequently in the region $x > b$, the curve representing $\psi(x)$, either above or below the x-axis, must approach the x-axis asymptotically. As a result, the wave function in this region has no nodal point. The situation is similar in the region $x < a$. On the other hand, in the region between a and b, the curve representing $\psi(x)$ is allowed to cross the x-axis and therefore nodal points may exist. Incidentally, points a and b are the points of inflection.

Since both ψ and $d\psi/dx$ must be continuous at a and b, we may not be able to construct an acceptable wave function for an arbitrary value of E. This is illustrated in Fig. 7–2 and explained below. When E is just above the minimum of $U(x)$, $g(x)$ between a and b is very small and the corresponding curvature of $\psi(x)$ is small. Thus the curve $\psi(x)$ is almost a straight line such as the curve PQ_1 in Fig. 7–2. If ψ and $d\psi/dx$ of the two sides are matched at P, they may not be matched at Q_1. As a result no acceptable wave function may be obtained for such an energy value. Now let E increase. A corresponding increase in $g(x)$ and consequently in the curvature will result. When the curvature has become so large that the slope of the curve just before $x = b$ turns out to be negative, it becomes possible to satisfy the continuity conditions at $x = b$. As E increases, sooner or later we shall come across a certain value of E at which the slope of ψ to the right of $x = b$ equals that to the left, and an acceptable wave function may be constructed. This corresponds to the first quantum state and the energy is the first eigenvalue. It may be seen in Fig. 7–2 that the first quantum state is represented by a curve which has no nodal point (curve PQ_2). If we increase the energy beyond the first eigenvalue, then the slopes of ψ to the left and to the right of $x = b$ will no longer be matched. As E increases the point Q will move downward and eventually will come up again. At Q_3 the curve PQ_3 may be joined smoothly with a curve in the region $x > b$ to give rise to an acceptable wave function. This corresponds to the second quantum state. It is seen in Fig. 7–2 that this wave function has one nodal point. As E increases further we may construct additional wave functions in a similar manner. Before the next acceptable wave function is reached, the curve ψ has to cross the x-axis once more. Thus the number of nodal points increases accordingly, the nth quantum state having $n - 1$ nodal points.

7–2 Approximate solution by the WKB method. It is usually difficult to solve the Schrödinger equation with an arbitrary potential. For a potential as simple as that of the linear harmonic oscillator, the solution of the corresponding Schrödinger equation in Section 4–2 is fairly complicated. However, an approximate method, known as the WKB method after its originators Wentzel,* Kramers,† and Brillouin,‡ is available, by which important information on energy eigenvalues and eigenfunctions may be obtained. This method will be described below.

Let us rewrite Eq. (7–3) as follows:

$$\ddot{\psi} + g\psi = 0. \tag{7–4}$$

If g is a constant, we may distinguish two cases: (1), if $g > 0$, ψ equals $\sin \sqrt{g}\, x$ or $\cos \sqrt{g}\, x$; (2), if $g < 0$, ψ equals $e^{-\sqrt{-g}\, x}$ or $e^{+\sqrt{-g}\, x}$. The argument or the exponent is a constant $\sqrt{|g|}$ multiplied by x; the amplitude is a constant. If $g(x)$ is not a constant but nearly a constant, the solution will be only slightly different from that just mentioned. It is thus a reasonable guess that $\sqrt{|g|}$ will be replaced by its average value and $\sqrt{|g|}\, x$ will be replaced by $\int^x \sqrt{|g(x)|}\, dx$; also that the amplitude will be slowly varying. These guesses will be shown to be correct in the following. Let us write

$$\psi(x) = e^{iZ(x)}. \tag{7–5}$$

As $Z(x)$ reduces to $\sqrt{g}\, x$ or $\sqrt{-g}\, x$ in the first approximation, $Z(x)$ is not far from a linear function of x. A linear function has the property that its second and higher derivatives vanish. Therefore we may expect the second and higher derivatives of $Z(x)$ to be small. Substituting Eq. (7–5) in Eq. (7–4) we have the differential equation for $Z(x)$,

$$\dot{Z}^2 - i\ddot{Z} = g(x). \tag{7–6}$$

---

* G. Wentzel, *Zeits. f. Physik*, **38,** 518 (1926).
† H. A. Kramers, *Zeits. f. Physik*, **39,** 828 (1926).
‡ L. Brillouin, *Comptes Rendus*, **183,** 24 (1926).

The first-approximation solution of this equation may be obtained by setting $\ddot{Z}$ equal to zero. Thus

$$\dot{Z} = \pm\sqrt{g(x)}. \tag{7-7}$$

The solution is

$$Z = \int^x \pm\sqrt{g(x)}\,dx. \tag{7-8}$$

The first approximation of ψ is thus what we had guessed. We now proceed to obtain the second approximation. Let

$$\dot{Z} = \pm\sqrt{g(x)} + \epsilon(x) \tag{7-9}$$

where $\epsilon(x)$ is a small correction. Since $\epsilon(x)$ is a small quantity, its derivatives are even smaller. Differentiating Eq. (7–9) and neglecting $\dot{\epsilon}(x)$, we have

$$\ddot{Z} = \pm\frac{\dot{g}(x)}{2\sqrt{g(x)}}. \tag{7-10}$$

Substituting Eq. (7–10) in Eq. (7–6) and neglecting ϵ^2, we have

$$\epsilon = \frac{i}{4}\frac{\dot{g}(x)}{g(x)}. \tag{7-11}$$

It follows that

$$\dot{Z} = \pm\sqrt{g(x)} + \frac{i}{4}\frac{\dot{g}(x)}{g(x)}. \tag{7-12}$$

Therefore

$$Z = \pm\int^x \sqrt{g(x)}\,dx + \frac{i}{4}\log g(x).$$

Equation (7–5) thus becomes

$$\psi(x) = \frac{1}{\sqrt[4]{g(x)}}\,e^{\pm i\int^x \sqrt{g(x)}\,dx}. \tag{7-13}$$

When $g > 0$, the general solution of Eq. (7–4) thus takes the following form:

$$\psi = \frac{A}{\sqrt[4]{g(x)}}\sin\left(\int^x \sqrt{g(x)}\,dx\right) + \frac{B}{\sqrt[4]{g(x)}}\cos\left(\int^x \sqrt{g(x)}\,dx\right), \tag{7-14}$$

where A and B are arbitrary constants. When $g < 0$, the general solution may be written as follows:

$$\psi = \frac{C}{\sqrt[4]{-g(x)}}\,e^{\int^x \sqrt{-g(x)}\,dx} + \frac{D}{\sqrt[4]{-g(x)}}\,e^{-\int^x \sqrt{-g(x)}\,dx}, \tag{7-15}$$

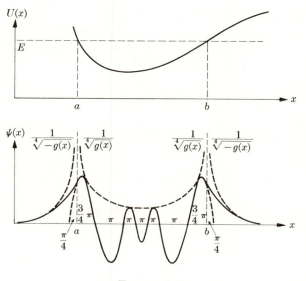

FIGURE 7-3

where C and D are arbitrary constants. Equation (7–14) applies to the region $a < x < b$ in Fig. 7–3; Eq. (7–15) applies to regions $x > b$ and $x < a$. A restriction on their applicability must be mentioned here. Both solutions fail in the immediate neighborhood of the points $x = a$ and $x = b$ because both $g(a)$ and $g(b)$ vanish and both Eqs. (7–14) and (7–15) diverge to infinity. The reason is that when g equals zero, the right-hand sides of Eqs. (7–10), (7–11), and (7–12) become infinite, and the assumption that $\ddot{Z}$ and ϵ are small is no longer valid. In the regions sufficiently remote from a and b, the assumption is valid and Eqs. (7–14) and (7–15) apply. We are thus confronted with two problems. First, what is the expression of ψ in the immediate neighborhood of the points $x = a$ and $x = b$? Second, how can the solutions in the regions to the left of a, to the right of a, and in the immediate neighborhood of a be matched to form a continuous solution (the same applies to the point $x = b$)? To solve the first problem we may approximate the potential function near the point a (or b) by a straight line and then shift the origin of the coordinate to a. Then $g(x)$ takes the form of x multiplied by a constant. The Schrödinger equation thus obtained may be solved in terms of the Bessel functions. Once this solution is obtained, the arbitrary constants A, B, C, D of the solutions in the regions to the right and to the left of a may be determined in such a way that these two solutions join smoothly with the solution in the immediate neighborhood of a. The procedure for joining the solutions is complicated; we shall state only the results, omitting the mathematical derivation. The results are that the constants of Eqs. (7–14) and (7–15) must be related in the following way (the solution in the immediate neighborhood of a (or b) will not be written down because it is seldom used in practice):

(1) The solution

$$\frac{\alpha}{\sqrt[4]{-g(x)}}\, e^{-\int_x^a \sqrt{-g(x)}\, dx}$$

to the left of a is related to the solution

$$\frac{2\alpha}{\sqrt[4]{g(x)}}\, \cos\left(\int_a^x \sqrt{g(x)}\, dx - \frac{\pi}{4}\right)$$

to the right of a, α being an arbitrary constant.

(2) The solution

$$\frac{\beta}{\sqrt[4]{-g(x)}}\, e^{+\int_x^a \sqrt{-g(x)}\, dx}$$

to the left of a is related to the solution

$$\frac{-\beta}{\sqrt[4]{g(x)}}\, \sin\left(\int_a^x \sqrt{g(x)}\, dx - \frac{\pi}{4}\right)$$

to the right of a, β being an arbitrary constant.

(3) The solution

$$\frac{\gamma}{\sqrt[4]{-g(x)}}\, e^{-\int_b^x \sqrt{-g(x)}\, dx}$$

to the right of b is related to the solution

$$\frac{2\gamma}{\sqrt[4]{g(x)}}\, \cos\left(\int_x^b \sqrt{g(x)}\, dx - \frac{\pi}{4}\right)$$

to the left of b, γ being an arbitrary constant.

(4) The solution

$$\frac{\delta}{\sqrt[4]{-g(x)}}\, e^{+\int_b^x \sqrt{-g(x)}\, dx}$$

to the right of b is related to the solution

$$\frac{-\delta}{\sqrt[4]{g(x)}}\, \sin\left(\int_x^b \sqrt{g(x)}\, dx - \frac{\pi}{4}\right)$$

to the left of b, δ being an arbitrary constant.

If the horizontal line representing energy E in Fig. 7–3 intersects the potential energy curve at two points a and b only, then of the two exponential solutions in the region $x > b$ or $x < a$, only the descending one may be acceptable because of the boundary condition at infinity. Thus only relations (1) and (3) need be considered. The solution ψ in the region $a < x < b$, joined smoothly with the others at a and b, may be determined by either relation (1) or (3). The two determinations must be identical. Hence

$$\frac{2\alpha}{\sqrt[4]{g(x)}} \cos\left(\int_a^x \sqrt{g(x)}\, dx - \frac{\pi}{4}\right) = \frac{2\gamma}{\sqrt[4]{g(x)}} \cos\left(\int_x^b \sqrt{g(x)}\, dx - \frac{\pi}{4}\right).$$

(7–16)

Equation (7–16) may be satisfied if

$$\left(\int_a^x \sqrt{g(x)}\, dx - \frac{\pi}{4}\right) + \left(\int_x^b \sqrt{g(x)}\, dx - \frac{\pi}{4}\right) = n\pi, \qquad n = 0, 1, 2, \ldots,$$

$$\gamma = (-1)^n \alpha.$$

(7–17)

The first condition may be rewritten as follows:

$$\int_a^b \sqrt{g(x)}\, dx = \left(n + \frac{1}{2}\right)\pi,$$

(7–18)

or

$$\frac{2\pi}{h} \int_a^b \sqrt{2M(E - U)}\, dx = \left(n + \frac{1}{2}\right)\pi.$$

(7–19)

An equivalent form is

$$\oint p_x\, dx = (n + \tfrac{1}{2})h.$$

(7–20)

The last equation resembles the Wilson-Sommerfeld quantum condition except for the appearance of the half-integer; it serves the same purpose of determining the permitted energy values. In order that the solution be joined smoothly at a and b and satisfy the boundary condition at infinity, the energy value E must satisfy Eq. (7–19) or its equivalent. Thus Eq. (7–19) is the condition determining the energy eigenvalues. It is now clear that the Wilson-Sommerfeld quantum condition is a rough approximation of the correct quantum-mechanical formula for energy quantization because it does not contain the half-integer. In practical applications, the integral of Eq. (7–19) may be evaluated by the method of contour integration in the complex plane; the energy eigenvalues may thus be determined.

Having obtained the energy eigenvalues and eigenfunctions we have solved the general one-dimensional problem. The condition for this approximate method to be valid is that $g(x)$ be slowing varying. This condition is usually satisfied in the high quantum number region where the wave function is nearly sinusoidal in a small section of the coordinate.

Some general properties of the eigenfunctions may be noted here. Between a and b the wave function in the high quantum number region is essentially a sine (or cosine) function the wavelength and amplitude of which are modulated according to the potential function. The wavelength is $2\pi/\sqrt{g(x)}$ in the neighborhood of x; the amplitude is proportional to $1/\sqrt[4]{g(x)}$. The probability distribution is represented by the square of the wave function. When there are many wave crests in a small region dx, the average value of the square of the wave function over dx is simply one-half of the square of the amplitude and thus is proportional to $1/\sqrt{g(x)}$. Since

$$\frac{1}{\sqrt{g(x)}} \sim \frac{1}{\sqrt{2M(E-U)}} \sim \frac{1}{v}, \qquad (7\text{--}21)$$

where v is the classical velocity, the quantum-mechanical probability distribution is inversely proportional to the classical velocity. Now consider the motion of a classical particle in the region between a and b. The time it spends in dx is dx/v. Thus the statistical distribution of the position of the particle in a long period of time is represented by a function inversely proportional to the velocity. Therefore the quantum-mechanical probability distribution is the same as the classical statistical distribution. A special case of this result, i.e., the linear harmonic oscillator, was discussed in Chapter 4. Its general validity is again expected from what we discussed in Section 2–7, the minor differences between the quantum-mechanical and the classical distributions i.e., the peculiar wavelike structure and the peculiar behavior at and beyond the endpoints a and b of the quantum-mechanical distribution being again attributable to the quantum force of Eq. (2–65).

Problems

7–1. Solve the problem of the linear harmonic oscillator by the WKB method and compare the results with those of Chapter 4.

7–2. Determine the energy eigenvalues by the WKB method when the potential of a linear harmonic oscillator is changed slightly by adding a term λx^3 where λ is a small constant:

$$U(x) = \tfrac{1}{2}kx^2 + \lambda x^3. \qquad (7\text{--}22)$$

Repeat for the potential

$$U(x) = \tfrac{1}{2}kx^2 + \lambda x^4. \qquad (7\text{--}23)$$

7–3. Is the WKB method applicable to the potential well problem? Work out some simple results and compare with those of Section 5–1. Explain.

7–4. Determine the energy eigenvalues of a particle in the following potential:

$$U(x) = kx^4. \tag{7-24}$$

7–5. When the potential energy curve has two minima, how may the Schrödinger equation be solved by the WKB method?

7–6. Calculate the probability current density of a wave function obtained by the WKB method. How is this result related to classical mechanics?

*7–7. Show that the phase of the time-dependent wave function (for a given energy) obtained by the WKB method, after being multiplied by $\hbar$, satisfies the Hamilton-Jacobi differential equation (2–59). The results of this and the last problem thus verify the relation between classical and quantum mechanics discussed in Section 2–7.

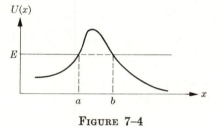

FIGURE 7–4

*7–8. Derive the expression for the transmission coefficient T of a particle penetrating through a potential barrier as shown in Fig. 7–4, the potential function $U(x)$ being such that the WKB method is applicable. Answer: $T \cong e^{-2\int_b^a \sqrt{g(x)}\,dx}$.

7–9. Show that Eq. (5–35) holds also for the above potential barrier.

---

* Indicates more difficult problems.

CHAPTER 8

THREE-DIMENSIONAL PROBLEMS

Having discussed one-dimensional problems in quantum mechanics, we now consider three-dimensional problems in this chapter. After a discussion of the space rotator, the central force problems will be considered. In particular the hydrogen atom will be dealt with in detail.

8–1 The space rotator. As mentioned earlier, the general solution of the time-dependent Schrödinger equation may be expressed as a Fourier series in time t, the coefficients of which, being functions of x, y, z, satisfy the time-independent Schrödinger equation,

$$\frac{\partial^2 \psi}{\partial x^2} + \frac{\partial^2 \psi}{\partial y^2} + \frac{\partial^2 \psi}{\partial z^2} + \frac{2M}{\hbar^2} (E - U)\psi = 0. \tag{8–1}$$

Equation (8–1) may be written in different coordinate systems. In spherical coordinates it takes the form

$$\frac{1}{r^2} \frac{\partial}{\partial r} \left(r^2 \frac{\partial \psi}{\partial r}\right) + \frac{1}{r^2} \frac{1}{\sin \theta} \frac{\partial}{\partial \theta} \left(\sin \theta \frac{\partial \psi}{\partial \theta}\right)$$
$$+ \frac{1}{r^2 \sin^2 \theta} \frac{\partial^2 \psi}{\partial \varphi^2} + \frac{2M}{\hbar^2} (E - U)\psi = 0. \tag{8–2}$$

In cylindrical coordinates it becomes

$$\frac{1}{\rho} \frac{\partial}{\partial \rho} \left(\rho \frac{\partial \psi}{\partial \rho}\right) + \frac{1}{\rho^2} \frac{\partial^2 \psi}{\partial \varphi^2} + \frac{\partial^2 \psi}{\partial z^2} + \frac{2M}{\hbar^2} (E - U)\psi = 0. \tag{8–3}$$

For a particular problem we shall choose a coordinate system in which the equation may be solved most easily.

The *space rotator* is defined as a particle constrained to move on a spherical surface without any other forces acting on it. Let the radius of the sphere be a. Clearly, the spherical coordinate system is the most convenient for treating the space rotator; Eq. (8–2) takes the following form:

$$\frac{1}{a^2 \sin \theta} \frac{\partial}{\partial \theta} \left(\sin \theta \frac{\partial \psi}{\partial \theta}\right) + \frac{1}{a^2 \sin^2 \theta} \frac{\partial^2 \psi}{\partial \varphi^2} + \frac{2M}{\hbar^2} E\psi = 0. \tag{8–4}$$

The wave function ψ is a function of θ and φ only. Solutions of Eq. (8–4) may be obtained by the method of separation of variables described below.

Assume a solution of the following form

$$\psi(\theta, \varphi) = \Theta(\theta)\Phi(\varphi) \tag{8–5}$$

where $\Theta(\theta)$ is a function of θ only and $\Phi(\varphi)$ is a function of φ only. Substituting in Eq. (8–4) we have

$$\Phi \frac{1}{a^2 \sin \theta} \frac{d}{d\theta} \left(\sin \theta \frac{d\Theta}{d\theta} \right) + \Theta \frac{1}{a^2 \sin^2 \theta} \frac{d^2\Phi}{d\varphi^2} + \frac{2M}{\hbar^2} E\Theta\Phi = 0 \tag{8–6}$$

or

$$\frac{1}{\Theta} \sin \theta \frac{d}{d\theta} \left(\sin \theta \frac{d\Theta}{d\theta} \right) + \frac{1}{\Phi} \frac{d^2\Phi}{d\varphi^2} + \frac{2M}{\hbar^2} Ea^2 \sin^2 \theta = 0. \tag{8–7}$$

Define a quantity m^2 by the following equation:

$$\frac{1}{\Phi} \frac{d^2\Phi}{d\varphi^2} = -m^2; \tag{8–8}$$

then we have

$$\frac{1}{\Theta} \sin \theta \frac{d}{d\theta} \left(\sin \theta \frac{d\Theta}{d\theta} \right) + \frac{2Ma^2}{\hbar^2} E \sin^2 \theta = m^2. \tag{8–9}$$

Equation (8–8) asserts that m^2 is independent of θ; on the other hand Eq. (8–9) requires that m^2 be independent of φ. The quantity m^2, being independent of both variables, thus must be a constant. The partial differential equation (8–4) is thus replaced by two ordinary differential equations, (8–8) and (8–9). Solutions of Eq. (8–4) may thus be obtained from the solutions Θ and Φ of Eqs. (8–8) and (8–9).

The solutions of Eq. (8–8) are readily obtained, i.e.,

$$\Phi = Ce^{\pm im\varphi} \tag{8–10}$$

where C is an arbitrary constant. Since the longitude angle φ has a period of 2π radians, the Φ function must also have a period of 2π; otherwise, Φ will not be a single-valued function and will not represent a single-valued probability distribution with respect to φ. This condition limits the value of m in Eq. (8–10). First of all, m cannot be an imaginary number. Among the real numbers, m can only be integers so that Φ may be a periodic function of period 2π. Thus

$$m = 0, \pm 1, \pm 2, \ldots . \tag{8–11}$$

The constant C is to be determined by normalization.

Equation (8–9) may be solved as follows: We first make the following substitution:

$$x = \cos \theta, \qquad \Theta(\theta) = y(x), \qquad \sin \theta \, \frac{d\Theta}{d\theta} = -(1 - x^2) \frac{dy}{dx}, \qquad (8\text{–}12)$$

$$\lambda = \frac{2Ma^2}{\hbar^2} E. \qquad (8\text{–}13)$$

The transformed equation is thus

$$(1 - x^2) \frac{d^2y}{dx^2} - 2x \frac{dy}{dx} + \left(\lambda - \frac{m^2}{1 - x^2} \right) y = 0. \qquad (8\text{–}14)$$

This differential equation is not ready to be solved by the power series method as we used in the linear harmonic oscillator. The reason is that it contains *poles* at the points $x = \pm 1$. A pole of a differential equation

$$\ddot{y} + p(x)\dot{y} + q(x)y = 0 \qquad (8\text{–}15)$$

at a point, say $x = c$, is defined as a singularity such that when $x \to c$,

$$
\begin{aligned}
p(x) &\to \infty \quad \text{but } (x - c)p(x) \to \text{finite value,} \\
q(x) &\to \infty \quad \text{but } (x - c)^2 q(x) \to \text{finite value.}
\end{aligned}
\qquad (8\text{–}16)
$$

When a pole exists at $x = c$, $y(x)$ must vanish at $x = c$. From Eqs. (8–15) and (8–16) we find that $y(x)$ may be written in the following form:

$$y(x) = (x - c)^\alpha v(x), \qquad (8\text{–}17)$$

where $v(x)$ satisfies a differential equation free from the pole at the point $x = c$. By substitution of Eq. (8–17) in Eq. (8–15) we may determine the value of α and the differential equation $v(x)$ has to satisfy; the latter, free from the pole, may be solved by expanding $v(x)$ into a power series. This procedure will be applied to remove the poles of Eq. (8–14). For the pole at $x = 1$, we rewrite Eq. (8–14) as follows:

$$(1 - x)^2 \ddot{y} - (1 - x) \frac{2x}{1 + x} \dot{y} + \left(\lambda - \frac{m^2}{1 - x^2} \right) \frac{1 - x}{1 + x} y = 0. \qquad (8\text{–}18)$$

Let

$$z = 1 - x; \qquad (8\text{–}19)$$

then

$$z^2 \ddot{y} + \frac{2z(1 - z)}{(2 - z)} \dot{y} - \left(\lambda + \frac{m^2}{z(z - 2)} \right) \frac{z}{z - 2} y = 0. \qquad (8\text{–}20)$$

Putting

$$y = z^\alpha v, \tag{8-21}$$

we have

$$\left[\alpha(\alpha - 1) + \frac{2(1 - z)}{2 - z} \alpha - \frac{\lambda z}{z - 2} - \frac{m^2}{(z - 2)^2} \right] v$$

$$+ \left[2\alpha + \frac{2(1 - z)}{2 - z} \right] z\dot{v} + z^2\ddot{v} = 0. \tag{8-22}$$

Equation (8–22) should be valid for $z = 0$, i.e.,

$$\alpha(\alpha - 1) + \alpha - \left(\frac{m}{2}\right)^2 = 0 \tag{8-23}$$

or

$$\alpha = \pm \frac{m}{2}. \tag{8-24}$$

Only the positive root is acceptable, as the negative root makes y diverge at $x = 1$. Thus we have

$$y = (1 - x)^{|m|/2} v(x). \tag{8-25}$$

The other pole at $x = -1$ may be treated in a similar manner. The result, together with Eq. (8–25), may be written

$$y = (1 - x)^{|m|/2}(1 + x)^{|m|/2} u(x) = (1 - x^2)^{|m|/2} u(x). \tag{8-26}$$

The differential equation $u(x)$ has to satisfy may be found by substitution of Eq. (8–26) in Eq. (8–18); the result is as follows:

$$(1 - x^2)\ddot{u} - 2(|m| + 1)x\dot{u} + (\lambda - |m| - m^2)u = 0. \tag{8-27}$$

Equation (8–27) may be solved by the power series method. Let

$$u(x) = \sum_\nu a_\nu x^\nu. \tag{8-28}$$

The recursion formula for the coefficients a_ν is found to be

$$a_{\nu+2} = \frac{\nu(\nu - 1) + 2(|m| + 1)\nu - \lambda + |m| + m^2}{(\nu + 2)(\nu + 1)} a_\nu. \tag{8-29}$$

For an arbitrary value of λ the series diverges at $x = \pm 1$. To obtain acceptable wave functions the series must terminate after a finite number of terms. The condition that the series terminates after the term $a_k x^k$ is

$$k(k-1) + 2(|m|+1)k - \lambda + |m| + m^2 = 0 \tag{8-30}$$

or

$$\begin{aligned}
\lambda &= k(k-1) + 2(|m|+1)k + |m| + |m|^2 \\
&= (|m|+k)(|m|+k+1) \\
&= l(l+1),
\end{aligned} \tag{8-31}$$

where

$$l = |m| + k.$$

Since

$$k = 0, 1, 2, \ldots, \tag{8-32}$$

and $|m|$ is an integer, the value of l is always a positive integer:

$$l = 0, 1, 2, \ldots. \tag{8-33}$$

Equation (8–31) thus determines the eigenvalues of Eq. (8–14); Eqs. (8–26) (8–28), and (8–29) determine the eigenfunctions. We have thus solved Eq. (8–9).

From the solutions Φ and Θ we may construct the solutions ψ of Eq. (8–4):

$$\psi_{lm}(\theta, \varphi) = N_{lm}\, e^{im\varphi} \sin^{|m|} \theta \sum_{\nu=0}^{k} a_\nu \cos^\nu \theta, \quad \begin{array}{l} m = 0, \pm 1, \pm 2, \ldots, \\ k = 0, 1, 2, \ldots, \\ l = |m| + k = 0, 1, 2, \ldots, \end{array} \tag{8-34}$$

where N_{lm} is a normalization constant which depends on l and m, and the coefficients a_ν are given by Eq. (8–29). The indices l and m are limited to the values listed above. Each set of values of l and m corresponds to a solution ψ_{lm}. The functions representing ψ_{lm} are called *spherical harmonics* in mathematics and are usually denoted by the notation $Y_{lm}(\theta, \varphi)$. The functions Θ are called *associated Legendre functions*, usually denoted by the notation $P_l^m (\cos \theta)$. Thus we may write:

$$\psi_{lm}(\theta, \varphi) = N_{lm} Y_{lm}(\theta, \varphi) = N_{lm}\, e^{im\varphi} P_l^m (\cos \theta). \tag{8-35}$$

The corresponding eigenvalue of ψ_{lm} is

$$E_l = \frac{\hbar^2}{2Ma^2}\, l(l+1), \quad l = 0, 1, 2, \ldots. \tag{8-36}$$

Introducing the moment of inertia $I = Ma^2$, we have

$$E_l = \frac{\hbar^2}{2I}\, l(l+1), \quad l = 0, 1, 2, \ldots. \tag{8-37}$$

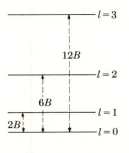

FIGURE 8–1

Thus the energy is quantized. The energy level diagram is shown in Fig. 8–1. The spacings of the energy levels have characteristic ratios. It was found experimentally that the spacings of energy levels of a diatomic molecule (within a band) actually exhibit these characteristic ratios. Such energy levels may thus be regarded as due to the rotational motion of the molecule as a space rotator. And the rule of energy quantization according to quantum mechanics is thus experimentally verified (Problem 8–1).

A few mathematical properties of the functions ψ_{lm} will be summarized below without proof.

1. The associated Legendre functions $P_l^m(x)$ are related to the *Legendre functions* $P_l(x)$ by the following equation:

$$P_l^m(x) = (1 - x^2)^{|m|/2} \frac{d^{|m|}}{dx^{|m|}} P_l(x) \tag{8–38}$$

(Problem 8–2), the Legendre functions being defined by the formula

$$P_l(x) = \frac{1}{2^l l!} \frac{d^l}{dx^l} (x^2 - 1)^l. \tag{8–39}$$

A few of them are listed below:

$$P_0(x) = 1,$$
$$P_1(x) = x,$$
$$P_2(x) = \tfrac{3}{2}x^2 - \tfrac{1}{2},$$
$$P_3(x) = \tfrac{5}{2}x^3 - \tfrac{3}{2}x,$$
$$P_4(x) = \tfrac{35}{8}x^4 - \tfrac{15}{2}x^2 + \tfrac{3}{8}.$$

2. The spherical harmonics are orthogonal to one another, i.e.,

$$\iint Y_{lm}(\theta, \varphi) Y_{l'm'}(\theta, \varphi) \sin \theta \, d\theta \, d\varphi = 0, \quad l \neq l', \quad m \neq m'. \tag{8–40}$$

3. The normalization constant is given by the expression

$$N_{lm} = \frac{1}{\sqrt{2\pi}} \sqrt{\frac{(2l+1)(l-m)!}{2(l+m)!}}. \tag{8-41}$$

4. Any function (well-behaved) of θ and φ over a sphere may be expanded in a series of spherical harmonics.

In the one-dimensional problems of the previous chapters we have only one quantum number. In this two-dimensional problem there appear two quantum numbers, m and l. We call m the *magnetic quantum number*[*] and l, the *azimuthal quantum number*. In two-dimensional problems the correspondence between eigenvalues and eigenfunctions may not be one-to-one. Equation (8–37) states that the eigenvalue is dependent only on the azimuthal quantum number l. For a given value of l, the magnetic quantum number m can take on any integral value from $-l$ to $+l$ as $l = |m| + k$. Thus there are $2l + 1$ possible values of m for a given l. Consequently there are $2l + 1$ wave functions ψ_{lm} corresponding to the same energy eigenvalue E_l. When there are several wave functions corresponding to the same energy eigenvalue, the wave functions are said to be *degenerate*.

We have found thus far a set of *particular solutions* of Eq. (8–4) by the method of separation of variables. Inasmuch as the set of solutions ψ_{lm} form an orthogonal set in terms of which any arbitrary function may be expanded, the *general solution* of Eq. (8–4) may be expressed by a series of ψ_{lm}:

$$\psi_E = \sum_{lm} a_{lm}\psi_{lm}. \tag{8-42}$$

Substituting Eq. (8–42) in Eq. (8–4), we obtain

$$\sum_{lm} a_{lm}\left[\frac{2Ma^2}{\hbar^2} E - l(l+1)\right]\psi_{lm} = 0. \tag{8-43}$$

In order to satisfy Eq. (8–43), all the coefficients of ψ_{lm} in Eq. (8–43) must vanish. To prove this, we may multiply Eq. (8–43) by a wave function $\psi_{l'm'}$ and then integrate with respect to θ and φ. Because of the orthogonal property of the spherical harmonics, all terms except that containing $\psi_{l'm'}$ drop out. Thus the coefficient of $\psi_{l'm'}$ must equal zero. Similarly all the other coefficients must also vanish. To satisfy these conditions E must be one of the values given by Eq. (8–36); say,

$$E_{l'} = \frac{\hbar^2}{2Ma^2} l'(l'+1), \tag{8-44}$$

---

[*] Because of the role it plays in the Zeeman effect (see Section 12–10).

and the coefficients a_{lm} for values of l other than l' must vanish, i.e.,

$$a_{lm} = 0, \qquad l \neq l'. \tag{8–45}$$

Consequently, an acceptable general solution of Eq. (8–4) exists only when E equals one of the energy eigenvalues given by Eq. (8–36). For a given eigenvalue specified by l' the general solution of Eq. (8–4) is a series of $\psi_{l'm}$ summed over m, i.e.,

$$\psi_{l'} = \sum_{m=-l}^{l} a_m \psi_{l'm}. \tag{8–46}$$

8–2 Central force problems. The central force problem is one in which the potential depends only on the distance between the particle and a fixed point in space (the force center). It is natural to take the force center as the origin and to employ the spherical coordinate system. The time-independent Schrödinger equation is thus

$$\frac{1}{r^2} \frac{\partial}{\partial r} \left(r^2 \frac{\partial \psi}{\partial r} \right) + \frac{1}{r^2 \sin \theta} \frac{\partial}{\partial \theta} \left(\sin \theta \frac{\partial \psi}{\partial \theta} \right)$$

$$+ \frac{1}{r^2 \sin^2 \theta} \frac{\partial^2 \psi}{\partial \varphi^2} + \frac{2M}{\hbar^2} [E - U(r)] \psi = 0. \tag{8–47}$$

Particular solutions of Eq. (8–47) may be obtained by the method of separation of variables which we have used in Section 8–1. General solutions may thus be obtained by superposition of particular solutions. On the other hand, we may also start from the general solution and obtain the same results. Consider the value of the general solution $\psi(r, \theta, \varphi)$ over a sphere of radius a. $\psi(a, \theta, \varphi)$, being a function of the variables θ and φ over a sphere, may be expanded in a series of the spherical harmonics. The coefficients of expansion are dependent on a:

$$\psi(a, \theta, \varphi) = \sum_{lm} C_{lm}(a) Y_{lm}(\theta, \varphi). \tag{8–48}$$

Considering all possible values of a, we may write

$$\psi(r, \theta, \varphi) = \sum_{lm} R_{lm}(r) Y_{lm}(\theta, \varphi), \tag{8–49}$$

where $R_{lm}(r)$ is a function of r. Substituting Eq. (8–49) in (8–47) we have

$$\sum_{lm} \left[\frac{d^2 R_{lm}}{dr^2} + \frac{2}{r} \frac{dR_{lm}}{dr} - \frac{1}{r^2} l(l+1) R_{lm} + \frac{2M}{\hbar^2} (E - U) R_{lm} \right] Y_{lm} = 0 \tag{8–50}$$

because

$$\frac{1}{\sin\theta}\frac{\partial}{\partial\theta}\left(\sin\theta\,\frac{\partial Y_{lm}}{\partial\theta}\right)+\frac{1}{\sin^2\theta}\frac{\partial^2 Y_{lm}}{\partial\varphi^2}=-l(l+1)Y_{lm}.\quad(8\text{--}51)$$

To satisfy Eq. (8–50) it is necessary that all coefficients of Y_{lm} vanish:

$$\frac{d^2 R_{lm}}{dr^2}+\frac{2}{r}\frac{dR_{lm}}{dr}-\frac{l(l+1)}{r^2}R_{lm}+\frac{2M}{\hbar^2}(E-U)R_{lm}=0.\quad(8\text{--}52)$$

Equation (8–52) specifies the condition R_{lm} has to satisfy; it is called the *radial equation*. Since the equation contains only the azimuthal quantum number l, not the magnetic quantum number m, R_{lm} is dependent on l only and we may drop the index m. This means that R_{lm}'s for Y_{33}, $Y_{32}, \ldots, Y_{3,-3}$, for example, are all the same, they being denoted by R_3. To solve Eq. (8–52), we introduce the transformation

$$y = rR_l \qquad\qquad (8\text{--}53)$$

Equation (8–52) thus takes the form

$$\frac{d^2 y}{dr^2}+\frac{2M}{\hbar^2}\left(E-U-\frac{\hbar^2}{2M}\frac{l(l+1)}{r^2}\right)y=0.\qquad(8\text{--}54)$$

This equation has the same form as a one-dimensional Schrödinger equation, the potential function of which is

$$U+\frac{\hbar^2}{2M}\frac{l(l+1)}{r^2}.$$

Thus the general methods for one-dimensional problems developed in Chapter 7 may be applied to solve Eq. (8–54). All qualitative and quantitative results of Chapter 7 remain valid provided the potential U is replaced by $U+(\hbar^2/2M)[l(l+1)/r^2]$. Once Eq. (8–54) is solved, the solutions ψ of Eq. (8–47) are readily obtained and the central force problem is thus solved. It may be interesting to look into the physical meaning of the additional term $(\hbar^2/2M)[l(l+1)/r^2]$. This additional potential represents a force $\hbar^2 l(l+1)/Mr^3$ obtained by differentiation of the potential with respect to r. We recall that in classical mechanics the central force problem may be solved with two equations, the radial and the angular equations. The radial equation is identical with the one-dimensional equation except that in this equation is added a centrifugal force p_θ^2/Mr^3 where p_θ is the angular momentum. It seems a reasonable guess that the force $\hbar^2 l(l+1)/Mr^3$ here may be identified with the centrifugal force. Later in Chapter 12 we shall identify $\hbar^2 l(l+1)$ with the square of the angular momentum [Eq. (12–62)]; the additional term is thus actually the centrifugal force.

Having obtained the general solution of the central force problem we shall briefly discuss in this section a few applications. In Chapter 5 we considered the one-dimensional potential well problem. The corresponding three-dimensional problem will now be discussed. The potential is specified by

$$U = \begin{cases} -V_0, & \text{for } r < a, \\ 0, & \text{for } r > a, \end{cases} \qquad (8\text{–}55)$$

where V_0 is a positive constant representing the well depth and a is the well radius. To solve Eq. (8–54) with this potential for negative energy values, we have to obtain separate solutions in the regions inside and outside the well. The interior solutions for $r < a$ may be expressed in terms of the *spherical Bessel functions*. The exterior solutions for $r > a$ may be expressed in terms of the *spherical Hankel functions*. Continuity conditions at $r = a$ may be satisfied only when the energy assumes certain selected values. Thus the energy is quantized; the eigenvalues may be obtained from the continuity conditions at $r = a$ (usually solved by numerical or graphical methods). The solution of this problem has important applications in nuclear physics, as the potential for a nucleon inside a nucleus may be approximated by a square well function. The energy eigenvalues represent the bound-state energy levels of the nucleon.

A second example concerns the motion of a particle confined by a spherical potential barrier of finite thickness. The motion of an α-particle inside a radioactive nucleus is a special case of this. The detailed solution of this problem is complicated and we will discuss only an approximate solution. The transmission coefficient in the three-dimensional problem is the ratio of $4\pi r^2 R_l^2$ outside the barrier to $4\pi r^2 R_l^2$ inside. This ratio is the same as that of y^2 outside to y^2 inside, which may be calculated from Eq. (8–54) in the same way as in the one-dimensional case. The only change we have to make in passing from the one- to the three-dimensional case is to include the centrifugal potential. In the one-dimensional case the transmission coefficient of a square barrier is approximately given by Eq. (5–37), i.e.,

$$T \approx e^{-2\sqrt{(2M/\hbar^2)(V_0-E)}\, a}. \qquad (8\text{–}56)$$

When the barrier is not square but curved such as that in Fig. 5–5 for the α-particle, $\sqrt{(2M/\hbar^2)(U-E)}$ in the exponent of Eq. (8–56) is not a constant. We may replace it approximately by its average value over the barrier. Thus we obtain the approximate formula

$$T \approx e^{-2\int_a^b \sqrt{(2M/\hbar^2)(U-E)}\, dx}. \qquad (8\text{–}57)$$

A rigorous calculation verifies this result. In the three-dimensional case the corresponding expression is

$$T \approx e^{-2\int_a^b \sqrt{(2M/\hbar^2)\{U+[\hbar^2/2M][l(l+1)/r^2]-E\}}\, dr}. \tag{8-58}$$

The potential function shown in Fig. 5–5 is known explicitly and the integral may be evaluated. Omitting the centrifugal potential for states of small values of l as an approximation, we obtain the *gamow formula*:

$$T \approx e^{-2\int_a^b \sqrt{(2M/\hbar^2)[(zZe^2/r)-E]}\, dr}$$

$$= e^{-2\sqrt{2MzZe^2b/\hbar^2}\,[\cos^{-1}\sqrt{a/b}-\sqrt{(a/b)-(a^2/b^2)}]}, \tag{8-59}$$

the parameter E being expressed in terms of b since $E = zZe^2/b$. Knowing the transmission coefficient we may proceed to determine the *radioactive constant* λ in α-decay which represents the probability of the α-particle's escaping the nucleus per unit time. Let v_{in} be the velocity of the α-particle inside the nucleus. The number of collisions the α-particle makes on the barrier wall per unit time is of the order of magnitude v_{in}/a where a is the nuclear radius. Upon each collision there is a probability T of escaping. The total probability of escaping per unit time is thus

$$\lambda \approx \frac{v_{in}}{a}\, e^{-2\sqrt{2MzZe^2b/\hbar^2}\,[\cos^{-1}\sqrt{a/b}-\sqrt{(a/b)-(a^2/b^2)}]}. \tag{8-60}$$

Equation (8–60) expresses the radioactive constant, or the reciprocal of the *mean life* τ, in terms of the nuclear parameters and the energy E of the escaping α-particle. In the example shown in Fig. 5–5, a/b is about $\frac{1}{6}$; when $a/b \to 0$, Eq. (8–60) leads to the following equation:

$$\frac{1}{\tau} \approx \frac{v_{in}}{a}\, e^{-(\pi zZe^2\sqrt{2M}/\hbar)/\sqrt{E}},$$

which expresses roughly the dependence of the mean life of an α-emitter on the energy of the α-particle emitted.

Having considered two potential barrier problems, we may include here another one, namely, the free particle in a rectangular box, although this is not a central force problem. A box of dimensions $a \times b \times c$ with nonpenetrable walls may be represented by a potential well with infinitely high barriers:

$$U = \begin{cases} 0, & \text{for } 0 < x < a,\ 0 < y < b,\ 0 < z < c, \\ \infty, & \text{for } 0 > x > a,\ 0 > y > b,\ 0 > z > c. \end{cases} \tag{8-61}$$

The time-independent Schrödinger equation for this three-dimensional problem may be solved most easily in the rectangular coordinate system,

$$\frac{\partial^2\psi}{\partial x^2} + \frac{\partial^2\psi}{\partial y^2} + \frac{\partial^2\psi}{\partial z^2} + \frac{2M}{\hbar^2}(E - U)\psi = 0. \tag{8–62}$$

Applying the method of separation of variables we let

$$\psi(x, y, z) = X(x)Y(y)Z(z), \tag{8–63}$$

$$E = E_x + E_y + E_z. \tag{8–64}$$

Equation (8–62) may thus be broken up in three equations, the first of which is

$$\frac{d^2X}{dx^2} + \frac{2M}{\hbar^2}[E_x - U(x)]X = 0, \tag{8–65}$$

where

$$U(x) = \begin{cases} 0, & \text{for } 0 < x < a, \\ \infty, & \text{for } 0 > x > a. \end{cases} \tag{8–66}$$

The other two equations may be obtained by replacing x by y and z. Each equation represents a one-dimensional problem which has been solved in Chapter 5. The eigenfunctions are sine (or cosine) functions with nodal points at the boundaries, i.e.,

$$X = C \sin \frac{n_x \pi}{a} x, \qquad n_x = 1, 2, 3, \dots. \tag{8–67}$$

The fact that the length a must contain an integral number of half-waves leads to the quantization of energy E_x, i.e.,

$$E_x = \frac{\hbar^2 \pi^2}{2Ma^2} n_x^2. \tag{8–68}$$

The product of the eigenfunctions of the three equations XYZ is the eigenfunction of Eq. (8–62); the sum of the eigenvalues of the three equations $E_x + E_y + E_z$ is the eigenvalue of the energy. Thus

$$\psi(x, y, z) = C \sin\left(\frac{n_x \pi}{a} x\right) \sin\left(\frac{n_y \pi}{b} y\right) \sin\left(\frac{n_z \pi}{c} z\right), \tag{8–69}$$

and

$$E = \frac{\hbar^2 \pi^2}{2M}\left(\frac{n_x^2}{a^2} + \frac{n_y^2}{b^2} + \frac{n_z^2}{c^2}\right), \tag{8–70}$$

where n_x, n_y, and n_z are positive integers. These results will be used later in Chapter 11.

8–3 The hydrogen atom. The hydrogen atom has one electron outside the nucleus. This electron moves under the influence of the Coulomb force of the nucleus. As the electronic motion is the main concern in atomic physics, we may regard the nucleus as a dimensionless point-charge of amount Ze, Z being the atomic number and e the absolute value of the electronic charge. As e is defined to be positive, the charge of an electron is $-e$. The potential of the electron in the field of the nucleus is thus

$$U = -\frac{Ze^2}{r}, \tag{8-71}$$

where r is the radius vector of the electron, the origin being at the center of the nucleus. For the hydrogen atom, Z equals unity. However, we shall carry Z in the formulation for the sake of generality. As the Coulomb force is a central force, we may apply the general theory developed in the last section. The time-independent Schrödinger equation becomes

$$\nabla^2\psi + \frac{2M}{\hbar^2}\left(E + \frac{Ze^2}{r}\right)\psi = 0. \tag{8-72}$$

Its general solution takes the form

$$\psi(r,\,\theta,\,\varphi) = \sum_{l,m} R_l(r)\,Y_{lm}(\theta,\,\varphi), \tag{8-73}$$

where $R_l(r)$, after the transformation $y = rR_l$, is to be determined by the radial equation

$$\frac{d^2y}{dr^2} + \frac{2M}{\hbar^2}\left[E + \frac{Ze^2}{r} - \frac{\hbar^2}{2M}\frac{l(l+1)}{r^2}\right]y = 0. \tag{8-74}$$

This differential equation may be solved exactly. Introducing a constant r_0 defined by

$$r_0 = \sqrt{\hbar^2/2M|E|}, \tag{8-75}$$

and changing the scale of the radius vector according to the equation

$$x = \frac{2}{r_0}r, \tag{8-76}$$

we may rewrite the radial equation in the following form:

$$\frac{d^2y}{dx^2} + \left[\pm\frac{1}{4} + \frac{A}{x} - \frac{l(l+1)}{x^2}\right]y = 0, \qquad \begin{array}{l} +\frac{1}{4} \text{ for } E > 0, \\ -\frac{1}{4} \text{ for } E < 0, \end{array} \tag{8-77}$$

where A is defined by

$$A = \frac{Ze^2}{2r_0|E|}. \tag{8-78}$$

In classical mechanics negative energies correspond to elliptical orbits while positive energies correspond to hyperbolic orbits. We shall see in the following that in quantum mechanics negative energies correspond to bound states and positive energies to unbound states in analogy with the results of classical mechanics. The case of negative energy will be considered in detail while that of positive energy will be discussed briefly at the end.

For $E < 0$, Eq. (8–77) becomes

$$\frac{d^2y}{dx^2} + \left[-\frac{1}{4} + \frac{A}{x} - \frac{l(l+1)}{x^2} \right] y = 0. \tag{8–79}$$

The asymptotic equation, when x approaches infinity, is

$$\frac{d^2y_\infty}{dx^2} - \frac{1}{4}\, y_\infty = 0, \tag{8–80}$$

which may be solved to obtain the asymptotic solution at infinity. This procedure has been employed in solving the differential equation for the linear harmonic oscillator. The solution of Eq. (8–80) is easily found to be

$$y_\infty = Ce^{\pm x/2}, \tag{8–81}$$

where C is an arbitrary constant. In order that the solution may be normalizable we have to discard the solution $e^{+x/2}$ which diverges at infinity. The solution of Eq. (8–79) may thus be written in the following form:

$$y = e^{-x/2}v(x), \tag{8–82}$$

where $v(x)$ is a function of x, presumably of a simpler structure than $y(x)$. By substitution, we find the differential equation $v(x)$ has to satisfy to be

$$\ddot{v} - \dot{v} + \left[\frac{A}{x} - \frac{l(l+1)}{x^2} \right] v = 0. \tag{8–83}$$

To remove the pole at $x = 0$ we let

$$v = x^\alpha u. \tag{8–84}$$

By substitution, Eq. (8–83) becomes

$$[\alpha(\alpha - 1) - l(l+1)]ux^{\alpha-2} + [\,\cdots\,]x^{\alpha-1} + [\,\cdots\,]x^\alpha = 0. \tag{8–85}$$

Equation (8–85) must be satisfied for all values of x, including $x = 0$, which requires, after dividing Eq. (8–85) by $x^{\alpha-2}$,

$$[\alpha(\alpha - 1) - l(l+1)] = 0. \tag{8–86}$$

The above equation has two roots:

$$\alpha = \begin{cases} l + 1, \\ -l. \end{cases} \tag{8-87}$$

The negative root is rejected because it makes $v(x)$, and consequently $\psi(x)$, diverge at $x = 0$. Thus we have

$$v = x^{l+1}u. \tag{8-88}$$

The differential equation for u may be obtained by substitution:

$$x\ddot{u} + [2(l + 1) - x]\dot{u} + (A - l - 1)u = 0. \tag{8-89}$$

This equation may be solved by the power series method. Let

$$u(x) = \sum_{k=0}^{\infty} a_k x^k. \tag{8-90}$$

The recursion formula for the coefficients a_k is found to be

$$a_{k+1} = \frac{k + l + 1 - A}{(k + 1)(k + 2l + 2)} a_k. \tag{8-91}$$

Consider the ratio of two successive coefficients. We find

$$\frac{a_{k+1}}{a_k} \to \frac{1}{k}, \qquad \text{when } k \to \infty. \tag{8-92}$$

For large values of k, the series $u(x)$ behaves like $\sum_k (1/k!)x^k$, which is the series expansion of e^x. Thus $R(x)$ behaves like $x^l e^{+x/2}$ which diverges at infinity. In order that the solution ψ may be normalizable it is necessary that the series terminate and become a polynomial. This condition will be fulfilled and the series will terminate after the term $a_k x^k$ when

$$A = k + l + 1. \tag{8-93}$$

Since A contains the energy $|E|$, Eq. (8-93) imposes a condition on the energy value. Therefore the energy is quantized and Eq. (8-93) determines the energy eigenvalues. As both k and l are positive integers, A must be a positive integer greater than 1. Let the sum $k + l + 1$ be denoted by n, which is called the *principal quantum number*,

$$n \equiv k + l + 1, \qquad \begin{aligned} k &= 0, 1, 2, \ldots, \\ l &= 0, 1, 2, \ldots. \end{aligned}$$

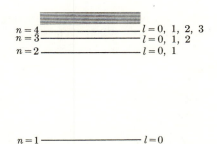

$$
\begin{array}{ll}
n = 4 & l = 0,\ 1,\ 2,\ 3 \\
n = 3 & l = 0,\ 1,\ 2 \\
n = 2 & l = 0,\ 1 \\[3em]
n = 1 & l = 0
\end{array}
$$

FIGURE 8–2

Equation (8–78) thus gives

$$
\frac{Ze^2\sqrt{2M|E|}}{2|E|\hbar} = n, \qquad n = 1, 2, \ldots, \tag{8–94}
$$

or

$$
E = -\frac{MZ^2e^4}{2\hbar^2(k + l + 1)^2} = -\frac{MZ^2e^4}{2\hbar^2n^2}, \qquad n = 1, 2, \ldots. \tag{8–95}
$$

Equation (8–95) gives the quantized energy values. We note that they are identical with those of Sommerfeld's theory if the number k is identified with the radial quantum number n_r of Eq. (1–34), and $l + 1$ is identified with the azimuthal quantum number n_θ of Eq. (1–35). Sommerfeld's n_θ takes integral values starting from 1. This is also the case for $l + 1$. The success of the old quantum theory in the energy quantization of the hydrogen atom is thus retained in quantum mechanics. The energy levels according to Eq. (8–95) are shown in Fig. 8–2.

We now return to Eq. (8–74). This equation is specified by one definite value of l. The energy E of this equation must be one of the values listed in Eq. (8–95) and thus is characterized by the quantum number n. Since $n = k + l + 1$ and k is not negative, we must require $n \geq l + 1$. On the other hand, once an energy value E_n is specified by an integer n, the general solution of Eq. (8–72) corresponding to E_n is a summation given by Eq. (8–73), each term of which leads to an equation of the form of Eq. (8–74). Since $n \geq l + 1$, the summation includes all terms whose values of l are smaller than or equal to $n - 1$. For each l, all possible values of m, i.e., $-l \leq m \leq l$, are to be included in the summation, since Eq. (8–74) is equally satisfied for all values of m. Thus the general solution ψ_n corresponding to E_n is

$$
\psi_n = \sum_{l=0}^{n-1}\sum_{m=-l}^{l} a_{klm}R_{kl}(r)\,Y_{lm}(\theta, \varphi), \qquad k = n - 1 - l, \tag{8–96}
$$

where $R_{kl}(r)$ is the radial solution containing a polynomial of the kth order and a_{klm} is an arbitrary constant. When we let a_{klm} equal unity for a particular set of values of k, l, and m but equal zero for all the others, we obtain a particular solution

$$\psi_{nlm} = R_{nl}(r)Y_{lm}(\theta, \varphi), \qquad l \leq n - 1, \qquad |m| \leq l. \qquad (8\text{-}97)$$

Here the index k is replaced by n since they are one-to-one related. From now on we shall always use the quantum numbers nlm to designate the eigenvalues and eigenfunctions. For $n = 3$, for example, the particular solutions are $\psi_{322}, \psi_{321}, \psi_{320}, \psi_{32-1}, \psi_{32-2}; \psi_{311}, \psi_{310}, \psi_{31-1}; \psi_{300}$. In general there are n^2 particular solutions for a given energy E_n, since

$$\sum_{l=0}^{n-1} (2l + 1) = n^2. \qquad (8\text{-}98)$$

These n^2 solutions are independent of one another since the spherical harmonics $Y_{lm}(\theta, \varphi)$ are mutually independent. As these n^2 independent solutions all correspond to the same energy value E_n, we have a degenerate case with a *multiplicity* equal to n^2. In later discussions the particular solutions ψ_{nlm} will be used more often than the general solution. It may be remarked here that the degeneracy with respect to l is accidental. It arises as a result of the peculiar nature of the Coulomb potential. In a general central force problem, different values of l usually correspond to different energy values. An analogy with the Sommerfeld theory will also be mentioned here. Solutions of different l but of the same energy E may be regarded as corresponding to the Sommerfeld elliptical orbits of different angular momenta but of the same energy (see Section 1–4). For a given l, solutions of different m may be regarded as corresponding to different spatial orientations of one elliptical orbit when spatial quantization is applied. The relation between the quantum number l and the angular momentum will be discussed in Chapter 12.

Turning our attention to the eigenfunctions, we write down the radial solution $R_{kl}(x)$:

$$\begin{aligned}
R_{kl}(x) &= \frac{y}{r} \\[2mm]
&= \frac{1}{x}\, e^{-x/2} v(x) \\[2mm]
&= \frac{1}{x}\, e^{-x/2} x^{l+1} u(x) \\[2mm]
&= e^{-x/2} x^l (a_0 + a_1 x + a_2 x^2 + \cdots + a_k x^k).
\end{aligned} \qquad (8\text{-}99)$$

The coefficients $a_1, \ldots, a_k$ may be expressed in terms of a_0 by successive application of Eq. (8–91). a_0 itself is to be determined by normalization. The radial solution given by Eq. (8–99) may be expressed in terms of the *associated Laguerre functions*, the properties of which will be briefly stated below without proof.

1. The *Laguerre polynomials* $L_k(x)$ are defined as follows:

$$L_k(x) = e^x \frac{d^k}{dx^k} (e^{-x} x^k). \qquad (8\text{–}100)$$

A few of them are listed below:

$$L_0(x) = 1,$$
$$L_1(x) = 1 - x,$$
$$L_2(x) = 2 - 4x + x^2,$$
$$L_3(x) = 6 - 18x + 9x^2 - x^3.$$

They satisfy the differential equation

$$x\ddot{L}_k + (1 - x)\dot{L}_k + kL_k = 0, \qquad (8\text{–}101)$$

and also the recursion formula

$$L_{k+1}(x) + (x - 1 - 2k)L_k(x) + k^2 L_{k-1}(x) = 0 \qquad (8\text{–}102)$$

(Problem 8–5).

2. The jth derivative of $L_k(x)$ is called the *associated Laguerre polynomial of degree $k - j$ and order j* which is denoted by $L_k^j(x)$; thus

$$L_k^j(x) = \frac{d^j}{dx^j} L_k(x). \qquad (8\text{–}103)$$

The differential equation satisfied by $L_k^j(x)$ may be obtained by differentiating Eq. (8–102) j times, the result being

$$x\ddot{L}_k^j + (j + 1 - x)\dot{L}_k^j + (k - j)L_k^j = 0 \qquad (8\text{–}104)$$

(Problem 8–5).

3. The solution $u(x)$ of Eq. (8–89) may be obtained by comparing Eq. (8–89) with Eq. (8–104). They become identical if we set

$$2(l + 1) = j + 1,$$
$$n - l - 1 = k - j. \qquad (8\text{–}105)$$

Solving Eq. (8–105) for k and j, we obtain

$$k = n + l,$$
$$j = 2l + 1. \tag{8–106}$$

Hence the solution of Eq. (8–89) is

$$u(x) = L_{n+l}^{2l+1}(x). \tag{8–107}$$

4. By the relations

$$x = \frac{2}{r_0} r = \frac{2}{\hbar} \sqrt{2M|E|}\, r = \frac{2}{\hbar} \frac{MZe^2}{\hbar n} r = \frac{2r}{na}, \tag{8–108}$$

where a is the Bohr radius of the first quantized orbit, i.e.,

$$a = \frac{\hbar^2}{ZMe^2} = (0.53 \times 10^{-8}) \frac{1}{Z}\ \text{cm}, \tag{8–109}$$

we may rewrite R_{nl} in terms of the associated Laguerre functions $e^{-x/2} x^l L_{n+l}^{2l+1}(x)$:

$$R_{nl}(r) = e^{-r/na} \left(\frac{2r}{na}\right)^l L_{n+l}^{2l+1}\left(\frac{2r}{na}\right). \tag{8–110}$$

The radial extension of the wave function, for $n = 1$, is thus of the order of a. This justifies the use of the Bohr radius for qualitative discussions in quantum mechanics.

5. The normalization constant of the radial function is

$$N_{nl} = \sqrt{\frac{4(n - l - 1)!}{a^3 n^4 [(n + l)!]^3}}. \tag{8–111}$$

Making use of these results we may write the eigenfunction ψ_{nlm} as follows:

$$\psi_{nlm} = \sqrt{\frac{4(n - l - 1)!}{a^3 n^4 [(n + l)!]^3}}\, e^{-r/na} \left(\frac{2r}{na}\right)^l L_{n+l}^{2l+1}\left(\frac{2r}{na}\right) Y_{lm}(\theta, \varphi), \tag{8–112}$$

where $Y_{lm}(\theta, \varphi)$ is the normalized spherical harmonic. The physical meaning of ψ_{nlm} will now be considered. According to Born's first assumption, the probability density distribution of the electron is to be represented by $\psi_{nlm}^* \psi_{nlm}$. The φ-dependence of this distribution is given by $e^{+im\varphi} \cdot e^{-im\varphi}$, which is a constant. Thus the probability density distribution of all ψ_{nlm} is symmetric with respect to the z-axis. The θ-dependence of this distribution is given by $\Theta^*\Theta$. For $l = 0$, Θ_{00} is a constant, and the distribution is spherically symmetric. For $l = 1$ there are three pos-

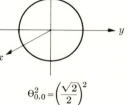

$$\Theta_{0,0}^2 = \left(\frac{\sqrt{2}}{2}\right)^2$$

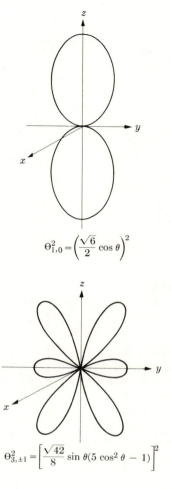

$$\Theta_{1,0}^2 = \left(\frac{\sqrt{6}}{2}\cos\theta\right)^2$$

$$\Theta_{1,\pm1}^2 = \left(\frac{\sqrt{3}}{2}\sin\theta\right)^2$$

$$\Theta_{3,\pm1}^2 = \left[\frac{\sqrt{42}}{8}\sin\theta(5\cos^2\theta - 1)\right]^2$$

FIGURE 8–3

sible values of m and two Θ functions. Θ_{10} equals $\cos\theta$ except for a constant factor and $\Theta_{1,\pm1}$ equals $\sin\theta$ except for a constant factor. The probability density distributions of these eigenfunctions are shown in Fig. 8–3. A slightly more complicated wave function $\Theta_{3,\pm1}$ is also shown in Fig. 8–3. Note that when $l - |m|$ is large, the curve representing Θ_{lm} consists of a large number of lobes because the θ-dependence of Θ has a wavelike structure. The r-dependence of the probability density distribution is illustrated in Fig. 8–4. The wavy structure of the distribution is demonstrated most clearly in the last curve. As $R_{kl}(r)$ contains a polynomial of the kth order, the curve representing it crosses the r-axis k times

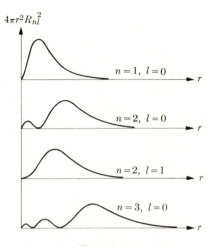

$4\pi r^2 R_{nl}^2$

$n = 1,\ l = 0$

$n = 2,\ l = 0$

$n = 2,\ l = 1$

$n = 3,\ l = 0$

FIGURE 8–4

in the region $0 < r < \infty$. For large values of k $(k = n - l - 1)$ the curve contains a large number of wave crests. Having discussed the probability density distribution we now turn our attention to the dynamical quantities. Born's second assumption states that the energy value in the state ψ_{nlm} is definitely known to be E_n. Once again, an exact knowledge of energy is accompanied with an uncertainty of position. States with the same value of n but different values of l and m are to have the same energy but, as will be noted in Chapter 12, different dynamical properties in other respects.

From the eigenfunctions we may derive another interesting quantity, namely, the probability current density given by Eq. (2–78),

$$\vec{j} = \frac{\hbar}{2Mi} \left(\Psi^* \nabla \Psi - \Psi \nabla \Psi^* \right). \qquad (8\text{–}113)$$

As we are using the spherical coordinates, the gradient ∇ must be expressed in this coordinate system, i.e.,

$$\nabla = \vec{r}\, \frac{\partial}{\partial r} + \vec{\theta}\, \frac{1}{r}\, \frac{\partial}{\partial \theta} + \vec{\varphi}\, \frac{1}{r \sin \theta}\, \frac{\partial}{\partial \varphi}, \qquad (8\text{–}114)$$

where $\vec{r}, \vec{\theta}, \vec{\varphi}$ are the three unit vectors in the spherical coordinate system. As the time-dependent wave function corresponding to the eigenfunction ψ_{nlm} is

$$\Psi_{nlm} = R_{nl}(r)\, \Theta_{lm}(\theta)\, \Phi_m(\varphi) e^{-i(E_n/\hbar)t}, \qquad (8\text{–}115)$$

the corresponding probability current density vector is

$$\vec{j} = \vec{r}\,\frac{\hbar}{2Mi}\,\Theta_{lm}^2\left(R_{nl}\frac{\partial R_{nl}}{\partial r} - R_{nl}\frac{\partial R_{nl}}{\partial r}\right)$$

$$+ \vec{\theta}\,\frac{\hbar}{2Mi}\,R_{nl}^2\left(\Theta_{lm}\frac{\partial \Theta_{lm}}{\partial \theta} - \Theta_{lm}\frac{\partial \Theta_{lm}}{\partial \theta}\right)$$

$$+ \vec{\varphi}\,\frac{\hbar}{2Mi}\,R_{nl}^2\Theta_{lm}^2\left(e^{-im\varphi}\frac{1}{r\sin\theta}\frac{\partial}{\partial\varphi}e^{im\varphi} - e^{im\varphi}\frac{1}{r\sin\theta}\frac{\partial}{\partial\varphi}e^{-im\varphi}\right)$$

$$= 0 + 0 + \vec{\varphi}\,R_{nl}^2\Theta_{lm}^2\,\frac{m\hbar}{Mr\sin\theta}\,\Phi_m^*\Phi_m$$

$$= \vec{\varphi}\,\frac{m\hbar\,\psi_{nlm}^*\,\psi_{nlm}}{Mr\sin\theta}. \tag{8–116}$$

It is zero along the radius vector and along the longitude, but not zero along the latitude. Furthermore, it is independent of time. Hence the current is stationary. This may be interpreted to mean that the probability elements circulate around the z-axis while maintaining the symmetry about the z-axis all the time. Since the electron carries a charge of $-e$, the probability current about the z-axis gives rise to an electric current which in turn produces a magnetic moment. The current ring passing through the area element $r\,dr\,d\theta$ has a magnetic moment

$$d\mathfrak{M} = \frac{1}{c}\,\pi(r\sin\theta)^2(-ej_\varphi)r\,dr\,d\theta.$$

The total magnetic moment is

$$\mathfrak{M} = -\frac{e\pi}{c}\iint r^3\sin^2\theta\,\frac{m\hbar\Psi^*\Psi}{Mr\sin\theta}\,dr\,d\theta$$

$$= -\frac{em\hbar}{2cM}\iint\psi^*\psi 2\pi r^2\sin\theta\,dr\,d\theta,$$

$$= -\frac{e\hbar}{2Mc}\,m,$$

which is m times the Bohr magneton $e\hbar/2Mc$ (Section 1–3). The quantum number m is thus seen to be related to the magnetic moment (hence the name magnetic quantum number). Incidentally, the same result was obtained in the old quantum theory, assuming space quantization.

The electronic motion specified by Ψ_{nlm} as described above is a kind quite foreign to our experience. The relation between this kind of quantum-mechanical description and the classical description based on the particle trajectory will now be discussed. The general solution of the time-dependent Schrödinger equation consists of a superposition of eigenfunctions each with a time factor $e^{-i(E_n/\hbar)t}$ attached. Since the wave functions Φ, Θ, R, all have a sort of wavelike structure, a superposition of Ψ_{nlm} in the high quantum number region may result in a localized wave packet which moves as a whole. Its center of gravity is to move on an elliptic orbit determined by classical mechanics (see Section 2–6). Accompanying this localization of the particle in the space and time there is a dispersion in the dynamical quantities. As a wave packet is made of many ψ_{nlm} of different n values, there is a probability distribution of the energy value (and other dynamical properties specified by l and m). For a wave packet representing a macroscopic object, both the dimension of the wave packet and the dispersion of the energy are small; both are beyond experimental detection. In such a case quantum mechanics is identical with classical mechanics for all practical purposes. We remark parenthetically that for a superposition of Ψ_{nlm}, the probability current density components along the radius vector and the longitude no longer equal zero and the current is no longer stationary. The results of Eq. (8–116) apply only to a single eigenfunction.

A few normalized eigenfunctions of the hydrogen atom are listed below as examples and for later use. For convenience r is changed to σ by the following transformation:

$$\sigma = \frac{n}{2}\, x = \frac{r}{a} = \frac{r}{\hbar^2/MZe^2}.$$

$$n = 1,\ l = 0, \quad \psi_{100} = \frac{1}{\sqrt{\pi}} \frac{1}{a^{3/2}} e^{-\sigma}$$

$$n = 2,\ l = 0, \quad \psi_{200} = \frac{1}{4\sqrt{2\pi}} \frac{1}{a^{3/2}} (2 - \sigma)e^{-\sigma/2},$$

$$n = 2,\ l = 1, \quad \psi_{211} = \frac{1}{8\sqrt{\pi}} \frac{1}{a^{3/2}} \sigma e^{-\sigma/2} \sin\theta\, e^{i\varphi}, \qquad (8\text{–}117)$$

$$\psi_{210} = \frac{1}{4\sqrt{2\pi}} \frac{1}{a^{3/2}} \sigma e^{-\sigma/2} \cos\theta,$$

$$\psi_{21-1} = \frac{1}{8\sqrt{\pi}} \frac{1}{a^{3/2}} \sigma e^{-\sigma/2} \sin\theta\, e^{-i\varphi},$$

$$n = 3,\ l = 0, \quad \psi_{300} = \frac{1}{81\sqrt{3\pi}} \frac{1}{a^{3/2}} (27 - 18\sigma + 2\sigma^2)e^{-\sigma/3},$$

$$n = 3, \, l = 1, \quad \psi_{311} = \frac{1}{81\sqrt{\pi}} \, \frac{1}{a^{3/2}} \, (6 - \sigma)\sigma e^{-\sigma/3} \sin \theta \, e^{i\varphi},$$

$$\psi_{310} = \frac{\sqrt{2}}{81\sqrt{\pi}} \, \frac{1}{a^{3/2}} \, (6 - \sigma)\sigma e^{-\sigma/3} \cos \theta,$$

$$\psi_{31-1} = \frac{1}{81\sqrt{\pi}} \, \frac{1}{a^{3/2}} \, (6 - \sigma)\sigma e^{-\sigma/3} \sin \theta \, e^{-i\varphi},$$

$$n = 3, \, l = 2, \quad \psi_{322} = \frac{1}{81\sqrt{4\pi}} \, \frac{1}{a^{3/2}} \, \sigma^2 e^{-\sigma/3} \sin^2 \theta \, e^{i2\varphi},$$

$$\psi_{321} = \frac{1}{81\sqrt{\pi}} \, \frac{1}{a^{3/2}} \, \sigma^2 e^{-\sigma/3} \sin \theta \cos \theta \, e^{i\varphi}, \tag{8-117 cont'd}$$

$$\psi_{320} = \frac{1}{81\sqrt{6\pi}} \, \frac{1}{a^{3/2}} \, \sigma^2 e^{-\sigma/3}(3 \cos^2 \theta - 1),$$

$$\psi_{32-1} = \frac{1}{81\sqrt{\pi}} \, \frac{1}{a^{3/2}} \, \sigma^2 e^{-\sigma/3} \sin \theta \cos \theta \, e^{-i\varphi},$$

$$\psi_{32-2} = \frac{1}{81\sqrt{4\pi}} \, \frac{1}{a^{3/2}} \, \sigma^2 e^{-\sigma/3} \sin^2 \theta \, e^{-i2\varphi}.$$

A remark concerning the spectrum of the hydrogen atom is in order here. The energy eigenvalues we have obtained, together with the Bohr frequency condition,* lead correctly to the Balmer formula. However, with better experimental resolution, a *fine structure* of the hydrogen lines was observed, the explanation of which is not included in the derivation of the Balmer formula. This phenomenon is attributed to the relativistic and spin effects (not to be discussed in this volume) and has been satisfactorily accounted for in quantum mechanics. It is interesting to note that in the old quantum theory, Sommerfeld obtained the experimentally correct formula by considering the relativistic effect only without the spin. However, this agreement is fortuitous. A recent development concerning the hydrogen spectrum is the discovery of the *Lamb shift* and its explanation by quantum electrodynamics. This topic is also beyond the scope of this book.

Up to this point we have limited our discussion to the solutions of negative energy values. These solutions correspond to bound states. Their

---

* The justification of applying the Bohr frequency condition, which is an assumption in the old quantum theory, in quantum mechanics will be discussed in the theory of radiation in Chapter 11.

wave functions are more or less localized in a finite region around the origin, as the radial wave function contains a declining exponential function. We now turn our attention to the positive energy solutions which correspond to unbound states. The radial equation (8–77) becomes

$$\frac{d^2y}{dx^2} + \left[\frac{1}{4} + \frac{A}{x} - \frac{l(l+1)}{x^2}\right]y = 0.$$

Its asymptotic equation

$$\ddot{y}_\infty + \tfrac{1}{4}y_\infty = 0 \qquad (8\text{–}118)$$

has solutions of the oscillatory type,

$$y_\infty = Ce^{\pm ix/2}. \qquad (8\text{–}119)$$

Unlike the exponential solutions, both solutions in Eq. (8–119) are compatible with the boundary condition at infinity, i.e., they may be normalized in the sense that a plane wave may be normalized. The general solution of y is thus

$$y = C_+ e^{+ix/2}v_+(x) + C_- e^{-ix/2}v_-(x), \qquad (8\text{–}120)$$

where C_+ and C_- are arbitrary constants, and $v_+(x)$ and $v_-(x)$ are two functions, presumably of a simpler structure than $y(x)$. The differential equation of $v_\pm$, found by substitution, is

$$\ddot{v}_\pm \pm i\dot{v}_\pm + \left[\frac{A}{x} - \frac{l(l+1)}{x^2}\right]v_\pm = 0. \qquad (8\text{–}121)$$

Removing the pole at $x = 0$ we obtain the result

$$v_\pm = x^{l+1}u_\pm. \qquad (8\text{–}122)$$

The differential equation of $u_\pm$ is found by substitution:

$$x\ddot{u}_\pm + [2(l+1) \pm ix]\dot{u}_\pm + [A \pm i(l+1)]u_\pm = 0. \qquad (8\text{–}123)$$

This equation may be solved by the power series method. The recursion formula is found to be

$$a_{k+1} = \frac{\mp i(k+l+1) - A}{(k+1)(k+2l+2)} a_k. \qquad (8\text{–}124)$$

This equation is similar to Eq. (8–9) except that a factor $\mp i$ is placed in front of $(k+l+1)$. For large values of k, the series behaves like $e^{\mp ix}$, which is normalizable at infinity. Consequently there is no need to terminate the series and the energy is not quantized. The energy eigenvalues of the hydrogen atom thus consist of a discrete spectrum in the negative energy region and a continuous spectrum in the positive energy region (as shown in Fig. 8–2). The continuous spectrum gives rise to a continuous band in the emission and absorption spectra beyond the series limit, which is observed experimentally. The eigenfunctions for positive energy values are mathematically complicated and we shall not discuss them further. Suffice it to say that they may be expressed in *confluent hypergeometric functions*, that the asymptotic behavior of the radial wave function is like a spherical wave $(1/x)e^{\mp(ix/2)}$, and that in the region near the nucleus the spherical wave is strongly distorted by the Coulomb field.

8–4 The positronium, mesic atoms, ionized atoms, alkali atoms, and exciton. The theory of the hydrogen atom may be applied to a number of hydrogenlike systems such as the positronium, mesic atoms, ionized atoms, alkali atoms, and exciton. These systems will be briefly discussed to show the wide applicability of quantum mechanics. They may be partly or wholly omitted without breaking the continuity of the book.

A *positron* is a positive counterpart of an electron, the two having the same mass but opposite charge. It *annihilates* with an electron, the mass of the two particles being converted to energy in the form of γ-rays. However, before the annihilation takes place, the electron may exist, for a very brief period, in the attractive force field of the positron. In this short time period the two charged particles form an "atom" which is identical with the hydrogen atom except that the nucleus of the atom is not a proton but a positron. Such an "atom" is called a *positronium*. As the theory of the hydrogen atom involves only the charge of the nucleus, which is the same for proton and positron, the theory is applicable to the positronium. On the other hand, the assumption in Section 8–3 that the force center is a fixed point in space is a good approximation in the case of the hydrogen atom (whose nucleus is 1836 times heavier than the electron) but is not for the positronium. Even in the hydrogen atom the nuclear mass is not infinite, and the theory has to be modified to account for the motion of the nucleus. This will be discussed in Chapter 12; the result (Section 12–8) is that the previous theory remains valid if the electronic mass is replaced by the *reduced mass* of the system $M_1 M_2/(M_1 + M_2)$, a conclusion also valid in classical mechanics. For the hydrogen atom the reduced mass is very close to the electron mass. The so-called *reduced-mass effect* is thus small. For the positronium, the reduced mass is just one-half of the electron mass. The theory of the hydrogen atom may thus be translated into that of the positronium if the electronic mass M is reduced by 2. The energy level scheme shown in Fig. 8–2 may be applied to the positronium if the energy scale shrinks by a factor of 2, since the energy eigenvalue E_n is proportional to the mass M. The wave functions may be taken over for the positronium if the scale of the coordinates is expanded by a factor of 2, as the Bohr radius is inversely proportional to M.

Another species of the transitory "atoms" is the *mesic atom*, formed by a negative *meson* and a regular atomic nucleus. The so-called μ-*meson* is a particle having a charge same as an electron but a mass 212 times heavier. The meson is eventually captured by the nucleus and its mass converted into energy. Before this happens, for a very brief period (about 10^{-6} sec) the meson and the nucleus form a hydrogenlike system. Since the μ-meson is 212 times heavier than the electron, the energy scale of the level scheme is to be expanded by a factor of 212 and the scale of the coordinates is to be reduced by a factor of 212. The "light" emitted by the mesic atom is thus of shorter wavelengths by a factor of 212. These rays are in the x-ray region and are called the *mesic x-rays*. On the other hand, the shrinking of the coordinate scale makes the meson closer to the nucleus. The meson thus can "see" the nucleus as an extended object instead of as a distant point as seen by the electron. (The first Bohr orbit of a μ-meson in a nucleus of $Z = 10$ has a radius about 2×10^{-12} cm, while the nuclear radius is about 0.5×10^{-12} cm; the two will be even closer for heavier atoms.) Inside the nucleus the charge is distributed in some fashion throughout the volume, not concentrated at a point. Consequently the electric field of the nucleus is not simply the Coulomb field of a point charge as assumed previously. This deviation of the field from that of a point charge causes a deviation of the energy levels from those of the Balmer formula. Mesic x-rays have been measured experimentally. Their deviations from the Balmer formula are also observed; from these deviations one may deduce information concerning the size and the charge distribution of the nucleus.

In atomic physics there are a number of hydrogenlike systems in which the results of the last section find immediate application. A singly ionized helium atom, containing one electron in a Coulomb field of two units of positive charge, is a hydrogenlike system with $Z = 2$. Since the Bohr radius a is inversely proportional to Z, the coordinate scale of the wave functions is to be reduced by a factor of 2. Similarly, as the energy eigenvalue is proportional to Z^2, the energy scale of the level scheme is to be expanded by a factor of 4. The doubly ionized lithium atom, which has one electron in a Coulomb field of a charge of $3e$, may be treated in a similar manner. So may be the triply ionized beryllium atom, etc.

The alkali atoms are approximately hydrogenlike. They all have one valence electron. Take the sodium atom as an example. It has 11 electrons outside the nucleus ($Z = 11$). We shall not discuss atomic structure systematically at this moment. It will suffice to say that 10 of the electrons are arranged in closed shells; and the eleventh, the valence electron, moves essentially outside the closed shells. So far as the electric field outside the closed shells is concerned, the nuclear charge $+11e$, shielded by the ten electrons carrying a charge of $-10e$, is equivalent to $+e$, the same as the hydrogen atom. Thus the valence electron is essentially in a hydrogenlike field. However, inside the closed shells the shielding effect of the 10 electrons becomes less and less as we approach the center, and eventually becomes zero in the region just outside the nucleus. Therefore the electrostatic potential increases from that of a hydrogenlike field $-e^2/r$ outside the closed shells to that of the full nuclear field $-Ze^2/r$ near the center. Such a potential may be approximated by the following formula:

$$U = -\frac{e^2}{r}\left(1 + \frac{\beta}{r}\right),\tag{8-125}$$

where β is a constant. The factor $[1 + (\beta/r)]$ accounts for the increase of the *effective nuclear charge* when r decreases. The radial equation for this potential is

$$\frac{d^2y}{dr^2} + \frac{2M}{\hbar^2}\left[E + \frac{e^2}{r} + \frac{e^2\beta}{r^2} - \frac{\hbar^2}{2M}\frac{l(l+1)}{r^2}\right]y = 0.\tag{8-126}$$

Define a number l' by the following equation:

$$l'(l'+1) = l(l+1) - \frac{2Me^2\beta}{\hbar^2}.\tag{8-127}$$

Equation (8–126) thus becomes

$$\frac{d^2y}{dr^2} + \frac{2M}{\hbar^2}\left[E + \frac{e^2}{r} - \frac{\hbar^2}{2M}\frac{l'(l'+1)}{r^2}\right]y = 0.\tag{8-128}$$

This equation is identical with Eq. (8–74) except that l' takes the place of l. Therefore it may be solved in a similar manner. The quantized energy values are, in analogy with Eq. (8–95),

$$E = -\frac{MZ^2e^4}{2\hbar^2(k+l'+1)^2}.\tag{8-129}$$

Since l' is less than l by virtue of Eq. (8–127) we may write

$$l' = l - \Delta_l,\tag{8-130}$$

where Δ_l is a positive number dependent on l, and is called the *quantum defect*. Equation (8–129) becomes

$$E = -\frac{MZ^2e^4}{2\hbar^2(n - \Delta_l)^2}, \qquad \begin{array}{l} n = 1, 2, \ldots, \\[4pt] l = 0, 1, \ldots, n-1. \end{array}\tag{8-131}$$

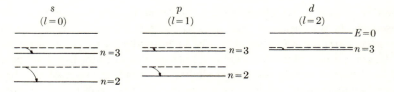

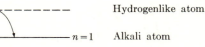

FIGURE 8-5

The energy levels are thus the same as before except that the quantum defect appears in the denominator. Since Δ_l depends on l, the energy values for different l's but the same n are no longer equal and the degeneracy is removed. We have mentioned that the degeneracy with respect to l in the hydrogen atom is accidental, being due to a peculiar property of the Coulomb field of a *point charge*. A slight change of this potential may remove the degeneracy; levels of the same n but different l values split apart among themselves. This is clearly illustrated by the alkali atoms. The level scheme corresponding to Eq. (8–131) is shown in Fig. 8–5. Levels of the same l are arranged in the same column. They are called s, p, d, f, g, . . . , levels for $l = 0, 1, 2, 3, 4, . . . ,$ respectively according to the spectroscopic convention. The hydrogen levels are marked out by dotted lines and the arrows indicate the shift caused by the quantum defects. The quantum defects depress the energy levels; different columns are depressed in different proportions, since Δ_l is dependent on l. Historically the quantum defects were first introduced in spectroscopy, on an empirical basis; it was also found empirically that Δ_l decreases with l. Quantum mechanics thus furnishes a natural explanation for the appearance of the quantum defects and enables us to calculate them on a theoretical basis. Equation (8–127) also implies that the quantum defect decreases with l.

The last example to be discussed here concerns an application in solid-state physics. The *semiconductor* has become increasingly important both in theoretical study and in practical application. The best-known semiconductor is germanium. Under ordinary conditions germanium is an insulator. However, when a beam of light is directed upon it, the resistance is immediately reduced and it becomes a conductor. This phenomenon is known as *photoconductivity*. We shall not discuss the theory of semiconductors here. Suffice it to say that the light energy is absorbed to liberate an electron from its regular position in the crystal lattice; this electron is able to move about freely inside the solid and thus conducts electrical current when a field is applied. However, conductivity is not entirely due to the electrons; the so-called *holes*, i.e., the vacant seats left by the electrons in the crystal lattice, also contribute to the conductivity. A hole migrates as the vacant seat is taken by a neighboring electron and thereby a new vacant seat is created. Since a hole represents the excess of one unit of positive charge, the migration of holes is equivalent to the flow of a stream of positively charged particles. Thus the holes also contribute to the conductivity. We now consider the following phenomenon: A crystal, when exposed to light within a certain frequency range, absorbs a number of discrete lines but fails to produce photoconductivity. The absorption lines have been analyzed and their pattern is found to resemble that of the hydrogen atom absorption lines. This phenomenon may be explained as follows: The electron may absorb an amount of energy which is insufficient to enable it to divorce itself completely from the hole. The positive hole and the negative electron thus form a bound "atomic" system similar to the hydrogen atom. The name *exciton* was given to this system. Actually an exciton may be regarded as a solid-state positronium. This bound system may exist in a number of states of quantized energies just as the hydrogen atom does, and the level scheme is to follow the same pattern. For light frequencies not high enough to excite electrons into the continuous energy levels (producing free electrons and holes), electrons may be excited to the discrete energy levels corresponding to bound states (producing excitons). As the electron and the hole are bound together to form a neutral system, they produce no conductivity. The frequencies of the absorption lines measured have been compared with those calculated from a Balmer-type formula and agreement with an accuracy of one part in one thousand was obtained except for the first line. A modification of the Balmer formula is necessary because of the fact that the Coulomb force between the electron and the hole is reduced by the dielectric medium by a factor K equal to the dielectric constant of germanium, which is about 16. This reduction may be incorporated into the previous formula by putting $Z = 1/K$. Since the energy eigenvalue is proportional to Z^2, the energy scale is to be reduced by a factor of K^2. As the Bohr radius is inversely proportional to Z, the coordinate scale of the wave function is to be expanded by a factor of K. For high quantum numbers the Bohr orbits of the electron around the hole embrace a large number of atoms, so that the above treatment of the effect of the dielectric medium is valid. On the other hand, the first orbit is not large enough to embrace a sufficient number of atoms so as to justify the use of the dielectric constant in determining the force field. Thus the energy level formula obtained by setting $Z = 1/K$ may be accurate for the excited levels but may not be correct for the first level. This accounts for the deviation from the Balmer-type formula of the first absorption line.

PROBLEMS

8–1. Determine the rotational energy levels of HCl, the interatomic distance being assumed to be a constant 1.27×10^{-8} cm and the two ions being regarded as mass-points (no internal structure).

8–2. Mathematical exercise: Show that the associated Legendre functions defined by Eq. (8–38) satisfy the differential equation (8–14).

8–3. Discuss the motion of the moon (or an earth satellite) about the earth according to quantum mechanics, both objects being considered as dimensionless mass-points. In particular, determine the order of magnitude of the quantum numbers of the moon's present orbit.

8–4. If the interatomic distances of HF, HCl, HBr, and HI are all assumed to be constant and equal to one another, what would be the differences among the rotational energy levels of the four molecules?

8–5. Mathematical exercise: Verify Eqs. (8–101), (8–102), and (8–104).

8–6. Discuss the energy eigenvalues and eigenfunctions of a π-mesic atom formed by a negative π-meson (276 times heavier than an electron) and an atomic nucleus. Calculate the wavelengths of the π-mesic x-rays for the $2p \rightarrow 1s$ transitions for the mesic atoms formed by Be, C, and O. For experimental results see: M. Camac *et al.*, *Phys. Rev.* **88,** 134 (1952), and M. B. Stearns *et al.*, *Phys. Rev.* **93,** 1123 (1954).

8–7. The alkali atoms Li, Na, K, Rb, and Cs are all hydrogenlike atoms, but their energy level schemes change from one to the other systematically. Describe and explain the trend of this change (omit the spin effect).

8–8. Calculate the wavelengths of the first five absorption lines of the exciton in germanium. How many germanium atoms are enclosed in the fifth orbit? And in the first orbit?

8–9. In Problem 8–1 the interatomic distance of HCl is now allowed to vary, and a restoring force proportional to the displacement from the equilibrium distance 1.27×10^{-8} cm is introduced. The force constant k is so large that the displacement is usually small compared with 1.27×10^{-8} cm. Determine the energy level of the molecule; the two ions are again regarded as mass points. *Note:* The solution provides the theoretical basis to analyze the spectra of diatomic molecules into rotational and vibrational energy levels.

8–10.[#] *Nuclear physics.* Study the possibility of muon induced nuclear fusion for energy generation. The mu-mesic atom has a dimension 200 times smaller than the hydrogen atom as the muon mass is 200 times larger than the electron. Substituting the muon for the electron in the *hydrogen molecular ion* would reduce the size of the latter 200 times and bring the two hydrogen nuclei closer to fuse. Study all aspects involved in generating fusion energy in this way.

8–11.[#] *Nuclear physics.* The potentials of Figs. 5–5 and 5–8 can be generalized to three dimensions to represent that of a proton generated and emitted in a nuclear reaction. Study the proton emitted, i.e., its energy spectrum and energy level width.

[#] Indicates new problems added to the expanded edition.

CHAPTER 9

THE THREE-DIMENSIONAL HARMONIC OSCILLATOR

The isotropic, three-dimensional harmonic oscillator is characterized by a potential which varies as the square of the radius vector. Though a central force problem, it may be solved not only in spherical coordinates but also in cylindrical coordinates and rectangular coordinates. We shall solve the problem in these respective coordinate systems and show that the results are equivalent. This also serves to illustrate the relations between the wave functions expressed in different coordinate systems.

9–1 Solution in rectangular coordinates. The potential of the isotropic, three-dimensional harmonic oscillator is expressed by

$$U = \tfrac{1}{2}Kr^2, \tag{9-1}$$

where K is the force constant and r is the distance of the oscillator from a fixed point in space taken as the origin of the coordinate system. In rectangular coordinates the potential may be expressed as follows:

$$U = \tfrac{1}{2}K(x^2 + y^2 + z^2). \tag{9-2}$$

The time-independent Schrödinger equation is thus

$$\frac{\partial^2 \psi}{\partial x^2} + \frac{\partial^2 \psi}{\partial y^2} + \frac{\partial^2 \psi}{\partial z^2} + \frac{2M}{\hbar^2}\left(E - \frac{1}{2}Kx^2 - \frac{1}{2}Ky^2 - \frac{1}{2}Kz^2\right)\psi = 0. \tag{9-3}$$

Applying the method of separation of variables, we assume a particular solution in which the variables are separated,

$$\psi(x, y, z) = X(x)Y(y)Z(z). \tag{9-4}$$

Substituting Eq. (9–4) in Eq. (9–3) and dividing by $\psi(x, y, z)$, we obtain

$$\frac{1}{X}\frac{d^2 X}{dx^2} + \frac{1}{Y}\frac{d^2 Y}{dy^2} + \frac{1}{Z}\frac{d^2 Z}{dz^2} + \frac{2M}{\hbar^2}\left(E - \frac{1}{2}Kx^2 - \frac{1}{2}Ky^2 - \frac{1}{2}Kz^2\right) = 0. \tag{9-5}$$

Define E_x by

$$\frac{1}{X}\frac{d^2 X}{dx^2} - \frac{2M}{\hbar^2}\frac{1}{2}Kx^2 = -\frac{2M}{\hbar^2}E_x. \tag{9-6}$$

Equation (9–5) thus becomes

$$\frac{1}{Y}\frac{d^2Y}{dy^2} + \frac{1}{Z}\frac{d^2Z}{dz^2} + \frac{2M}{\hbar^2}\left(E - \frac{1}{2}Ky^2 - \frac{1}{2}Kz^2\right) = \frac{2M}{\hbar^2}E_x. \quad (9\text{–}7)$$

Equation (9–6) implies that E_x is independent of y and z, while Eq. (9–7) asserts that E_x is independent of x. Therefore E_x must be a constant. Equation (9–6) may be written as follows:

$$\frac{d^2X}{dx^2} + \frac{2M}{\hbar^2}\left(E_x - \frac{1}{2}Kx^2\right)X = 0. \quad (9\text{–}8)$$

Equation (9–8) is an ordinary differential equation for the function $X(x)$. In a similar manner the variables y and z in Eq. (9–7) may be separated. The equations thus obtained are:

$$\frac{d^2Y}{dy^2} + \frac{2M}{\hbar^2}\left(E_y - \frac{1}{2}Ky^2\right)Y = 0, \quad (9\text{–}9)$$

$$\frac{d^2Z}{dz^2} + \frac{2M}{\hbar^2}\left(E_z - \frac{1}{2}Kz^2\right)Z = 0, \quad (9\text{–}10)$$

where E_y and E_z are constants subject to the relation

$$E = E_x + E_y + E_z. \quad (9\text{–}11)$$

The partial differential equation of three variables is thus separated into three ordinary differential equations. The wave function $\psi(x, y, z)$ is the product of the eigenfunctions of the three equations and the energy E is the sum of the eigenvalues E_x, E_y, E_z of the same.

Equations (9–8), (9–9), and (9–10) are identical in form with the equation of the one-dimensional harmonic oscillator which has been solved in Chapter 4. Hence their eigenvalues and eigenfunctions may be written according to the results of Chapter 4 as follows:

$E_x = \hbar\omega(n_x + \frac{1}{2})$,

$$X_{n_x}(x) = \left(\frac{M\omega}{\hbar}\right)^{1/4}\frac{1}{(\sqrt{\pi}\,2^{n_x}n_x!)^{1/2}}H_{n_x}\left(\sqrt{\frac{M\omega}{\hbar}}\,x\right)e^{-(M\omega/2\hbar)x^2},$$

$$n_x = 0, 1, 2, \ldots, \quad (9\text{–}12)$$

$E_y = \hbar\omega(n_y + \frac{1}{2})$,

$$Y_{n_y}(y) = \left(\frac{M\omega}{\hbar}\right)^{1/4}\frac{1}{(\sqrt{\pi}\,2^{n_y}n_y!)^{1/2}}H_{n_y}\left(\sqrt{\frac{M\omega}{\hbar}}\,y\right)e^{-(M\omega/2\hbar)y^2},$$

$$n_y = 0, 1, 2, \ldots, \quad (9\text{–}13)$$

$E_z = \hbar\omega(n_z + \frac{1}{2})$,

$$Z_{n_z}(z) = \left(\frac{M\omega}{\hbar}\right)^{1/4} \frac{1}{(\sqrt{\pi}\, 2^{n_z} n_z!)^{1/2}} H_{n_z}\left(\sqrt{\frac{M\omega}{\hbar}}\, z\right) e^{-(M\omega/2\hbar)z^2},$$

$$n_z = 0, 1, 2, \ldots, \qquad (9\text{--}14)$$

where n_x, n_y, and n_z are three quantum numbers having integral values. Out of these solutions we construct solutions of Eq. (9–3):

$$\psi_{n_x n_y n_z}(x, y, z) = X_{n_x}(x) Y_{n_y}(y) Z_{n_z}(z). \qquad (9\text{--}15)$$

The corresponding energy eigenvalues are

$$E_{n_x n_y n_z} = \hbar w(n_x + n_y + n_z + \tfrac{3}{2}), \quad \left.\begin{matrix} n_x \\ n_y \\ n_z \end{matrix}\right\} = 0, 1, 2, \ldots. \qquad (9\text{--}16)$$

It may be noted that the energy eigenvalue depends only on the sum $n_x + n_y + n_z$. There exist a number of combinations of n_x, n_y, and n_z which give the same sum. For example, the following sets of values, $(2, 0, 0)$, $(0, 2, 0)$, $(0, 0, 2)$, $(1, 1, 0)$, $(1, 0, 1)$, $(0, 1, 1)$, all give the same sum 2. Consequently the eigenfunctions ψ_{200}, ψ_{020}, ψ_{002}, ψ_{110}, ψ_{101}, ψ_{011} all correspond to the same energy eigenvalue $\hbar\omega(2 + \frac{3}{2})$, and we have a case of degeneracy. Defining a quantum number n by

$$n \equiv n_x + n_y + n_z, \qquad n = 0, 1, 2, \ldots,$$

we may write the energy eigenvalues as follows:

$$E_n = \hbar\omega(n + \tfrac{3}{2}), \qquad n = 0, 1, 2, \ldots. \qquad (9\text{--}17)$$

The energy levels are shown in Fig. 9–1.

The method of separation of variables enables us to obtain particular solutions of the time-independent Schrödinger equation. Since the par-

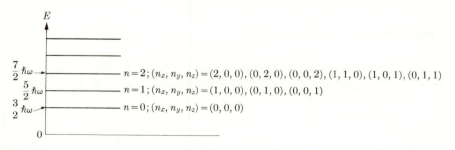

FIGURE 9–1

ticular solutions, Eqs. (9–12)–(9–14), form complete orthonormal sets, an arbitrary function (well-behaved) $F(x, y, z)$ may be expanded in a series of these functions. First, consider $F(x, y, z)$ as a function of x for fixed y and z; it may be expanded in a series of $\psi_{n_x}(x)$. The coefficients of expansion, being functions of y and z, may be expanded again first in terms of y and then in terms of z. $F(x, y, z)$ may thus be expressed by the eigenfunctions as follows:

$$F(x, y, z) = \sum_{n_x n_y n_z} a_{n_x n_y n_z} \psi_{n_x}(x) \psi_{n_y}(y) \psi_{n_z}(z). \qquad (9\text{–}18)$$

In particular, the general solution $\psi(x, y, z)$ of Eq. (9–3) may be expressed in the series form of Eq. (9–18). Substituting the series in Eq. (9–3) and making use of the properties of $\psi_{n_x}(x)$, $\psi_{n_y}(y)$, and $\psi_{n_z}(z)$, we obtain

$$\sum_{n_x n_y n_z} a_{n_x n_y n_z}(E - E_{n_x n_y n_z}) \psi_{n_x}(x) \psi_{n_y}(y) \psi_{n_z}(z) = 0. \qquad (9\text{–}19)$$

The above equation may be satisfied identically for all values of x, y, z only when

$$a_{n_x n_y n_z} = 0, \qquad \text{for} \qquad E \neq E_{n_x n_y n_z}. \qquad (9\text{–}20)$$

Thus solutions of Eq. (9–3) exist only when E equals one of the eigenvalues E_n given by Eq. (9–17). For each E_n, the corresponding general solution consists of all terms for which $n_x + n_y + n_z = n$.

Remembering that $r^2 = x^2 + y^2 + z^2$ and $n = n_x + n_y + n_z$, we may write the particular solution $\psi_{n_x}(x) \psi_{n_y}(y) \psi_{n_z}(z)$ as follows:

$$\psi_{n_x n_y n_z}(x, y, z) = \left(\frac{M\omega}{\hbar}\right)^{3/4} \frac{1}{(\pi^{3/2} 2^n n_x! n_y! n_z!)^{1/2}}$$

$$\times H_{n_x}\left(\sqrt{\frac{M\omega}{\hbar}}\, x\right) H_{n_y}\left(\sqrt{\frac{M\omega}{\hbar}}\, y\right) H_{n_z}\left(\sqrt{\frac{M\omega}{\hbar}}\, z\right) e^{-(M\omega/2\hbar)r^2}. \qquad (9\text{–}21)$$

This expression will be used later.

9–2 Solution in cylindrical coordinates. The cylindrical coordinates (ρ, φ, z) are related to the rectangular coordinates (x, y, z) by the following equations of transformation:

$$x = \rho \cos \varphi,$$
$$y = \rho \sin \varphi, \qquad (9\text{–}22)$$
$$z = z.$$

The Schrödinger equation of the oscillator, when transformed to the cylindrical coordinates, takes the following form:

$$\frac{1}{\rho}\frac{\partial}{\partial\rho}\left(\rho\frac{\partial\psi}{\partial\rho}\right) + \frac{1}{\rho^2}\frac{\partial^2\psi}{\partial\varphi^2} + \frac{\partial^2\psi}{\partial z^2} + \frac{2M}{\hbar^2}\left(E - \frac{1}{2}K\rho^2 - \frac{1}{2}Kz^2\right)\psi = 0. \quad (9\text{–}23)$$

Applying the method of separation of variables again, we let

$$\psi(\rho,\varphi,z) = P(\rho)\Phi(\varphi)Z(z). \quad (9\text{–}24)$$

Equation (9–23) may be transformed to the following by substitution:

$$\frac{1}{P(\rho)\rho}\frac{d}{d\rho}\left(\rho\frac{dP(\rho)}{d\rho}\right) + \frac{1}{\Phi(\varphi)\rho^2}\frac{d^2\Phi(\varphi)}{d\varphi^2} + \frac{1}{Z(z)}\frac{d^2Z(z)}{dz^2}$$

$$+ \frac{2M}{\hbar^2}(E - \tfrac{1}{2}K\rho^2 - \tfrac{1}{2}Kz^2) = 0. \quad (9\text{–}25)$$

$Z(z)$ may be separated from the above equation as in Section 9–1. Thus

$$\frac{d^2Z(z)}{dz^2} + \frac{2M}{\hbar^2}\left(E_z - \frac{1}{2}Kz^2\right)Z(z) = 0. \quad (9\text{–}26)$$

The solution of this equation is already written in Eq. (9–14). After having separated $Z(z)$, the remaining equation may be separated by a procedure similar to that in Section 8–1. The equation for Φ is thus

$$\frac{d^2\Phi(\varphi)}{d\varphi^2} + m^2\Phi(\varphi) = 0, \quad (9\text{–}27)$$

where m^2 is the separation constant. The normalized solutions are

$$\Phi(\varphi) = \frac{1}{\sqrt{2\pi}}e^{im\varphi}, \qquad m = 0, \pm 1, \pm 2, \ldots, \quad (9\text{–}28)$$

according to the results of Section 8–1. The remaining equation for $P(\rho)$ is

$$\frac{d^2P(\rho)}{d\rho^2} + \frac{1}{\rho}\frac{dP(\rho)}{d\rho} + \frac{2M}{\hbar^2}\left(E_{\rho\varphi} - \frac{1}{2}K\rho^2 - \frac{\hbar^2}{2M}\frac{m^2}{\rho^2}\right)P(\rho) = 0, \quad (9\text{–}29)$$

where

$$E_{\rho\varphi} = E - E_z. \quad (9\text{–}30)$$

In analogy with the procedure in Chapter 4 we find the asymptotic equation of Eq. (9–29) to be

$$\frac{d^2P_\infty}{d\rho^2} - \frac{MK}{\hbar^2}\rho^2 P_\infty = 0, \quad (9\text{–}31)$$

which has two solutions, i.e.,

$$P_\infty = e^{\pm \sqrt{MK/\hbar^2}\,\rho^2/2} \equiv e^{\pm(M\omega/\hbar)\rho^2/2}, \qquad (9\text{-}32)$$

ω being defined by Eq. (4–4). The positive solution is ruled out because of the boundary condition at infinity. Following a procedure similar to that in Chapter 4, we let

$$P(\rho) = e^{-(M\omega/\hbar)\rho^2/2} v(\rho). \qquad (9\text{-}33)$$

The differential equation of $v(\rho)$ is obtained by substitution,

$$\ddot{v} + \left(\frac{1}{\rho} - 2\,\frac{M\omega}{\hbar}\,\rho\right)\dot{v} + \left(\frac{2ME_{\rho\varphi}}{\hbar^2} - \frac{m^2}{\rho^2} - 2\,\frac{M\omega}{\hbar}\right)v = 0. \qquad (9\text{-}34)$$

Introducing a variable ξ by the following equation:

$$\xi = \sqrt{M\omega/\hbar}\,\rho, \qquad (9\text{-}35)$$

and letting

$$\lambda = \frac{2E_{\rho\varphi}}{\hbar\omega} \qquad (9\text{-}36)$$

as in Chapter 4, we obtain

$$\frac{d^2 v}{d\xi^2} + \left(\frac{1}{\xi} - 2\xi\right)\frac{dv}{d\xi} + \left(\lambda - 2 - \frac{m^2}{\xi^2}\right)v = 0. \qquad (9\text{-}37)$$

This equation has a pole at $\xi = 0$ which may be removed by the method discussed in Chapter 8. Let

$$v(\xi) = \xi^\alpha u(\xi), \qquad u(0) \neq 0. \qquad (9\text{-}38)$$

Equation (9–37) thus becomes

$$u[\alpha(\alpha - 1) + \alpha - m^2]\xi^{\alpha-2} + [2\alpha\dot{u} + \dot{u}]\xi^{\alpha-1}$$
$$+ [\ddot{u} + (\lambda - 2\alpha - 2)u]\xi^\alpha - [2\dot{u}]\xi^{\alpha+1} = 0. \qquad (9\text{-}39)$$

Dividing through $\xi^{\alpha-2}$ and letting $\xi \to 0$, we obtain the equation

$$\alpha^2 - m^2 = 0. \qquad (9\text{-}40)$$

In order that v be finite at $\xi = 0$, α must be positive. Thus

$$\alpha = |m|. \qquad (9\text{-}41)$$

Equation (9–39), after substitution by Eq. (9–40), becomes the following differential equation for $u(\xi)$,

$$\xi\ddot{u} + (2|m| + 1 - 2\xi^2)\dot{u} + \xi(\lambda - 2|m| - 2)u = 0, \qquad (9\text{–}42)$$

which may be solved by the power series method. Let

$$u(\xi) = \sum_k a_k \xi^k. \qquad (9\text{–}43)$$

The relations between the coefficients are found to be:

$$(2|m| + 1)a_1 = 0,$$
$$\cdot \ \cdot \ \cdot \ \cdot \ \cdot \ \cdot \ \cdot \ \cdot \ \cdot \ \cdot \ \cdot \ \cdot \ \cdot \ \cdot \ \cdot \ \cdot \ \cdot \ \cdot \ , $$
$$a_{k+2} = \frac{2k + 2|m| - \lambda + 2}{(k + 2)(k + 1) + (k + 2)(2|m| + 1)}\, a_k. \qquad (9\text{–}44)$$

The series contains even powers of ξ only, as the odd terms vanish because of the condition $a_1 = 0$ and the recursion formula. Since

$$\frac{a_{k+2}}{a_k} \to \frac{2}{k}, \qquad \text{as} \quad k \to \infty, \qquad (9\text{–}45)$$

the series diverges like the function e^{ξ^2} (see Eq. (4–23) in Chapter 4). Thus P diverges like $e^{\xi^2/2}$. In order that the solution be normalizable, it is necessary that the series break off after a finite number of terms. The condition that the even series terminate after the term $a_k\xi^k$ is

$$\lambda = 2(k + |m| + 1), \qquad k = 0, 2, 4, \ldots, \qquad (9\text{–}46)$$

or

$$E_{\rho\varphi} = \hbar\omega(k + |m| + 1), \qquad k = 0, 2, 4, \ldots. \qquad (9\text{–}47)$$

Equation (9–47) gives the energy eigenvalues $E_{\rho\varphi}$ of Eq. (9–29). The corresponding eigenfunctions are

$$P(\xi) = N_\xi e^{-\xi^2/2}\xi^{|m|}\sum_{\nu=0}^{k} a_\nu \xi^\nu, \qquad (9\text{–}48)$$

where N_ξ is a normalization constant. From the solutions $P(\rho)$, $\Phi(\varphi)$ and $Z(z)$ we obtain the eigenvalues and eigenfunctions of Eq. (9–23). The expression for the eigenvalues is

$$E = \hbar\omega(k + |m| + n_z + \tfrac{3}{2}), \quad \begin{aligned} k &= 0, 2, 4, \ldots, \\ m &= 0, \pm 1, \pm 2, \ldots, \\ n_z &= 0, 1, 2, \ldots, \end{aligned}$$

or

$$E = \hbar\omega(n + \tfrac{3}{2}), \qquad n = 0, 1, 2, \ldots; \qquad (9\text{–}49)$$

the expression for the eigenfunctions is

$$\psi_{kmn_z}(\xi, \varphi, z) = N_\xi e^{-\xi^2/2} \xi^{|m|} \left(\sum_{\nu=0}^{k} a_\nu \xi^\nu \right) \frac{1}{\sqrt{2\pi}} e^{im\varphi} N_z H_{n_z}(\xi_z) e^{-\xi_z^2/2}, \quad (9\text{-}50)$$

where N_ξ and N_z are normalization constants and $\xi_z \equiv \sqrt{(M\omega/\hbar)}\, z$. Here the quantum numbers are k, m, and n_z. The energy eigenvalues given by Eq. (9-49) are identical with those given by Eq. (9-16) obtained by using the rectangular coordinate system. The relation between the two sets of eigenfunctions will be considered presently, after which the multiplicity of the degenerate energy levels will be discussed.

In Section 9-1 the general solution of the Schrödinger equation for an oscillator corresponding to an eigenvalue E_n is found to be a linear superposition of terms $\psi_{n_x}(x)\psi_{n_y}(y)\psi_{n_z}(z)$ subject to the condition $n_x + n_y + n_z = n$. For a set of quantum numbers k, m, n_z such that $k + |m| + n_z = n$, the solution $\psi_{kmn_z}(\rho, \varphi, z)$ is one corresponding to energy E_n and thus should be expressible as a superposition of terms $\psi_{n_x}(x)\psi_{n_y}(y)\psi_{n_z}(z)$. This will be verified as follows: First of all, $\psi_{n_z}(z)$ is contained in both solutions and may be ignored for the present purpose. We need only show that $\psi_{km}(\rho, \varphi)$ may be expressed in terms of $\psi_{n_x}(x)\psi_{n_y}(y)$. The latter may be written explicitly as follows:

$$\psi_{n_x}(\xi_x)\psi_{n_y}(\xi_y) \sim e^{-\xi_x^2/2} H_{n_x}(\xi_x) e^{-\xi_y^2/2} H_{n_y}(\xi_y). \quad (9\text{-}51)$$

The combined exponential factor is exactly the same as the exponential factor of ψ_{km}, since $x^2 + y^2 = \rho^2$. The product of $H_{n_x}(\xi_x)$ and $H_{n_y}(\xi_y)$ is a polynomial of x and y, the order of which is $n_x + n_y$. A superposition of terms $\psi_{n_x}(x)\psi_{n_y}(y)$ subject to the condition $n_x + n_y = |m| + k$, leads to a polynomial of x and y of the $(|m| + k)$th order. When changed to polar coordinates this expression becomes a polynomial of ρ of the $(|m| + k)$th order. Now ψ_{km} actually contains a polynomial of ρ of the $(|m| + k)$th order. Take a concrete example in which $m = 1$ and $k = 2$.

$$\psi_{21}(\xi, \varphi) = N_\xi e^{-\xi^2/2} \xi \left(a_0 - \frac{a_0}{2} \xi^2 \right) \frac{1}{\sqrt{2\pi}} e^{i\varphi}$$

$$= N_\xi e^{-(\xi_x^2 + \xi_y^2)/2} \frac{a_0}{\sqrt{2\pi}} (\xi_x + i\xi_y) \left(1 - \frac{\xi_x^2 + \xi_y^2}{2} \right) \quad (9\text{-}52)$$

$$\sim e^{-(\xi_x^2 + \xi_y^2)/2} (\xi_x^3 + i\xi_y^3 + i\xi_x^2\xi_y + \xi_x\xi_y^2 - 2\xi_x - 2i\xi_y).$$

The last expression may be shown to be representable in terms of $\psi_3(\xi_x)\psi_0(\xi_y)$, $\psi_2(\xi_x)\psi_1(\xi_y)$, $\psi_1(\xi_x)\psi_2(\xi_y)$, and $\psi_0(\xi_x)\psi_3(\xi_y)$. From the expression of the Hermite polynomials, Eq. (4–26), the following result may be verified,

$$H_3(\xi_x)H_0(\xi_y) + iH_2(\xi_x)H_1(\xi_y) + H_1(\xi_x)H_2(\xi_y) + iH_0(\xi_x)H_3(\xi_y)$$
$$= 8(\xi_x^3 + i\xi_y^3 + i\xi_x^2\xi_y + \xi_x\xi_y^2 - 2\xi_x - 2i\xi_y). \tag{9–53}$$

Therefore

$$\psi_{21}(\rho, \varphi) \sim \psi_3(x)\psi_0(y) + i\psi_2(x)\psi_1(y) + \psi_1(x)\psi_2(y) + i\psi_0(x)\psi_3(y). \tag{9–54}$$

The relation between the two sets of eigenfunctions in different coordinate systems is thus established.

Let the sum of n_x and n_y be n. As there are $n + 1$ possible combinations of the two quantum numbers n_x and n_y which lead to the same sum n, the energy level E_n in rectangular coordinates, omitting the z-component as before, is $(n + 1)$-fold degenerate. On the other hand, in cylindrical coordinates $n = k + |m|$, where $k = 0, 2, 4, \ldots$ and $m = 0, \pm 1, \pm 2, \ldots$. When n is even there are $(n/2) + 1$ possible values of k. For $k = n$, there is only one possible value of m, i.e., zero. For $k = 0, 2, \ldots, n - 2$, each k is accompanied by two possible values of m, i.e., $\pm(n - k)$. Thus the total number of combinations of the two quantum numbers k and m for a fixed, even value of n is $2 \times (n/2) + 1 = n + 1$. When n is odd there are $(n + 1)/2$ possible values of k and each is accompanied by two possible values of m. The total number of combinations is thus also $n + 1$. Since each combination of quantum numbers represents one independent eigenfunction, the energy level in the cylindrical coordinates, ignoring the z-component, is also $(n + 1)$-fold degenerate. Comparing the results in the two coordinate systems we conclude that the *multiplicity* of a degenerate level is independent of the coordinate system. The quantum numbers of the degenerate levels according to both coordinate systems are listed in Fig. 9–2.

The conclusion of Section 9–1 that the general solution of the time-independent Schrödinger equation corresponding to energy E_n is a linear superposition of terms $\psi_{n_x n_y n_z}$ subject to the condition $n_x + n_y + n_z = n$ may be restated as follows: The solutions corresponding to energy E_n form a *linear manifold* with the functions $\psi_{n_x n_y n_z}$ $(n_x + n_y + n_z = n)$ as *base functions*. A linear manifold is defined as a group of functions generated from a set of base functions $u_1, u_2, \ldots, u_s$ by linear superposition, i.e.,

$$u = a_1 u_1 + a_2 u_2 + \cdots a_s u_s, \tag{9–55}$$

where $a_1, a_2, \ldots, a_s$ are arbitrary constants. The base functions are assumed to be mutually independent. Since there are infinitely many possible values for each constant, the total number of functions in the manifold is ∞^s, i.e., an infinity of the sth order. The theory of linear transformation

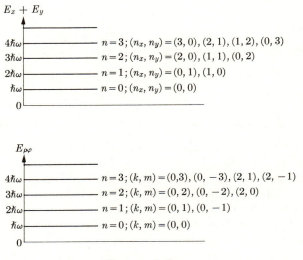

FIGURE 9–2

enables us to change from one set of base functions to another, but the number of independent base functions is always the same, that is, s. The above-discussed relation between eigenfunctions in two different coordinate systems may now be restated as follows: The eigenfunctions in one coordinate system may be regarded as a particular set of base functions by which the manifold of degenerate solutions may be expressed. Different coordinate systems lead to different sets of base functions, the number of base functions and so the multiplicity remaining the same. Different sets of base functions are related by linear transformations. They are equivalent to one another in representing the same manifold.

9–3 Solution in spherical coordinates. As the potential is a function of the radius vector only, it represents a central force and the problem may be solved in spherical coordinates according to the general method of Chapter 8. The time-independent Schrödinger equation now takes the following form:

$$\frac{1}{r^2}\frac{\partial}{\partial r}\left(r^2\frac{\partial\psi}{\partial r}\right) + \frac{1}{r^2\sin\theta}\frac{\partial}{\partial\theta}\left(\sin\theta\frac{\partial\psi}{\partial\theta}\right)$$
$$+ \frac{1}{r^2\sin^2\theta}\frac{\partial^2\psi}{\partial\varphi^2} + \frac{2M}{\hbar^2}\left(E - \frac{1}{2}Kr^2\right)\psi = 0. \qquad (9\text{–}56)$$

Its general solution corresponding to a given energy E is

$$\psi = \sum_{lm} R_l(r)\,Y_{lm}(\theta, \varphi), \qquad (9\text{–}57)$$

where $R_l(r)$ satisfies the radial equation

$$\frac{d^2 R_l}{dr^2} + \frac{2}{r}\frac{dR_l}{dr} + \frac{2M}{\hbar^2}\left(E - \frac{1}{2}Kr^2 - \frac{\hbar^2}{2M}\frac{l(l+1)}{r^2}\right)R_l = 0. \quad (9\text{–}58)$$

The asymptotic equation and its solution are the same as Eqs. (9–31) and (9–32). Thus we let

$$R_l(r) = e^{-\sqrt{MK/\hbar^2}\,r^2/2}v(r) = e^{-(M\omega/\hbar)r^2/2}v(r) \equiv e^{-\xi^2/2}v(\xi). \quad (9\text{–}59)$$

The differential equation of $v(\xi)$ may be obtained by substitution,

$$\ddot{v} + \left(\frac{2}{\xi} - 2\xi\right)\dot{v} + \left(\lambda - 3 - \frac{l(l+1)}{\xi^2}\right)v = 0, \quad (9\text{–}60)$$

where again

$$\lambda = \frac{2E}{\hbar\omega}. \quad (9\text{–}61)$$

Equation (9–60) is similar to Eq. (9–37) and may be solved analogously. Let

$$v(\xi) = \xi^\alpha u(\xi), \qquad u(0) \neq 0. \quad (9\text{–}62)$$

Equation (9–60) becomes

$$u[\alpha(\alpha - 1) + 2\alpha - l(l+1)]\xi^{\alpha-2} + [2\alpha\dot{u} + 2\dot{u}]\xi^{\alpha-1}$$
$$+ [\ddot{u} - 2\alpha u + (\lambda - 3)u]\xi^\alpha + [-2\dot{u}]\xi^{\alpha+1} = 0. \quad (9\text{–}63)$$

Dividing this equation by $\xi^{\alpha-2}$ and letting $\xi \to 0$, we obtain

$$\alpha(\alpha - 1) + 2\alpha - l(l+1) = 0. \quad (9\text{–}64)$$

The solutions of this equation are

$$\alpha = \begin{cases} l, \\ -(l+1). \end{cases} \quad (9\text{–}65)$$

As before the negative root is rejected on account of the condition of normalization. After substitution of $\alpha = l$ in Eq. (9–63), the latter becomes the differential equation of $u(\xi)$,

$$\xi\ddot{u} + (2l + 2 - 2\xi^2)\dot{u} + (\lambda - 3 - 2l)\xi u = 0. \quad (9\text{–}66)$$

Equation (9–66) may be solved by the power series method. Assume

$$u = \sum_k a_k\xi^k.$$

The relations between the coefficients are, analogous to those in Section 9–2,

$$(2l + 2)a_1 = 0,$$
$$. \quad . \quad . \quad . \quad . \quad . \quad . \quad . \quad . \quad . \quad . \quad . \quad . \quad . \quad . \quad , \tag{9–67}$$
$$a_{k+2} = \frac{2k + 3 + 2l - \lambda}{(k + 2)(k + 1) + (2l + 2)(k + 2)} \, a_k.$$

Thus the odd terms vanish because of the first equation $a_1 = 0$ and the recursion formula. The even terms form a series which diverges like the series of e^{ξ^2}. Thus $R_l(\xi)$ diverges like the function $e^{+\xi^2/2}$. Acceptable solutions do not exist unless the series breaks off after a finite number of terms. The condition that the series terminate after the kth term $a_k \xi^k$ is

$$2k + 3 + 2l - \lambda = 0. \tag{9–68}$$

Equation (9–61) thus leads to

$$E = \hbar\omega(k + l + \tfrac{3}{2}), \quad \begin{array}{l} l = 0, 1, 2, \ldots, \\ k = 0, 2, 4, \ldots, \end{array}$$

or
$$\tag{9–69}$$
$$E = \hbar\omega(n + \tfrac{3}{2}), \quad n = 0, 1, 2, \ldots,$$

where $n = k + l$. The energy eigenvalues are thus exactly the same as those obtained in the two previous sections. The eigenfunctions corresponding to quantum numbers klm are

$$\psi_{klm}(r, \theta, \varphi) = N_k e^{-\xi^2/2} \xi^l \left(\sum_{\nu=0}^{k} a_\nu \xi^\nu \right) Y_{lm}(\theta, \varphi), \tag{9–70}$$

where N_k is a normalization constant and $\xi = \sqrt{(M\omega/\hbar)} \, r$. As in Section 9–2, the eigenfunctions in spherical coordinates may be expressed as linear combinations of those in the other coordinate systems. This may be demonstrated by an example. Consider the eigenfunction of $k = 2$, $l = 1$, $m = 0$. It may be transformed as follows:

$$\psi_{210}(r, \theta, \varphi)$$
$$= N_k e^{-\xi^2/2} \xi(a_0 - \tfrac{2}{5}a_0\xi^2)(\sqrt{6}/2) \cos\theta$$
$$\sim e^{-\xi^2/2}(5 - 2\xi_x^2 - 2\xi_y^2 - 2\xi_z^2)\xi_z$$
$$= e^{-\xi^2/2}[(1 - 2\xi_x^2)\xi_z + (1 - 2\xi_y^2)\xi_z + (3\xi_z - 2\xi_z^3)] \tag{9–71}$$
$$= e^{-\xi^2/2}[-\tfrac{1}{4}H_2(\xi_x)H_1(\xi_z) - \tfrac{1}{4}H_2(\xi_y)H_1(\xi_z) - \tfrac{1}{4}H_3(\xi_z)]$$
$$\sim -\psi_2(\xi_x)\psi_0(\xi_y)\psi_1(\xi_z) - \psi_0(\xi_x)\psi_2(\xi_y)\psi_1(\xi_z) - \psi_0(\xi_x)\psi_0(\xi_y)\psi_3(\xi_z).$$

The last expression gives the linear combination of eigenfunctions in the rectangular coordinate system. Since $k + l = 3$, $\psi_{210}(r, \theta, \varphi)$ belongs to the energy eigenvalue E_3. The eigenfunctions in the last line of Eq. (9–71) all belong to energy eigenvalue E_3, since the sums of the three indices are all 3. We can show that the multiplicity of the degenerate level E_n is the same as in the two other coordinate systems. In rectangular coordinates the multiplicity of an energy level E_n is the number of combinations of n_x, n_y, n_z the sum of which is the same n. First there are $n + 1$ possible values of n_x. Once n_x is fixed, the amount $n - n_x$ may be divided in $(n - n_x + 1)$ ways between n_y and n_z. The total number of combinations (and thus the multiplicity) is given by

$$\text{multiplicity} = \sum_{n_x=0}^{n} (n - n_x + 1) = (n + 1)\,\frac{n + 1 + 1}{2}$$

$$= \frac{1}{2}\,(n + 1)(n + 2). \tag{9–72}$$

In spherical coordinates the multiplicity of E_n may be obtained as follows: When n is even, there are $n/2 + 1$ possible values of k. For each value of k the value of l is $n - k$ and there are $2l + 1$ possible values of m. The total number of combinations (and thus the multiplicity) is given by

$$\text{multiplicity} = \sum_{k/2=0}^{n/2} [2(n - k) + 1] = \left(\frac{n}{2} + 1\right)\frac{(2n + 1) + 1}{2}$$

$$= \frac{1}{2}(n + 1)(n + 2). \tag{9–73}$$

When n is odd there are $(n + 1)/2$ possible values of k. For each k the multiplicity due to m is the same as before. Thus

$$\text{multiplicity} = \sum_{k/2=0}^{(n-1)/2} [2(n - k) + 1] = \left(\frac{n + 1}{2}\right)\frac{(2n + 1) + 3}{2}$$

$$= \frac{1}{2}\,(n + 1)(n + 2), \tag{9–74}$$

the same as in the even n case. Therefore the multiplicity of the degenerate level E_n in spherical coordinates is exactly the same as in rectangular coordinates. According to the discussion at the end of Section 9–2, the set of degenerate eigenfunctions corresponding to energy E_n in spherical coordinates represents a new set of base functions for the same manifold of degenerate solutions. The energy level diagram and the quantum numbers are shown in Fig. 9–3. Comparing it with Fig. 9–1 we find that the energy values and the multiplicities are all the same, as they should be.

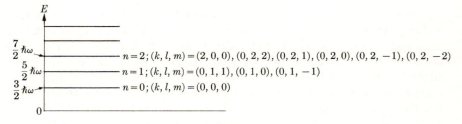

FIGURE 9-3

The energy levels in a *general* central force problem may be labeled according to the following system: The spectroscopic notations s, p, d, f, ..., are used to indicate the value of l. All states differing by m only are not distinguished and are considered as one level since their energies are all the same. Thus m will not appear in the notation designating a level. A number is to be attached in front of the spectroscopic notation to indicate the ordinal number of this level in the column of levels of the same l. Thus $3p$ indicates the third level of $l = 1$ counted from the bottom of the energy scale. It is understood that this level has a multiplicity of $2l + 1$ due to the magnetic quantum number m. The physical meaning of this number placed before the spectroscopic notation is that it is equal to the number of radial nodes plus one. In the present case, the harmonic oscillator, this number equals $\frac{1}{2}k + 1$ or $\frac{1}{2}(n - l) + 1$. This system should not be confused with the spectroscopic system in which the number before the spectroscopic notation is the principal quantum number n. The designation of the energy levels of the harmonic oscillator according to this system is shown in the upper left part of Fig. 9–4. Due to a peculiar property of the potential, some of the levels, e.g., $1g$, $2d$, and $3s$, happen to coincide. If the potential is changed slightly, these levels will split apart. The situation is similar to that of the Coulomb potential in which a number of levels also happen to coincide and will split apart when the potential is changed slightly. The levels of the Coulomb potential are labeled according to the present system and are shown in the upper right part of Fig. 9–4. Among them the levels $1g$, $2f$, and $3d$, for example, coincide. If we change the potential gradually starting from the oscillator potential (lower left of Fig. 9–4) and finally end up at the Coulomb potential (lower right of Fig. 9–4), the energy levels will change their positions accordingly: starting from the scheme in the upper left part of Fig. 9–4, passing through a succession of intermediate stages where the degenerate levels split apart, and finally ending up in the scheme in the upper right part of Fig. 9–4 where the levels are regrouped together in a new degenerate pattern. This discussion shows that no matter how the potential changes, the number of levels remains the same. The positions of the levels may

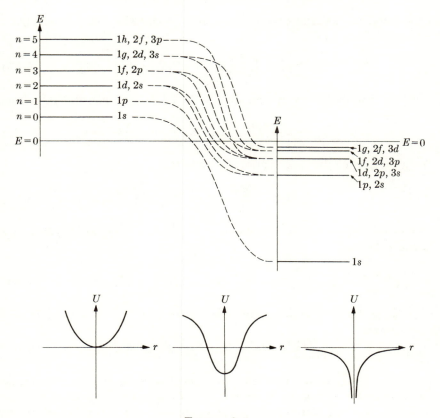

FIGURE 9–4

change but the quantum numbers and the characteristics associated with the quantum numbers remain the same. These observations have a bearing in the discussion of the nucleonic energy levels in nuclear physics. Incidentally, the present system of labeling is used in nuclear physics.

9–4 Orthogonality of the wave functions. We have proved or demonstrated the orthogonality of wave functions in a number of special cases. It will now be shown that this is a general property of all eigenfunctions of the three-dimensional Schrödinger equation. Let ψ_1 and ψ_2 be two solutions for energies E_1 and E_2 respectively,

$$\nabla^2\psi_1 + \frac{2M}{\hbar^2}\,(E_1 - U)\psi_1 = 0, \tag{9–75}$$

$$\nabla^2\psi_2 + \frac{2M}{\hbar^2}\,(E_2 - U)\psi_2 = 0. \tag{9–76}$$

In general, the solutions are complex functions. The complex conjugate equation of Eq. (9–75) is

$$\nabla^2 \psi_1^* + \frac{2M}{\hbar^2} (E_1 - U)\psi_1^* = 0. \tag{9–77}$$

Multiply Eq. (9–76) by ψ_1^* and Eq. (9–77) by ψ_2 and take the difference,

$$\psi_1^* \nabla^2 \psi_2 - \psi_2 \nabla^2 \psi_1^* + \frac{2M}{\hbar^2} (E_2 - E_1)\psi_1^* \psi_2 = 0. \tag{9–78}$$

Integrating over the space variables, we obtain

$$\iiint (\psi_1^* \nabla^2 \psi_2 - \psi_2 \nabla^2 \psi_1^*) \, d\tau + \frac{2M}{\hbar^2} (E_2 - E_1) \iiint \psi_1^* \psi_2 \, d\tau = 0. \tag{9–79}$$

The first integral may be transformed to a surface integral at infinity by Green's theorem,

$$\iint \left(\psi_1^* \frac{\partial \psi_2}{\partial n} - \psi_2 \frac{\partial \psi_1^*}{\partial n} \right) d\sigma + \frac{2M}{\hbar^2} (E_2 - E_1) \iiint \psi_1^* \psi_2 \, d\tau = 0. \tag{9–80}$$

Since ψ_1 and ψ_2 vanish at infinity, the first term of Eq. (9–80) equals zero and we have

$$\iiint \psi_1^* \psi_2 \, d\tau = 0, \quad \text{if} \quad E_1 \neq E_2. \tag{9–81}$$

Equation (9–81) shows that any two eigenfunctions of different energy eigenvalues are orthogonal to each other. Incidentally, this proof may be presented without introducing the complex conjugate equation when the Schrödinger equation is real. The complex conjugate equation becomes necessary when the Schrödinger equation contains an imaginary term, e.g., $ia(\partial \psi/\partial z)$ (where i is the imaginary number $\sqrt{-1}$ and a is a constant); such a situation arises in the treatment of the Zeeman effect (Chapter 12). The reader may verify that Eq. (9–81) remains valid after the inclusion of such a term.

When there are several wave functions $\psi_1, \psi_2, \ldots$, corresponding to the same energy E, the orthogonality relation expressed by Eq. (9–81) in general does not hold for them. In spite of this it is always possible, as will be shown presently, to find a set of base functions which are orthogonal to one another. Moreover there exist many such sets. Let ψ_a and ψ_b be two wave functions of the same eigenvalue E_n not orthogonal to each other. From them a new set of base functions φ_a and φ_b may be obtained by the transformation

$$\varphi_a = a_{11}\psi_a + a_{12}\psi_b,$$
$$\varphi_b = a_{21}\psi_a + a_{22}\psi_b, \tag{9–82}$$

provided that

$$\begin{vmatrix} a_{11} & a_{12} \\ a_{21} & a_{22} \end{vmatrix} \neq 0.$$

One of the many transformations leading to orthogonal sets is as follows:

$$\varphi_a = \psi_a,$$
$$\varphi_b = \psi_a + k\psi_b, \tag{9–83}$$

where

$$k = -\frac{\iiint \psi_a^* \psi_a \, d\tau}{\iiint \psi_a^* \psi_b \, d\tau}, \tag{9–84}$$

because

$$\iiint \varphi_a^* \varphi_b \, d\tau = \iiint \psi_a^* (\psi_a + k\psi_b) \, d\tau$$
$$= \iiint \psi_a^* \psi_a \, d\tau + k \iiint \psi_a^* \psi_b \, d\tau \tag{9–85}$$
$$= 0.$$

For threefold degeneracy, we may construct φ_a, φ_b, φ_c out of ψ_a, ψ_b, ψ_c as follows:

$$\varphi_a = \psi_a,$$
$$\varphi_b = \psi_a + k_1\psi_b, \tag{9–86}$$
$$\varphi_c = \psi_a + k_2\psi_b + k_3\psi_c,$$

where k_1 is to be determined by Eq. (9–84) so that φ_a and φ_b are orthogonal to each other, k_2 and k_3 are to be determined similarly by the two conditions that φ_c is orthogonal to both φ_a and φ_b. This procedure* may be generalized to the 4, 5, . . . , and n-fold degeneracy.

In later applications we shall always use orthogonal sets of eigenfunctions whenever degeneracy occurs. We also have them normalized. These properties may be expressed in one single equation

$$\iiint \psi_i^* \psi_j \, d\tau = \delta_{ij}, \tag{9–87}$$

where δ_{ij} is the *Kronecker δ symbol*, equal to unity when $i = j$ and zero for all others.

---

* Known as the Gram-Schmidt orthogonalization process.

PROBLEMS

9–1. Solve the anisotropic, three-dimensional harmonic oscillator problem the potential of which is

$$U = \tfrac{1}{2}K_x x^2 + \tfrac{1}{2}K_y y^2 + \tfrac{1}{2}K_z z^2. \tag{9–88}$$

Correlate the energy levels with those in Section 9–1 by letting $K_x \to K$, $K_y \to K$, $K_z \to K$.

*9–2. Solve the hypothetical problem of an isotropic, four-dimensional harmonic oscillator, the potential of which is

$$U = \tfrac{1}{2}K(x^2 + y^2 + z^2 + w^2). \tag{9–89}$$

The Schrödinger equation is

$$\frac{\partial^2 \psi}{\partial x^2} + \frac{\partial^2 \psi}{\partial y^2} + \frac{\partial^2 \psi}{\partial z^2} + \frac{\partial^2 \psi}{\partial w^2} + \frac{2M}{\hbar^2}(E - U)\psi = 0. \tag{9–90}$$

Approach the problem by using different coordinate systems. Discuss the degeneracy of energy levels and the correlation of eigenfunctions in different coordinate systems. *Note:* Schrödinger's equation in a space of more than three dimensions will be introduced in Chapter 12.

9–3. Correlate the energy levels of the three-dimensional harmonic oscillator with those of the three-dimensional square-well potential. *Remark:* The study of the nuclear energy levels may be approximated by a potential intermediate between these two potentials.

9–4. Let $\psi_1 = e^{-x^2}$, $\psi_2 = e^{-2x^2}$, $\psi_3 = e^{-3x^2}$ be three eigenfunctions corresponding to the same energy eigenvalue. Find an orthonormal set of eigenfunctions by the Gram-Schmidt orthogonalization process.

---

* Indicates more difficult problems.

CHAPTER 10

TIME-INDEPENDENT PERTURBATION THEORY

The major mathematical problem in quantum mechanics is to solve the Schrödinger equation. However, even for a simple problem this partial differential equation is usually difficult to solve. In the applications we have had so far, the potential functions are comparatively simple, so that exact, analytic solutions may be obtained. Such cases are rare. As a matter of fact, in most applications, the Schrödinger equation is solved by approximate or numerical methods. The most important of these is the perturbation method, which is applicable to cases where the potential function is only slightly different from that of a well-solved problem. Its basic idea is taken from the perturbation theory in classical mechanics. Take an example in celestial mechanics. The motion of a planet in the solar system is essentially determined by the gravitational force of the sun. However, the gravitational attraction of the other planets, though small, also influences the planetary motion. As a result, the orbit is not a closed ellipse, but a slowly precessing "ellipse". The perturbation theory enables us to calculate the changes of the elliptical orbit due to a perturbing potential, such as the advance of perihelion. Similar situations arise in quantum mechanics and a perturbation theory is developed along the same line. In this chapter, we consider the perturbation theory for the solution of the time-independent Schrödinger equation, which may thus be called the time-independent (or stationary) perturbation theory.* The time-dependent perturbation theory will be discussed in the next chapter.

10–1 Perturbation theory for nondegenerate levels. Suppose that we know the solution of a time-independent Schrödinger equation the potential of which is U. The purpose of the perturbation theory is to determine how the eigenvalues and eigenfunctions of this equation change when the potential U is changed to a small amount.

Let us rewrite the Schrödinger equation in the following form:

$$-\frac{\hbar^2}{2M}\left(\frac{\partial^2}{\partial x^2} + \frac{\partial^2}{\partial y^2} + \frac{\partial^2}{\partial z^2}\right)\psi + U\psi = E\psi. \qquad (10\text{–}1)$$

Equation (10–1) resembles the classical equation for conservation of energy,

$$\frac{1}{2M}(p_x^2 + p_y^2 + p_z^2) + U = E. \qquad (10\text{–}2)$$

---

* E. Schrödinger, *Ann. d. Physik* **80**, 437 (1926).

If we replace p_x, p_y, and p_z by the *differential operators*

$$\frac{\hbar}{i}\frac{\partial}{\partial x}, \qquad \frac{\hbar}{i}\frac{\partial}{\partial y}, \qquad \text{and} \qquad \frac{\hbar}{i}\frac{\partial}{\partial z},$$

that is,

$$p_x \rightarrow \frac{\hbar}{i}\frac{\partial}{\partial x}, \qquad p_y \rightarrow \frac{\hbar}{i}\frac{\partial}{\partial y}, \qquad p_z \rightarrow \frac{\hbar}{i}\frac{\partial}{\partial z}, \qquad (10\text{--}3)$$

Eq. (10–2) becomes an operator equation. (The student is expected to know differential operators from his study of vector analysis.) Let both sides of Eq. (10–2) operate on the function ψ; as a result Eq. (10–1) is reproduced. Until we discuss the important implications of Eq. (10–3) in Chapter 12, Eq. (10–3) may be regarded as a mnemonic aid for writing the Schrödinger equation. Since the left-hand side of Eq. (10–2) is the classical Hamiltonian $H(p_x, p_y, p_z, x, y, z)$ the time-independent Schrödinger equation may be written as follows:

$$H\left(\frac{\hbar}{i}\frac{\partial}{\partial x}, \frac{\hbar}{i}\frac{\partial}{\partial y}, \frac{\hbar}{i}\frac{\partial}{\partial z}, x, y, z\right)\psi = E\psi,$$

or simply

$$H\psi = E\psi \qquad (10\text{--}4)$$

where H is understood to be an operator.

When the potential changes slightly, the Hamiltonian changes by the same amount. The problem to be solved by the perturbation theory may be restated as follows: Given the eigenvalues and eigenfunctions of the *unperturbed problem*,

$$H^{(0)}\psi^{(0)} = E^{(0)}\psi^{(0)}, \qquad (10\text{--}5)$$

find the eigenvalues and eigenfunctions of the *perturbed problem*,

$$(H^{(0)} + \lambda H^{(1)})\psi = E\psi, \qquad (10\text{--}6)$$

in which the Hamiltonian differs from the unperturbed problem by a small amount $\lambda H^{(1)}$ (usually due to a change of the potential). λ denotes an infinitesimal parameter and is separated from the finite part $H^{(1)}$ for the purpose of determining the order of infinitesimals in power-series expansion. This point will be made clear in later discussions.

Let the eigenvalues and eigenfunctions of the unperturbed problem be $E_1^{(0)}, E_2^{(0)}, \ldots, E_n^{(0)}, \ldots,$ and $\psi_1^{(0)}, \psi_2^{(0)}, \ldots \psi_n^{(0)} \ldots,$ respectively. Thus,

$$H^{(0)}\psi_n^{(0)} = E_n^{(0)}\psi_n^{(0)}, \qquad n = 1, 2, \ldots n, \ldots. \qquad (10\text{--}7)$$

In the nondegenerate case each eigenvalue corresponds to one eigenfunction, and the eigenfunctions are orthogonal to one another. They may be normalized to form an orthonormal set, which satisfies

$$\int \psi_i^{(0)*} \psi_j^{(0)} \, d\tau = \delta_{ij}. \tag{10-8}$$

This will always be assumed throughout the later discussions. When the perturbation term is small, the eigenvalues and eigenfunctions of Eq. (10-6) should not differ greatly from those of Eq. (10-5). Their first-order changes will be of the order of magnitude of λ; their second-order changes will be of the order of λ^2, etc. Thus we may write the eigenvalues and eigenfunctions of the perturbed equation as follows:

$$E_n = E_n^{(0)} + \lambda E_n^{(1)} + \lambda^2 E_n^{(2)} + \cdots, \tag{10-9}$$

$$\psi_n = \psi_n^{(0)} + \lambda \psi_n^{(1)} + \lambda^2 \psi_n^{(2)} + \cdots, \tag{10-10}$$

where $E_n^{(1)}, E_n^{(2)}, \ldots$, and $\psi_n^{(1)}, \psi_n^{(2)}, \ldots$, are quantities to be determined. For mathematical convenience we may expand $\psi_n^{(1)}, \psi_n^{(2)}, \ldots$, in series of the unperturbed eigenfunctions $\psi_n^{(0)}$, as any well-behaved function may be expanded in a series of a complete set of orthonormal functions;

$$\psi_n^{(1)} = \sum_m C_{nm}^{(1)} \psi_m^{(0)}, \tag{10-11}$$

$$\psi_n^{(2)} = \sum_m C_{nm}^{(2)} \psi_m^{(0)}, \tag{10-12}$$

$$\cdot \quad \cdot \quad \cdot \quad \cdot \quad \cdot \quad \cdot \quad ,$$

where $C_{nm}^{(1)}, C_{nm}^{(2)}, \ldots$, are coefficients of expansion. Substituting Eqs. (10-9) and (10-10) in Eq. (10-6), we have

$$(H^{(0)} + \lambda H^{(1)})(\psi_n^{(0)} + \lambda \psi_n^{(1)} + \lambda^2 \psi_n^{(2)} + \cdots)$$
$$= (E_n^{(0)} + \lambda E_n^{(1)} + \lambda^2 E_n^{(2)} + \cdots)(\psi_n^{(0)} + \lambda \psi_n^{(1)} + \lambda^2 \psi_n^{(2)} + \cdots) \cdot \tag{10-13}$$

Equation (10-13) may be changed to

$$H^{(0)} \psi_n^{(0)} + (H^{(0)} \psi_n^{(1)} + H^{(1)} \psi_n^{(0)}) \lambda + (H^{(0)} \psi_n^{(2)} + H^{(1)} \psi_n^{(1)}) \lambda^2 + \cdots$$
$$= E_n^{(0)} \psi_n^{(0)} + (E_n^{(0)} \psi_n^{(1)} + E_n^{(1)} \psi_n^{(0)}) \lambda$$
$$+ (E_n^{(0)} \psi_n^{(2)} + E_n^{(1)} \psi_n^{(1)} + E_n^{(2)} \psi_n^{(0)}) \lambda^2 + \cdots. \tag{10-14}$$

In order that Eq. (10-14) may be satisfied for all values of λ, the coefficients of the same powers of λ on both sides of Eq. (10–14) must equal each other. Thus we have

$$H^{(0)}\psi_n^{(0)} = E_n^{(0)}\psi_n^{(0)}, \tag{10–15}$$

$$H^{(0)}\psi_n^{(1)} + H^{(1)}\psi_n^{(0)} = E_n^{(0)}\psi_n^{(1)} + E_n^{(1)}\psi_n^{(0)}, \tag{10–16}$$

$$H^{(0)}\psi_n^{(2)} + H^{(1)}\psi_n^{(1)} = E_n^{(0)}\psi_n^{(2)} + E_n^{(1)}\psi_n^{(1)} + E_n^{(2)}\psi_n^{(0)}, \tag{10–17}$$

$$\cdots \cdots \cdots$$

These equations specify the conditions $E_n^{(i)}$'s and $\psi_n^{(i)}$'s have to satisfy and may be solved to obtain $E_n^{(i)}$'s and $\psi_n^{(i)}$'s. Equation (10–15) is satisfied automatically by virtue of Eq. (10–7); Eq. (10–16) may be solved to determine $E_n^{(1)}$ and $\psi_n^{(1)}$; Eq. (10–17) may be solved with the help of the solutions of Eq. (10–16) to obtain $E_n^{(2)}$ and $\psi_n^{(2)}$, etc. The solution of the perturbed problem (E_n and ψ_n) may thus be obtained; the series may be determined to any order of λ as we desire.

Making use of the series expansion of $\psi_n^{(1)}$ in terms of $\psi_n^{(0)}$'s, we solve Eq. (10–16) to determine the perturbation of the first order. Substituting Eq. (10–11) in Eq. (10–16), we have

$$\sum_m C_{nm}^{(1)} H^{(0)}\psi_m^{(0)} + H^{(1)}\psi_n^{(0)} = \sum_m C_{nm}^{(1)} E_n^{(0)}\psi_m^{(0)} + E_n^{(1)}\psi_n^{(0)}. \tag{10–18}$$

By virtue of Eq. (10–15), the last equation may be changed to

$$\sum_m C_{nm}^{(1)} (E_m^{(0)} - E_n^{(0)})\psi_m^{(0)} + H^{(1)}\psi_n^{(0)} = E_n^{(1)}\psi_n^{(0)}. \tag{10–19}$$

Multiplying Eq. (10–19) by $\psi_n^{(0)*}$ and integrating over the space variables, we obtain the following result by making use of Eq. (10–8):

$$E_n^{(1)} = \int \psi_n^{(0)*} H^{(1)} \psi_n^{(0)} \, d\tau. \tag{10–20}$$

This equation gives the first-order perturbation of the energy eigenvalue. Multiplying Eq. (10–19) by $\psi_k^{(0)*} (k \neq n)$ and integrating as before, we obtain the following equation with the help of the orthonormal property of $\psi_n^{(0)}$'s:

$$C_{nk}^{(1)} (E_k^{(0)} - E_n^{(0)}) + \int \psi_k^{(0)*} H^{(1)} \psi_n^{(0)} \, d\tau = 0. \tag{10–21}$$

Since k may take any value of the running index m except n, we may replace k by m. Thus,

$$C_{nm}^{(1)} = \frac{\int \psi_m^{(0)*} H^{(1)} \psi_n^{(0)} \, d\tau}{E_n^{(0)} - E_m^{(0)}}, \qquad m \neq n. \tag{10–22}$$

This equation gives the coefficients of expansion $C_{nm}^{(1)}$ of the first-order perturbation of the eigenfunction for all values of m except $m = n$. For $m = n$ this method does not enable us to determine the coefficient $C_{nn}^{(1)}$. Actually $C_{nn}^{(1)}$ may take any value and still satisfy Eq. (10–16). Since $C_{nn}^{(1)}$ introduces in $\psi_n^{(1)}$ a term of $\psi_n^{(0)}$ which is the same as the unperturbed eigenfunction, we may combine this term with the unperturbed eigenfunction. Thus

$$\psi_n = (1 + \lambda C_{nn}^{(1)})\psi_n^{(0)} + \lambda \sum_m{}' C_{nm}^{(1)}\psi_m^{(0)} + \cdots. \qquad (10\text{–}23)$$

The symbol $\sum_m'$ denotes a summation over all values of m except $m = n$. This notation will be adopted throughout the book. The coefficient $C_{nn}^{(1)}$ is eventually determined by normalization which requires $C_{nn}^{(1)}$ to be zero. Therefore we may forget about the term $C_{nn}^{(1)}$ in the expansion of $\psi_n^{(1)}$. Equations (10–20) and (10–22) thus give us the complete solution of the first-order perturbation.

The reader may wonder how it is possible to determine the infinitely many coefficients $C_{n1}^{(1)}, C_{n2}^{(1)}, \ldots C_{nm}^{(1)}, \ldots$, out of one single equation (10–16). The reason is that Eq. (10–16) is an identical equation which holds for infinitely many points (x, y, z) in space. Thus it implies infinitely many equations.

The second-order perturbation may be determined by Eq. (10–17). After substitution of Eq. (10–12), Eq. (10–17) takes the form

$$\sum_m C_{nm}^{(2)}H^{(0)}\psi_m^{(0)} + \sum_m{}' C_{nm}^{(1)}H^{(1)}\psi_m^{(0)}$$

$$= \sum_m C_{nm}^{(2)}E_n^{(0)}\psi_m^{(0)} + \sum_m{}' C_{nm}^{(1)}E_n^{(1)}\psi_m^{(0)} + E_n^{(2)}\psi_n^{(0)} \qquad (10\text{–}24)$$

or

$$\sum_m C_{nm}^{(2)}(E_m^{(0)} - E_n^{(0)})\psi_m^{(0)} = \sum_m{}' C_{nm}^{(1)}(E_n^{(1)} - H^{(1)})\psi_m^{(0)} + E_n^{(2)}\psi_n^{(0)}. \qquad (10\text{–}25)$$

The second-order perturbation of the eigenvalue and eigenfunction may be obtained from Eq. (10–25) in a manner similar to that used for the first-order perturbation. Thus $E_n^{(2)}$ may be obtained from Eq. (10–25) by multiplying it by $\psi_n^{(0)*}$ and integrating over the space variables:

$$E_n^{(2)} = \sum_m{}' C_{nm}^{(1)} \int \psi_n^{(0)*}H^{(1)}\psi_m^{(0)}\, d\tau. \qquad (10\text{–}26)$$

Putting in the values of the coefficients $C_{nm}^{(1)}$, we have

$$E_n^{(2)} = \sum_m{}' \frac{\int \psi_m^{(0)*}H^{(1)}\psi_n^{(0)}\, d\tau \cdot \int \psi_n^{(0)*}H^{(1)}\psi_m^{(0)}\, d\tau}{E_n^{(0)} - E_m^{(0)}}. \qquad (10\text{–}27)$$

Multiplying by $\psi_k^{(0)*}$ instead of $\psi_n^{(0)*}(k \neq n)$ we obtain in a similar manner

$$C_{nk}^{(2)} = \sum_m{}' \frac{C_{nm}^{(1)}}{E_k^{(0)} - E_n^{(0)}} \left(E_n^{(1)} \delta_{km} - \int \psi_k^{(0)*} H^{(1)} \psi_m^{(0)} \, d\tau \right), \qquad k \neq n. \tag{10–28}$$

For convenience we define the *matrix element* of H with respect to two wave functions ψ_n and ψ_m by

$$H_{nm} = \int \psi_n^* H \psi_m \, d\tau. \tag{10–29}$$

Care must be taken with regard to the order of the two indices, because H_{mn} is not necessarily equal to H_{nm} as may be shown readily by simple examples. Also we insist upon writing H between ψ_n^* and ψ_m, for H may contain operators. The results of the first-order perturbation may be rewritten thus:

$$E_n^{(1)} = H_{nn}^{(1)}, \tag{10–30}$$

$$C_{nm}^{(1)} = \frac{H_{mn}^{(1)}}{E_n^{(0)} - E_m^{(0)}}. \tag{10–31}$$

Likewise we rewrite the results of the second-order perturbation, the *dummy index* m in Eqs. (10–27) and (10–28) being replaced by another letter, s (for summation), and the *running index* k being replaced by m which is now reserved for this particular purpose:

$$E_n^{(2)} = \sum_s{}' \frac{H_{ns}^{(1)} H_{sn}^{(1)}}{E_n^{(0)} - E_s^{(0)}}, \tag{10–32}$$

$$C_{nm}^{(2)} = \frac{1}{E_n^{(0)} - E_m^{(0)}} \sum_s{}' \frac{(H_{ms}^{(1)} - H_{nn}^{(1)} \delta_{ms}) H_{sn}^{(1)}}{E_n^{(0)} - E_s^{(0)}}. \tag{10–33}$$

(Note that $\psi_n^{(i)}$ is not needed for the calculation of $E_n^{(i)}$ but is needed for $E_n^{(i+1)}$.)

To illustrate the application of the perturbation theory, let us consider a charged linear harmonic oscillator placed in a *weak*, uniform electric field F. The potential function of the electric field is $-Fx$. Therefore the potential energy of the oscillator is $\frac{1}{2}Kx^2$ plus a small quantity $-qFx$ where q is the charge of the oscillator. The Schrödinger equation is thus

$$\frac{d^2\psi}{dx^2} + \frac{2M}{\hbar^2} \left(E - \frac{1}{2} Kx^2 + qFx \right) \psi = 0. \tag{10–34}$$

As the field is weak (F is small), the additional term $-qFx$ of the potential may be regarded as a perturbation. Thus we let

$$\lambda H^{(1)} = -qFx.$$

Since F is a small quantity, we may identify $-qF$ with λ,

$$\lambda = -qF. \tag{10-35}$$

Hence

$$H^{(1)} = x. \tag{10-36}$$

The unperturbed problem was solved in Chapter 4. The eigenvalues and eigenfunctions expressed by Eqs. (4-25) and (4-34) are now the unperturbed solutions, $E_n^{(0)}$ and $\psi_n^{(0)}$. The first- and second-order perturbations of the eigenvalues and eigenfunctions may be obtained by Eqs. (10-30) to (10-33). When F is small, the power series, Eqs. (10-9) and (10-10), converge rapidly and the approximate solutions obtained by the perturbation method will be accurate enough for practical purposes. To complete the calculation we first evaluate the matrix elements of $H^{(1)}$. From the wave functions of Eq. (4-34) and $H^{(1)}$ of Eq. (10-36), we have

$$H_{mn}^{(1)} = \int_{-\infty}^{\infty} \left(\frac{M\omega}{\hbar}\right)^{1/2} \frac{1}{(\sqrt{\pi}\, 2^m m!)^{1/2}} \frac{1}{(\sqrt{\pi}\, 2^n n!)^{1/2}}$$

$$\times H_m\left(\sqrt{\frac{M\omega}{\hbar}}\, x\right) H_n\left(\sqrt{\frac{M\omega}{\hbar}}\, x\right) e^{-(M\omega/\hbar)x^2} x\, dx \tag{10-37}$$

or, in terms of the variable ξ,

$$H_{mn}^{(1)} = \sqrt{\frac{\hbar}{M\omega}} \frac{1}{(\pi 2^{m+n} m! n!)^{1/2}} \int_{-\infty}^{\infty} H_m(\xi)\xi H_n(\xi) e^{-\xi^2}\, d\xi.$$

Replacing $\xi H_n(\xi)$ according to Eq. (4-32), we obtain

$$H_{mn}^{(1)} = \sqrt{\frac{\hbar}{M\omega}} \frac{1}{(\pi 2^{m+n} m! n!)^{1/2}} \int_{-\infty}^{\infty} H_m(\xi)\left[\frac{1}{2} H_{n+1}(\xi) + n H_{n-1}(\xi)\right] e^{-\xi^2}\, d\xi. \tag{10-38}$$

The right-hand side does not vanish only when m equals either $n+1$ or $n-1$ because of the orthogonal relation. These nonvanishing values may be obtained with the aid of the normalization integral, Eq. (4-33). For $m = n+1$, we have

$$\int_{-\infty}^{\infty} [H_{n+1}(\xi)e^{-\xi^2/2}]^2 \, d\xi = \sqrt{\pi}\, 2^{n+1}(n+1)!. \qquad (10\text{-}39)$$

Thus

$$H_{n+1,n}^{(1)} = \sqrt{\frac{\hbar}{M\omega}}\, \frac{1}{[\pi 2^{2n+1}(n+1)!n!]^{1/2}}\, \frac{1}{2}\, \sqrt{\pi}\, 2^{n+1}(n+1)!$$

$$= \sqrt{\hbar/M\omega}\, \sqrt{(n+1)/2}. \qquad (10\text{-}40)$$

For $m = n - 1$, we have similarly

$$H_{n-1,n}^{(1)} = \sqrt{\frac{\hbar}{M\omega}}\, \frac{1}{[\pi 2^{2n-1}(n-1)!n!]^{1/2}}\, n\sqrt{\pi}\, 2^{n-1}(n-1)!$$

$$= \sqrt{\hbar/M\omega}\, \sqrt{n/2}. \qquad (10\text{-}41)$$

Thus,

$$H_{m,n}^{(1)} = \begin{cases} \sqrt{\hbar/M\omega}\, \sqrt{(n+1)/2}, & \text{for } m = n+1, \\ 0, & \text{for } m \neq n \pm 1, \qquad (10\text{-}42) \\ \sqrt{\hbar/M\omega}\, \sqrt{n/2}, & \text{for } m = n-1. \end{cases}$$

Equations (10–30) to (10–33) thus lead to the following results:

$$E_n^{(1)} = 0, \qquad (10\text{-}43)$$

$$C_{nm}^{(1)} = \begin{cases} -1/\hbar\omega\sqrt{\hbar/M\omega}\, \sqrt{(n+1)/2}, & m = n+1, \\ 0, & m \neq n \pm 1, \qquad (10\text{-}44) \\ 1/\hbar\omega\sqrt{\hbar/M\omega}\, \sqrt{n/2}, & m = n-1, \end{cases}$$

$$E_n^{(2)} = \frac{-(\hbar/M\omega)[(n+1)/2] + (\hbar/M\omega)(n/2)}{\hbar\omega} = -\frac{1}{2M\omega^2}, \qquad (10\text{-}45)$$

$$C_{nr}^{(2)} = \begin{cases} (1/4M\omega^2)(1/\hbar\omega)\sqrt{(n+1)(n+2)}, & r = n+2, \\ 0, & r \neq n \pm 2, \qquad (10\text{-}46) \\ (1/4M\omega^2)(1/\hbar\omega)\sqrt{n(n-1)}, & r = n-2. \end{cases}$$

With these results we write the eigenvalues and eigenfunctions of the perturbed problem, accurate to the second order, as follows (remembering $K = M\omega^2$),

$$E_n = \hbar\omega\left(n + \frac{1}{2}\right) - \frac{q^2 F^2}{2K}, \tag{10–47}$$

$$\psi_n = \psi_n^{(0)} + qF\sqrt{\frac{n+1}{2\hbar\omega K}}\,\psi_{n+1}^{(0)} - qF\sqrt{\frac{n}{2\hbar\omega K}}\,\psi_{n-1}^{(0)}$$

$$+ \frac{q^2 F^2\sqrt{(n+1)(n+2)}}{4\hbar\omega K}\,\psi_{n+2}^{(0)} + \frac{q^2 F^2\sqrt{n(n-1)}}{4\hbar\omega K}\,\psi_{n-2}^{(0)}, \tag{10–48}$$

where $\psi_n^{(0)}$, $\psi_{n+1}^{(0)}$, . . . , are the unperturbed solutions given by Eq. (4–34). We notice that the first-order energy perturbation is zero. This result is expected in classical mechanics, since the gain of energy when the oscillator is on the $-x$ side is compensated by the loss of energy when the oscillator is on the $+x$ side. The second-order energy perturbation is a negative quantity. This is also expected in the classical theory. Because of the deformation of the orbit by the positive force qF, the oscillator will spend a little more time on the $+x$ side than on the $-x$ side. Therefore the average of potential energy will favor the $+x$ side, i.e., a negative amount. This argument also illustrates the previous mathematical result that the first-order eigenfunction affects the second-order eigenvalue. It is also natural that this second-order energy change should be inversely proportional to the force constant K, since a small value of K corresponds to an orbit easily deformable and the energy change will be large. We notice from Eq. (10–48) that the change of the eigenfunction in the first order involves only eigenfunctions with quantum numbers differing from n by 1, and that in the second order it involves only those with quantum numbers differing by 2. These results have important applications later.

The first-order perturbation of energy, $\int \psi_n^{(0)*}\lambda H^{(1)}\psi_n^{(0)}\,d\tau$, has a simple physical interpretation. It is just the average value of the perturbation Hamiltonian $\lambda H^{(1)}$ over the unperturbed state $\psi_n^{(0)}$. This result is analogous to one in the perturbation theory of classical mechanics, i.e., the perturbation energy is the time average of the perturbing potential over the unperturbed orbit.

10–2 Perturbation theory for degenerate levels. Equations (10–30) to (10–33), which summarize the results of the last section, contain the energy differences $E_n^{(0)} - E_m^{(0)}$ in the denominators. When several energy levels are very close to one another, the energy differences among them are small and the right-hand side of Eqs. (10–31) to (10–33) may be very large. If these levels actually coincide, they become a degenerate level. The energy differences $E_n^{(0)} - E_m^{(0)}$ among them equal zero and the previous equations no longer give valid solutions because the right-hand side becomes infinite. Therefore a new method is needed for the perturbation theory of a cluster of closely spaced energy levels which includes the degenerate level as a special case.

Let the energy eigenvalues of the unperturbed Schrödinger equation be arranged in such an order that the cluster of g closely spaced levels be labeled from $E_1^{(0)}$ to $E_g^{(0)}$; all others be labeled from $E_{g+1}^{(0)}$ on. The eigenvalues $E_1^{(0)}, \ldots, E_g^{(0)}$ are nearly equal to one another. For the special case of a degenerate level, $E_1^{(0)}, \ldots, E_g^{(0)}$ are all the same. The eigenfunctions are denoted by

$$\psi_1^{(0)}, \psi_2^{(0)}, \ldots, \psi_g^{(0)} \mid \psi_{g+1}^{(0)}, \psi_{g+2}^{(0)}, \ldots,$$

corresponding to eigenvalues

$$E_1^{(0)}, E_2^{(0)}, \ldots, E_g^{(0)} \mid E_{g+1}^{(0)}, E_{g+2}^{(0)}, \ldots,$$

respectively. Let them be an orthonormal set. For the degenerate case we have shown in Section 9–4 that it is always possible to find an orthogonal set of g functions by making linear combinations of the g degenerate eigenfunctions; such a set is to be used as $\psi_1^{(0)}, \psi_2^{(0)}, \ldots, \psi_g^{(0)}$. The perturbed eigenfunctions of $\psi_1^{(0)}, \psi_2^{(0)}, \ldots, \psi_g^{(0)}$, designated by ψ_n where $1 \leq n \leq g$ (the index n will be used in this sense exclusively in this section), may be expanded in series of the unperturbed eigenfunctions. Thus

$$\psi_n = C_{n1}\psi_1^{(0)} + \cdots + C_{ng}\psi_g^{(0)} + C_{n,g+1}\psi_{g+1}^{(0)} + \cdots, \quad 1 \leq n \leq g.$$
$$(10\text{–}49)$$

When the values of $E_1^{(0)}, \ldots, E_g^{(0)}$ are nearly equal but not exactly equal, the results of Section 10–1, Eq. (10–31), tell us that the coefficients $C_{n1}, \ldots, C_{ng}$ are large while the other coefficients $C_{n,g+1}, \ldots,$ are small. Thus in the degenerate case it is a reasonable guess that $C_{n1}, \ldots, C_{ng}$ are finite numbers, while $C_{n,g+1}, \ldots,$ are small quantities at least of the first order. Thus the power series expansion of ψ_n may be written in the form

$$\psi_n = \psi_n^{\{0\}} + \lambda\psi_n^{\{1\}} + \lambda^2\psi_n^{\{2\}} + \cdots,$$
$$\equiv \sum_{l=1}^{g} C_{nl}^{(0)}\psi_l^{(0)} + \lambda \sum_{m=g+1}^{\infty} C_{nm}^{(1)}\psi_m^{(0)} + \lambda^2 \sum_{m=g+1}^{\infty} C_{nm}^{(2)}\psi_m^{(0)} + \cdots.$$
$$(10\text{–}50)$$

The series expansion of E_n is again

$$E_n = E_n^{(0)} + \lambda E_n^{(1)} + \lambda^2 E_n^{(2)} + \cdots. \qquad (10\text{–}51)$$

Substituting Eqs. (10–50) and (10–51) in the perturbed Schrödinger equation, we have

$$(H^{(0)} + \lambda H^{(1)}) \left(\sum_{l=1}^{g} C_{nl}^{(0)} \psi_l^{(0)} + \lambda \sum_{m=g+1}^{\infty} C_{nm}^{(1)} \psi_m^{(0)} + \lambda^2 \sum_{m=g+1}^{\infty} C_{nm}^{(2)} \psi_m^{(0)} + \cdots \right)$$

$$= (E_n^{(0)} + \lambda E_n^{(1)} + \lambda^2 E_n^{(2)} + \cdots)$$

$$\times \left(\sum_{l=1}^{g} C_{nl}^{(0)} \psi_l^{(0)} + \lambda \sum_{m=g+1}^{\infty} C_{nm}^{(1)} \psi_m^{(0)} + \lambda^2 \sum_{m=g+1}^{\infty} C_{nm}^{(2)} \psi_m^{(0)} + \cdots \right). \quad (10\text{–}52)$$

Equating terms of the same power of λ on both sides and making use of the unperturbed Schrödinger equation, we obtain equations for various orders of perturbation:

$$\sum_{l=1}^{g} C_{nl}^{(0)} E_l^{(0)} \psi_l^{(0)} = \sum_{l=1}^{g} C_{nl}^{(0)} E_n^{(0)} \psi_l^{(0)}, \quad (10\text{–}53)$$

$$\sum_{m=g+1}^{\infty} C_{nm}^{(1)} E_m^{(0)} \psi_m^{(0)} + \sum_{l=1}^{g} C_{nl}^{(0)} H^{(1)} \psi_l^{(0)} = \sum_{m=g+1}^{\infty} C_{nm}^{(1)} E_n^{(0)} \psi_m^{(0)}$$

$$+ \sum_{l=1}^{g} C_{nl}^{(0)} E_n^{(1)} \psi_l^{(0)}, \quad (10\text{–}54)$$

$$\sum_{m=g+1}^{\infty} C_{nm}^{(2)} E_m^{(0)} \psi_m^{(0)} + \sum_{m=g+1}^{\infty} C_{nm}^{(1)} H^{(1)} \psi_m^{(0)} = \sum_{m=g+1}^{\infty} C_{nm}^{(2)} E_n^{(0)} \psi_m^{(0)}$$

$$+ \sum_{m=g+1}^{\infty} C_{nm}^{(1)} E_n^{(1)} \psi_m^{(0)} + \sum_{l=1}^{g} C_{nl}^{(0)} E_n^{(2)} \psi_l^{(0)}, \quad (10\text{–}55)$$

$$\cdot \quad \cdot \quad \cdot \quad \cdot \quad \cdot \quad \cdot$$

Equation (10–53) is the zero-th order equation which is satisfied identically in the degenerate case. Equation (10–54) is the first-order equation from which we may determine the perturbation energy $E_n^{(1)}$ and the coefficients $C_{nl}^{(0)}$ and $C_{nm}^{(1)}$. For the degenerate case we shall start by solving Eq. (10–54), since Eq. (10–53) is already satisfied. Multiplying Eq. (10–54) by $\psi_s^{(0)*}(1 \leq s \leq g)$ and integrating over the space variables, we have

$$\sum_{l=1}^{g} (H_{sl}^{(1)} - E_n^{(1)} \delta_{sl}) C_{nl}^{(0)} = 0, \quad 1 \leq s \leq g. \quad (10\text{–}56)$$

Let s run through all numbers from 1 to g; we write down the g equations explicitly as follows:

$$(H_{11}^{(1)} - E_n^{(1)})C_{n1}^{(0)} + H_{12}^{(1)}C_{n2}^{(0)} + \cdots + H_{1g}^{(1)}C_{ng}^{(0)} = 0,$$

$$H_{21}^{(1)}C_{n1}^{(0)} + (H_{22}^{(1)} - E_n^{(1)})C_{n2}^{(0)} + \cdots + H_{2g}^{(1)}C_{ng}^{(0)} = 0, \quad (10\text{--}57)$$

$$\cdots \cdots \cdots \cdots \cdots \cdots$$

$$H_{g1}^{(1)}C_{n1}^{(0)} + H_{g2}^{(1)}C_{n2}^{(0)} + \cdots + (H_{gg}^{(1)} - E_n^{(1)})C_{ng}^{(0)} = 0.$$

This set of homogeneous linear equations may have a set of nonvanishing solutions $C_{nl}^{(0)}, \ldots, C_{ng}^{(0)}$ provided that the determinant of the coefficients vanishes, i.e.,

$$\begin{vmatrix} H_{11}^{(1)} - E_n^{(1)} & H_{12}^{(1)} & \cdots & H_{1g}^{(1)} \\ H_{21}^{(1)} & H_{22}^{(1)} - E_n^{(1)} & \cdots & H_{2g}^{(1)} \\ \vdots & \vdots & & \vdots \\ H_{g1}^{(1)} & H_{g2}^{(1)} & \cdots & H_{gg}^{(1)} - E_n^{(1)} \end{vmatrix} = 0. \quad (10\text{--}58)$$

This equation gives the condition which the first-order perturbation $E_n^{(1)}$ has to satisfy. Equation (10–58) is an algebraic equation of the gth order in the variable $E_n^{(1)}$, therefore it has g roots. The first-order perturbation $E_n^{(1)}$ thus has g possible values. Each root of Eq. (10–58) gives rise to a set of solutions of Eq. (10–57), i.e., a set of ratios of $C_{n1}^{(0)}, \ldots, C_{ng}^{(0)}$ which determines a zero-th order eigenfunction $\psi_n^{\{0\}}$ (the normalization condition determines the magnitudes of $C_{n1}^{(0)}, \ldots, C_{ng}^{(0)}$ completely). Thus we have g eigenvalues and g eigenfunctions corresponding to g levels. In other words, the g-fold degenerate level is split into g levels. From Eq. (10–57) it may be shown that the g roots of Eq. (10–58) may be expressed in terms of the g unperturbed eigenfunctions $\psi_n^{\{0\}}$ as follows (see Problem 10–6):

$$E_n^{(1)} = \int \psi_n^{\{0\}*} H^{(1)} \psi_n^{\{0\}} \, d\tau.$$

This equation is identical with Eq. (10–30) of the nondegenerate case except for the substitution of $\psi_n^{\{0\}}$ for $\psi_n^{(0)}$.

Having determined the values of $E_n^{(1)}$ and $C_{nl}^{(0)}$, we turn our attention to the coefficients $C_{nm}^{(1)}$. These coefficients may be determined from Eq. (10–54) by multiplying it by $\psi_m^{(0)*}(m > g)$ and integrating over the space variables. Thus

$$C_{nm}^{(1)}(E_m^{(0)} - E_n^{(0)}) + \int \psi_m^{(0)*} H^{(1)} \left(\sum_{l=1}^{g} C_{nl}^{(0)} \psi_l^{(0)} \right) d\tau = 0. \quad (10\text{--}59)$$

Since $\sum_{l=1}^{q} C_{nl}^{(0)} \psi_l^{(0)}$ is the zero-th order eigenfunction $\psi_n^{(0)}$, we may denote the integral in Eq. (10–59) by $H_{mn}^{(1)}$ as n is used exclusively for $\psi_n^{(0)}$ $(n = 1, 2, \ldots, g)$, that is,

$$H_{mn}^{(1)} = \int \psi_m^{(0)*} H^{(1)} \psi_n^{(0)} \, d\tau, \qquad m > g, \; n = 1, 2, \ldots, g.$$

Equation (10–59) thus leads to

$$C_{nm}^{(1)} = \frac{H_{mn}^{(1)}}{E_n^{(0)} - E_m^{(0)}}, \qquad m > g, \; n = 1, 2, \ldots, g. \qquad (10\text{–}60)$$

Equation (10–60) has the same form as Eq. (10–31). This result is expected because, after the degeneracy has been removed by perturbation, each zero-th order eigenfunction acts like a nondegenerate eigenfunction. Our task of determining the perturbation energy $E_n^{(1)}$ and the expansion coefficients $C_{nl}^{(0)}$, and $C_{nm}^{(1)}$ from Eq. (10–54) is thus accomplished.

Equation (10–55), the second-order equation, may now be solved to determine the second-order perturbation of energy $E_n^{(2)}$ and the expansion coefficients $C_{nm}^{(2)}$ Multiplying it by $\psi_n^{(0)*}$ and integrating over the space variables, we have

$$E_n^{(2)} = \sum_{m=g+1}^{\infty} C_{nm}^{(1)} H_{nm}^{(1)}, \qquad (10\text{–}61)$$

provided that $\psi_n^{(0)}$ is normalized, which we shall always assume. By Eq. (10–60) we have

$$E_n^{(2)} = \sum_{m=g+1}^{\infty} \frac{H_{mn}^{(1)} H_{nm}^{(1)}}{E_n^{(0)} - E_m^{(0)}}. \qquad (10\text{–}62)$$

Multiplying Eq. (10–55) by $\psi_r^{(0)*} (r > g)$ and integrating, we obtain

$$C_{nr}^{(2)} E_r^{(0)} + \sum_{m=g+1}^{\infty} C_{nm}^{(1)} H_{rm}^{(1)} = C_{nr}^{(2)} E_n^{(0)} + C_{nr}^{(1)} E_n^{(1)}, \qquad (10\text{–}63)$$

or

$$C_{nr}^{(2)} = \frac{\sum_{m=g+1}^{\infty} C_{nm}^{(1)} (E_n^{(1)} \delta_{rm} - H_{rm}^{(1)})}{E_r^{(0)} - E_n^{(0)}}, \qquad (10\text{–}64)$$

or

$$C_{nr}^{(2)} = \frac{1}{E_r^{(0)} - E_n^{(0)}} \sum_{m=g+1}^{\infty} \frac{H_{mn}^{(1)} (E_n^{(1)} \delta_{rm} - H_{rm}^{(1)})}{E_n^{(0)} - E_m^{(0)}}. \qquad (10\text{–}65)$$

Equations (10–62) and (10–65) give the second-order eigenvalues and eigenfunctions. They are identical with Eqs. (10–32) and (10–33) except that $\sum_{m=g+1}^{\infty}$ takes the place of $\sum_m'$. This result is also expected for the reason mentioned previously.

The results for the degenerate levels may be summarized in the following statement: they are the same as the results for the nondegenerate levels, i.e., Eqs. (10–30, 31, 32, 33), provided that $\psi_n^{(0)}$ is replaced by $\psi_n^{\{0\}}$ and $\sum'$ is replaced by $\sum_{m=g+1}^{\infty}$.

The above theory applies to the degenerate case in which the g levels of eigenvalues $E_1^{(0)}, \ldots, E_g^{(0)}$ coincide. If the g levels are not exactly the same but very close to one another, then the solution will be slightly different. Equation (10–53) will not be satisfied identically but will be satisfied to the zero-th order provided that the differences among $E_1^{(0)}, \ldots,$ $E_g^{(0)}$ are of the first order. The first-order residue of Eq. (10–53) may now be absorbed in the first-order equation (10–54). The only change in the previous solution that has to be made in order to include this effect is to modify the secular equation (10–58) as follows:

$$
\begin{vmatrix}
E_1^{(0)}-E_n^{(0)}+H_{11}^{(1)}-E_n^{(1)} & H_{12}^{(1)} & \cdots & H_{1g}^{(1)} \\
H_{21}^{(1)} & E_2^{(0)}-E_n^{(0)}+H_{22}^{(1)}-E_n^{(1)} \cdots & & H_{2g}^{(1)} \\
\vdots & \vdots & & \vdots \\
H_{g1}^{(1)} & H_{g2}^{(1)} & \cdots E_g^{(0)}-E_n^{(0)}+H_{gg}^{(1)}-E_n^{(1)}
\end{vmatrix} = 0.
$$

(10–66)

Others remain the same. Equation (10–66) gives g solutions of $(E_n^{(0)} + E_n^{(1)})$. These correspond to the perturbed energies (to the first order) of the g states in the cluster.

If the g roots of Eq. (10–58) are all distinct, the perturbed energies of the g levels are distinct and the degeneracy is completely removed. If Eq. (10–58) contains multiple roots, then the degeneracy is only partially removed. The multiple root corresponds to several levels still coinciding with one another so far as the first-order perturbation is concerned. However, the remaining degeneracy may be removed when higher orders of perturbation are considered. To illustrate this situation we consider a special case in which the g roots of Eq. (10–58) are all the same. In this case the matrix elements in Eq. (10–58) are such that

$$
H_{11}^{(1)} = H_{22}^{(1)} = \cdots = H_{gg}^{(1)},
$$

and all others are zero. The g-fold degeneracy thus remains in spite of the first-order perturbation. As all coefficients in Eq. (10–57) vanish, the values of $C_{nl}^{(0)}(l = 1, 2, \ldots, g)$ become indeterminate. We may remove the degeneracy and determine the $C_{nl}^{(0)}(l = 1, 2, \ldots, g)$ by considering the next order of perturbation. Multiplying Eq. (10–55) by $\psi_s^{(0)}$ $(1 \leq s \leq g)$ and integrating, we obtain

$$
\sum_{m=g+1}^{\infty} C_{nm}^{(1)} H_{sm}^{(1)} = \sum_{l=1}^{g} C_{nl}^{(0)} E_n^{(2)} \, \delta_{sl}.
$$

Equation (10–60) may be rewritten as follows:

$$C_{nm}^{(1)} = \sum_{l=1}^{g} C_{nl}^{(0)} \frac{H_{ml}^{(1)}}{E_n^{(0)} - E_m^{(0)}}.$$

Substituting the values of $C_{nm}^{(1)}$ in the previous equation, we have

$$\sum_{l=1}^{g} \sum_{m=g+1}^{\infty} C_{nl}^{(0)} \frac{H_{sm}^{(1)} H_{ml}^{(1)}}{E_n^{(0)} - E_m^{(0)}} - \sum_{l=1}^{g} C_{nl}^{(0)} E_n^{(2)} \delta_{sl} = 0$$

or

$$\sum_{l=1}^{g} C_{nl}^{(0)} \left[\sum_{m=g+1}^{\infty} \frac{H_{sm}^{(1)} H_{ml}^{(1)}}{E_n^{(0)} - E_m^{(0)}} - E_n^{(2)} \delta_{sl} \right] = 0.$$

Letting $s = 1, 2, \ldots, g$ we obtain g equations which $C_{n1}^{(0)}, C_{n2}^{(0)} \ldots, C_{ng}^{(0)}$ must satisfy. The secular equation for this set of simultaneous linear equations is

$$\begin{vmatrix} \sum_{m=g+1}^{\infty} \frac{H_{1m}^{(1)} H_{m1}^{(1)}}{E_n^{(0)} - E_m^{(0)}} - E_n^{(2)} & \sum_{m=g+1}^{\infty} \frac{H_{1m}^{(1)} H_{m2}^{(1)}}{E_n^{(0)} - E_m^{(0)}} & \cdots \\ \sum_{m=g+1}^{\infty} \frac{H_{2m}^{(1)} H_{m1}^{(1)}}{E_n^{(0)} - E_m^{(0)}} & \sum_{m=g+1}^{\infty} \frac{H_{2m}^{(1)} H_{m2}^{(1)}}{E_n^{(0)} - E_m^{(0)}} - E_n^{(2)} & \cdots \\ \cdots & \cdots & \cdots \\ \cdots & \cdots & \sum_{m=g+1}^{\infty} \frac{H_{gm}^{(1)} H_{mg}^{(1)}}{E_n^{(0)} - E_m^{(0)}} - E_n^{(2)} \end{vmatrix} = 0.$$

The g roots of $E_n^{(2)}$ specify the second-order perturbation energies. If they are distinct, the degeneracy is removed; each root leads to a set of coefficients $C_{n1}^{(0)}, C_{n2}^{(0)}, \ldots, C_{ng}^{(0)}$, and thus to a zero-th order wave function $\psi_n^{[0]}$.

10–3 A generalized perturbation theory. The perturbation theory developed in the last two sections may be shown to be special cases of a more general perturbation theory* which will be discussed below.

The purpose of this theory is to solve the eigenvalue problem

$$H\psi(x) = E\psi(x), \tag{10–67}$$

with the help of a complete set of orthonormal functions, $F_1(x), F_2(x), \ldots$, not necessarily limited to the unperturbed wave functions. The orthogonal

---

* P. S. Epstein, *Phys. Rev.* **28**, 695 (1926).

property may be of a more general type defined by

$$\int F_m^*(x)F_n(x)\rho(x)\,dx = \delta_{mn}, \tag{10–68}$$

where δ_{mn} is the Kronecker δ-symbol, and $\rho(x)$ is a function of x which may be called the density function. $\rho(x)\,dx$ may be, but is not necessarily limited to, the volume element $d\tau$ when generalized coordinates are used. Hence the integral of $F_m^*(x)F_n(x)$ over the volume may take other values than δ_{mn}, which may be denoted by γ_{mn}. Thus

$$\int F_m^*(x)F_n(x)\,d\tau = \gamma_{mn}. \tag{10–69}$$

The solution $\psi(x)$ of Eq. (10–67) may be expanded in a series of $F_n(x)$'s:

$$\psi(x) = \sum_n a_n F_n(x).$$

By substitution, Eq. (10–67) becomes

$$\sum_n a_n(H - E)F_n(x) = 0.$$

Multiplying by $F_m^*(x)\,d\tau$ and integrating over the space variables we obtain

$$\sum_n a_n(H_{mn} - E\gamma_{mn}) = 0, \tag{10–70}$$

where H_{mn} is the matrix element of the Hamiltonian operator with respect to $F_m(x)$ and $F_n(x)$ defined by

$$H_{mn} \equiv \int F_m^*(x)HF_n(x)\,d\tau. \tag{10–71}$$

In order that a set of nonvanishing solutions of a_n exist, the determinant of the coefficients of Eqs. (10–70) must equal zero, i.e.,

$$\begin{vmatrix} H_{11} - E\gamma_{11} & H_{12} - E\gamma_{12} & \dots & H_{1n} - E\gamma_{1n} & \dots \\ H_{21} - E\gamma_{21} & H_{22} - E\gamma_{22} & \dots & H_{2n} - E\gamma_{2n} & \dots \\ \vdots & \vdots & & \vdots & \\ H_{n1} - E\gamma_{n1} & H_{n2} - E\gamma_{n2} & \dots & H_{nn} - E\gamma_{nn} & \dots \\ \vdots & \vdots & & \vdots & \end{vmatrix} = 0. \tag{10–72}$$

If the set of functions $F_n(x)$ is orthonormal in the conventional sense, then $\rho(x) = 1$ and $\gamma_{mn} = \delta_{mn}$; the above equation becomes

$$\begin{vmatrix} H_{11} - E & H_{12} & \ldots & H_{1n} & \ldots \\ H_{21} & H_{22} - E & \ldots & H_{2n} & \ldots \\ \vdots & \vdots & & \vdots & \\ H_{n1} & H_{n2} & \ldots & H_{nn} - E & \ldots \\ \vdots & \vdots & & \vdots & \end{vmatrix} = 0. \quad (10\text{–}73)$$

The roots of Eq. (10–73) are the eigenvalues of Eq. (10–67). For each eigenvalue, the corresponding solutions $a_1, a_2, \ldots, a_n, \ldots$, of Eq. (10–70) determine the corresponding eigenfunction.

We shall show that this general theory reduces to the previous theories when certain approximations are introduced. First, a trivial case may be verified. If the functions $F_n(x)$ are the eigenfunctions themselves, then $\rho(x) = 1$ and the off-diagonal elements of Eq. (10–73) vanish because of the general orthogonal property of the eigenfunctions. Equation (10–73) thus reduces to

$$(H_{11} - E)(H_{22} - E) \ldots (H_{nn} - E) \ldots = 0. \quad (10\text{–}74)$$

The energy eigenvalues are

$$E = H_{11}, H_{22}, \ldots H_{nn}, \ldots.$$

This result is expected, since

$$H_{nn} = \int \psi_n^*(x) H \psi_n(x) \rho(x) \, dx$$

$$= E_n \int \psi_n^*(x) \psi_n(x) \rho(x) \, dx$$

$$= E_n.$$

The eigenfunction corresponding to E_n, given by Eq. (10–69), is found to contain only one term, i.e., $\psi_n(x)$ itself.

Next, let the functions $F_n(x)$ be the eigenfunctions of the unperturbed problem [again $\rho(x) = 1$],

$$F_n(x) = \psi_n^{(0)}(x),$$

$$H^{(0)}\psi_n^{(0)} = E_n^{(0)}\psi_n^{(0)},$$

$$H = H^{(0)} + \lambda H^{(1)}.$$

The off-diagonal elements are small quantities of the first order, since

$$H_{mn} = 0 + \lambda H_{mn}^{(1)}, \quad m \neq n;$$

they may be neglected in a first approximation. Equation (10–73) thus reduces to Eq. (10–74), and the energy eigenvalues are

$$
\begin{aligned}
E_n &= H_{nn} \\
&= H_{nn}^{(0)} + \lambda H_{nn}^{(1)} \\
&= E_n^{(0)} + \lambda \int \psi_n^{(0)*} H^{(1)} \psi_n^{(0)} \, d\tau.
\end{aligned}
$$

The above equation is the same as the result of the first-order perturbation for nondegenerate levels.

More accurate eigenvalues may be obtained by successive approximations. To find the second approximation of E_n, Eq. (10–73) may be kept to the second order. Among the diagonal elements, $H_{nn} - E_n$ is a small quantity of the second order while the others, $H_{mm} - E_n$, are finite. The off-diagonal elements are small quantities of the first order. The second-order approximation of Eq. (10–73) may be obtained by: (1), replacing E in Eq. (10–73) with H_{nn} except for the element $H_{nn} - E$, and (2), retaining the off-diagonal elements $H_{1n}, H_{2n}, \ldots,$ and $H_{n1}, H_{n2}, \ldots$. Thus we have

$$
\begin{vmatrix}
H_{11}-H_{nn} & 0 & 0 & \cdots & H_{1n} & 0 & \cdots \\
0 & H_{22}-H_{nn} & 0 & \cdots & H_{2n} & 0 & \cdots \\
0 & 0 & H_{33}-H_{nn} & \cdots & H_{3n} & 0 & \cdots \\
\vdots & \vdots & \vdots & & \vdots & \vdots & \\
H_{n1} & H_{n2} & H_{n3} & \cdots & H_{nn}-E & H_{n,n+1} & \cdots \\
0 & 0 & 0 & \cdots & H_{n+1,n} & H_{n+1,n+1}-H_{nn} & \cdots \\
\vdots & \vdots & \vdots & & \vdots & \vdots &
\end{vmatrix} = 0.
$$

$$(10\text{–}75)$$

Equation (10–75) may be reduced to the following expression:

$$
-\frac{H_{n1}H_{1n}}{H_{11} - H_{nn}} \prod_{m}' (H_{mm} - H_{nn}) - \frac{H_{n2}H_{2n}}{H_{22} - H_{nn}} \prod_{m}' (H_{mm} - H_{nn}) + \cdots
$$

$$
+ (H_{nn} - E)\prod_{m}' (H_{mm} - H_{nn}) + \cdots = 0, \qquad (10\text{–}76)
$$

where Π_m' is a product of factors of all values of m except $m = n$. It follows that

$$
E = H_{nn} + \sum_{m}' \frac{H_{nm}H_{mn}}{H_{nn} - H_{mm}} \qquad (10\text{–}77)
$$

$$
= E_n^{(0)} + \lambda H_{nn}^{(1)} + \lambda^2 \sum_{m}' \frac{H_{nm}^{(1)}H_{mn}^{(1)}}{E_n^{(0)} - E_m^{(0)}} + \cdots. \qquad (10\text{–}78)
$$

Thus the second-order perturbation of the nondegenerate level is reproduced.

If the unperturbed equation has degenerate solutions $u_1(x), \ldots, u_g(x)$ then the off-diagonal elements H_{mn} for the degenerate solutions $u_1(x), \ldots, u_g(x)$ do not vanish in the zero-th order. The first-order approximation of Eq. (10–73) is thus

$$\begin{vmatrix} H_{11}-E & H_{12} & \ldots & H_{1g} & 0 & 0 & \ldots \\ H_{21} & H_{22}-E & \ldots & H_{2g} & 0 & 0 & \ldots \\ \vdots & \vdots & & \vdots & \vdots & \vdots & \\ H_{g1} & H_{g2} & \ldots & H_{gg}-E & 0 & 0 & \ldots \\ 0 & 0 & \ldots & 0 & H_{g+1,g+1}-E & 0 & \ldots \\ 0 & 0 & \ldots & 0 & 0 & H_{g+2\,g+2}-E & \ldots \\ \vdots & \vdots & & \vdots & \vdots & \vdots & \end{vmatrix} = 0.$$

$$(10\text{–}79)$$

The energy eigenvalues of the nondegenerate levels (for which $n > g$) are thus the same as before, while those of the degenerate levels ($n \leq g$) are given by

$$\begin{vmatrix} H_{11} - E & H_{12} & \ldots & H_{1g} \\ H_{21} & H_{22} - E & \ldots & H_{2g} \\ \vdots & \vdots & & \vdots \\ H_{g1} & H_{g2} & \ldots & H_{gg} - E \end{vmatrix} = 0. \qquad (10\text{–}80)$$

The first-order perturbation of the degenerate level is thus reproduced.

This general theory may be carried out to any degree of accuracy by including in Eq. (10–72) the proper nondiagonal elements. The more nondiagonal elements we include, the more accurate, in general, the solution may be. When all nondiagonal elements are included, we obtain the exact solution.

10–4 The Stark effect. When a radiating atom is placed in an electrostatic field, each of its spectral lines splits into a number of components. This phenomenon is known as the Stark effect, named after its discoverer who in 1913 observed the splitting of the Balmer lines in a field of 100,000 volts/cm. Since a very strong electric field is required to produce an observable effect, it was discovered much later than the discovery of the Zeeman effect, the splitting of spectral lines due to a magnetic field (1897). The normal Zeeman effect was explained immediately after its discovery by the classical theory—the Lorentz theory of the electron. (It may be mentioned that in the same year (1897) Thomson discovered the electron.) The Stark effect, on the contrary, cannot be explained by the classical theory. In the same year (1913) of its discovery, Bohr initiated the development of the old quantum theory. This theory was able to explain the Stark effect as well as the Zeeman effect; it was considered as a major triumph of the old quantum theory. Yet a complete account of these effects would have to await the formulation of quantum mechanics around 1925. From the spectroscopic point of view the Zeeman effect is more important; it will be discussed in Chapter 12. The Stark effect will be discussed here as an example of the perturbation theories developed in Sections 10–1, 10–2, and 10–3. For this purpose we are interested in one aspect of the Stark effect only, i.e., the quantization of energy. The other aspects, including the selection rules, intensity, and polarization of the component lines, belong to the theory of radiation, and so will not be discussed here in spite of the fact that they may be treated on the basis of the correspondence principle.

(A) *The old quantum theory.* For historical, as well as mathematical, interest, the old quantum theory treatment of the Stark effect will be briefly sketched here. The electron in a hydrogenlike atom, when exposed to an electrostatic field F in the positive z-direction, will have an additional amount of potential energy $+eFz$ where e is the numerical value of the electronic charge:

$$e = +4.80 \times 10^{-10} \text{ esu.}$$

The total energy is thus

$$E = \frac{1}{2M} (p_x^2 + p_y^2 + p_z^2) - \frac{Ze^2}{r} + eFz. \qquad (10\text{–}81)$$

The parabolic coordinates ξ, η, φ are introduced by the following equations of transformation:

$$x = \sqrt{\xi\eta} \cos \varphi,$$
$$y = \sqrt{\xi\eta} \sin \varphi, \qquad (10\text{–}82)$$
$$z = \frac{\xi - \eta}{2}.$$

The *Lagrangian** is then transformed in parabolic coordinates:

$$L = \frac{M}{8} \left(\frac{\eta}{\xi} \dot{\xi}^2 + \frac{\xi}{\eta} \dot{\eta}^2 + 4\xi\eta\dot{\varphi}^2 + \dot{\xi}^2 + \dot{\eta}^2 \right) + \frac{Ze^2}{r} - eFz. \qquad (10\text{–}83)$$

---

* The Lagrangian is a function *defined* by $L = \frac{1}{2}m(\dot{x}^2 + \dot{y}^2 + \dot{z}^2) - U(x, y, z)$, where $U(x, y, z)$ is the potential energy; from the Lagrangian the generalized momenta may be obtained by the *definition* Eq. (10–84).

The generalized momenta are

$$p_\xi = \frac{\partial L}{\partial \dot\xi} = \frac{M}{4}\left(1 + \frac{\eta}{\xi}\right)\dot\xi,$$

$$p_\eta = \frac{\partial L}{\partial \dot\eta} = \frac{M}{4}\left(1 + \frac{\xi}{\eta}\right)\dot\eta, \tag{10–84}$$

$$p_\varphi = \frac{\partial L}{\partial \dot\varphi} = M\xi\eta\dot\varphi.$$

Equation (10–81) thus becomes

$$E = \frac{1}{2M(\xi + \eta)}\left[4\xi p_\xi^2 + 4\eta p_\eta^2 + \left(\frac{1}{\xi} + \frac{1}{\eta}\right)p_\varphi^2\right] - \frac{Ze^2}{(\xi + \eta)/2} + eF\,\frac{\xi - \eta}{2}. \tag{10–85}$$

The right-hand side of Eq. (10–85) is actually the Hamiltonian. The Wilson-Sommerfeld quantum conditions assert that

$$\oint p_\xi\,d\xi = n_1 h,$$

$$\oint p_\eta\,d\eta = n_2 h, \tag{10–86}$$

$$\oint p_\varphi\,d\varphi = n_3 h.$$

Rewrite Eq. (10–85):

$$4\xi p_\xi^2 + 4\eta p_\eta^2 + \left(\frac{1}{\xi} + \frac{1}{\eta}\right)p_\varphi^2 - 4MZe^2 + eMF(\xi^2 - \eta^2) - 2M(\xi + \eta)E = 0. \tag{10–87}$$

This equation may be satisfied if p_ξ, p_η, p_φ satisfy the following equations respectively:*

$$p_\varphi = \alpha, \tag{10–88}$$

$$4\xi p_\xi^2 + \frac{1}{\xi}\alpha^2 + eMF\xi^2 - 2M\xi E = \beta, \tag{10–89}$$

$$4\eta p_\eta^2 + \frac{1}{\eta}\alpha^2 - 4MZe^2 - eMF\eta^2 - 2M\eta E = -\beta, \tag{10–90}$$

---

* In solving the Hamilton-Jacobi differential equation we use the method of separation of variables to obtain a complete integral containing three arbitrary constants E, α, β. A similar procedure is adopted here. Physically p_φ is a constant α because φ is a cyclic variable.

where α and β are constants. This may be verified by straightforward substitution. Solve Eqs. (10–88), (10–89), and (10–90) for p_φ, p_ξ, and p_η in terms of φ, ξ, η respectively, substitute the results in Eqs. (10–86), and carry out the integration. The three equations so obtained express the three constants α, β, and E in terms of the three quantum numbers n_1, n_2, and n_3. The result for E is

$$E = -\frac{MZ^2e^4}{2\hbar^2n^2} - \frac{3\hbar^2F}{2MZe}\,n(n_2 - n_1), \qquad n = n_1 + n_2 + n_3. \qquad (10\text{–}91)$$

It shows that the energy is quantized. In addition to the Balmer term value, there is now a second term, proportional to the field F, representing the additional energy due to the electric field.

(B) *The nondegenerate perturbation method.* The quantum-mechanical treatment of the Stark effect* may be carried out in parabolic coordinates by using the techniques of the nondegenerate perturbation theory of Section 10–1; it may also be carried out in spherical coordinates as a degenerate problem by the method of Section 10–2. Since our purpose is to illustrate the perturbation methods, both approaches will be presented as examples of the nondegenerate and degenerate perturbation theories. Finally, the general theory of Section 10–3 will also be applied.

The time-independent Schrödinger equation for this problem is

$$\nabla^2\psi + \frac{2M}{\hbar^2}\left(E + \frac{Ze^2}{r} - eFz\right)\psi = 0. \qquad (10\text{–}92)$$

By writing the Laplacian in parabolic coordinates, Eq. (10–92) may be rewritten as follows:

$$\frac{\partial}{\partial\xi}\left(\xi\frac{\partial\psi}{\partial\xi}\right) + \frac{\partial}{\partial\eta}\left(\eta\frac{\partial\psi}{\partial\eta}\right) + \frac{1}{4}\left(\frac{1}{\xi} + \frac{1}{\eta}\right)\frac{\partial^2\psi}{\partial\varphi^2}$$

$$+ \frac{M}{2\hbar^2}\left\{(\xi + \eta)E + 2Ze^2 - \frac{eF}{2}\,(\xi^2 - \eta^2)\right\}\psi = 0. \qquad (10\text{–}93)$$

The similarity between Eq. (10–93) and Eq. (10–87) may be noted; the procedures of separating variables in solving these two equations are analogous. Let

$$\psi(\xi, \eta, \varphi) = X(\xi)Y(\eta)\Phi(\varphi). \qquad (10\text{–}94)$$

---

* P. S. Epstein, *Phys. Rev.* **28,** 695 (1926).

The separated equations are thus

$$\frac{d^2\Phi}{d\varphi^2} = -m^2\Phi, \tag{10–95}$$

$$\frac{d}{d\xi}\left(\xi\frac{dX}{d\xi}\right) - \frac{m^2}{4\xi}X + \frac{M}{2\hbar^2}\left\{\xi E + Z_1 - \frac{eF}{2}\xi^2\right\}X = 0, \tag{10–96}$$

$$\frac{d}{d\eta}\left(\eta\frac{dY}{d\eta}\right) - \frac{m^2}{4\eta}Y + \frac{M}{2\hbar^2}\left\{\eta E + Z_2 + \frac{eF}{2}\eta^2\right\}Y = 0, \tag{10–97}$$

where m, Z_1, Z_2 are constants and

$$Z_1 + Z_2 = 2Ze^2. \tag{10–98}$$

The solution of the Φ equation is well known and the constant m must be an integer. By the substitution,

$$X = \frac{u}{\sqrt{\xi}}, \tag{10–99}$$

$$Y = \frac{v}{\sqrt{\eta}}; \tag{10–100}$$

Eqs. (10–96) and (10–97) may be rewritten as follows:

$$\frac{d^2u}{d\xi^2} + \left\{\frac{M}{2\hbar^2}\left(E + \frac{Z_1}{\xi}\right) + \frac{1-m^2}{4\xi^2}\right\}u = \frac{eMF}{4\hbar^2}\xi u, \tag{10–101}$$

$$\frac{d^2v}{d\eta^2} + \left\{\frac{M}{2\hbar^2}\left(E + \frac{Z_2}{\eta}\right) + \frac{1-m^2}{4\eta^2}\right\}v = -\frac{eMF}{4\hbar^2}\eta v. \tag{10–102}$$

Since Eq. (10–102) may be reproduced from Eq. (10–101) by replacing F by $-F$ we need to solve only one of them, say Eq. (10–101). We shall consider the solution of this equation as a one-dimensional problem, which is nondegenerate, and treat the term on the right-hand side as a perturbation. The unperturbed equation

$$\frac{d^2u^{(0)}}{d\xi^2} + \left\{\frac{M}{2\hbar^2}\left(E^{(0)} + \frac{Z_1}{\xi}\right) + \frac{1-m^2}{4\xi^2}\right\}u^{(0)} = 0 \tag{10–103}$$

has a form similar to the radial equation of the hydrogen atom, Eq. (8–74), except that the factor $\frac{1}{4}(1-m^2)$ takes the place of $-l(l+1)$ in Eq. (8–74). Therefore it may be solved in a similar manner. The asymptotic solution is to be obtained first. The pole is then removed and the remaining equation may be solved by the power series method. In order that the solutions may be acceptable, the series must terminate. Thus the energy is quantized. The quantum number thus introduced is denoted by n_1. We shall omit the details and merely state the results. The eigenfunctions are given by

$$u_{n_1}^{(0)}(\xi) = Nx^{(m+1)/2}e^{-x/2}L_{n_1+m}^m(x), \qquad n_1 = 0, 1, 2, \ldots, \qquad (10\text{-}104)$$

where

$$x = \sqrt{-[2ME^{(0)}/\hbar^2]}\; \xi, \qquad (10\text{-}105)$$

and $L_{n_1+m}^m(x)$ is the associated Laguerre polynomial. The eigenvalues are determined by the equation

$$\frac{Z_1}{2}\sqrt{-\frac{M}{2\hbar^2E^{(0)}}} = 1 + \frac{m}{2} - \frac{1}{2} + n_1, \qquad n_1 = 0, 1, 2, \ldots. \qquad (10\text{-}106)$$

The unperturbed solution of Eq. (10-102) may be obtained by replacing ξ by η and Z_1 by Z_2. The quantum number of Eq. (10-102) is denoted by n_2. Thus

$$\frac{Z_2}{2}\sqrt{-\frac{M}{2\hbar^2E^{(0)}}} = 1 + \frac{m}{2} - \frac{1}{2} + n_2, \qquad n_2 = 0, 1, 2, \ldots. \qquad (10\text{-}107)$$

Equation (10-98) thus leads to the following result:

$$Ze^2\sqrt{-M/[2\hbar^2E^{(0)}]} = 1 + m + n_1 + n_2. \qquad (10\text{-}108)$$

Consequently, the unperturbed energy eigenvalues are

$$E^{(0)} = -\frac{MZ^2e^4}{2\hbar^2(n_1 + n_2 + m + 1)^2} \qquad (10\text{-}109)$$

$$= -\frac{MZ^2e^4}{2\hbar^2n^2}, \qquad n = n_1 + n_2 + m + 1 = 1, 2, 3, \ldots.$$

The above equation is identical with Eq. (8-95), which gives energy eigenvalues of the hydrogen atom, except that a different but equivalent set of quantum numbers is used. This result is expected, since the unperturbed problem is the hydrogen atom itself expressed in a different coordinate system. We have seen in Chapter 9 that different coordinate systems lead to the same results. After having solved the unperturbed problem, we turn to the first-order perturbation. Equation (10-101) may be rewritten as follows:

$$\frac{2\hbar^2}{M}\left\{-\frac{d^2u}{d\xi^2} - \frac{1 - m^2}{4\xi^2}u - \frac{M}{2\hbar^2}\frac{Z_1}{\xi}u\right\} + \frac{1}{2}eF\xi u = Eu. \qquad (10\text{-}110)$$

The perturbation term is thus

$$\lambda H^{(1)} = \tfrac{1}{2} e F \xi. \tag{10-111}$$

Since the unperturbed solutions are nondegenerate, the first-order perturbation of energy is

$$\lambda E_{n_1}^{(1)} = \tfrac{1}{2} e F \int_0^\infty u_{n_1}^{(0)}(\xi) \xi u_{n_1}^{(0)}(\xi) \, d\xi. \tag{10-112}$$

The evaluation of the integral is complicated; we shall merely state the results in the following:

$$\lambda E_{n_1}^{(1)} = \frac{1}{2} \, e F \, \frac{\hbar^2}{M Z_1} \, [6(n_1 + m)^2 + 6(1 - m)(n_1 + m) + (m - 1)(m - 2)]. \tag{10-113}$$

On the other hand, a similar equation may be obtained from Eq. (10–102), the only difference being the sign of F. Thus

$$\lambda E_{n_2}^{(1)} = -\, \frac{1}{2} \, e F \, \frac{\hbar^2}{M Z_2}$$
$$\times \, [6(n_2 + m)^2 + 6(1 - m)(n_2 + m) + (m - 1)(m - 2)]. \tag{10-114}$$

When $u(\xi)$ and $v(\eta)$ combine to form a complete solution, the total energy E appears in both Eqs. (10–101) and (10–102). Therefore the two perturbation energies calculated from the two equations must equal each other, i.e.,

$$\lambda E_{n_1}^{(1)} = \lambda E_{n_2}^{(1)} \equiv \lambda E_n^{(1)}. \tag{10-115}$$

From Eqs. (10–113) and (10–114) we obtain

$$(Z_1 + Z_2) \lambda E_n^{(1)} = \frac{1}{2} \, e F \, \frac{\hbar^2}{M} \, 6(n_1 - n_2)(n_1 + n_2 + m + 1). \tag{10-116}$$

Substitution by Eq. (10–98) leads to

$$\lambda E_n^{(1)} = \frac{3}{2} \, \frac{F \hbar^2}{M Z e} \, (n_1 - n_2) n. \tag{10-117}$$

The first-order perturbation energy is thus obtained. It is the same as that obtained by the old quantum theory, Eq. (10–91).

(C) *The degenerate perturbation method.* We now approach the same problem by a different method. The Schrödinger equation for the Stark effect, Eq. (10–92), may be solved in spherical coordinates, the term eFz being treated as a perturbation. The unperturbed problem is simply the hydrogen atom problem which has been solved in spherical coordinates. The solutions are degenerate. Accordingly, the perturbed energy values may be obtained by solving Eq. (10–58). The perturbation term is

$$\lambda H^{(1)} = eFz = eFr \cos \theta. \tag{10–118}$$

Let

$$\lambda = eF, \qquad H^{(1)} = r \cos \theta. \tag{10–119}$$

The matrix elements are

$$H^{(1)}_{nlm,n'l'm'} = \iiint \psi^*_{nlm} r \cos \theta \psi_{n'l'm'} r^2 \sin \theta \, dr \, d\theta \, d\varphi, \tag{10–120}$$

where ψ_{nlm} are the hydrogen atom wave functions, some of which are listed in Eq. (8–117). For the ground state, $n = 1$, $l = 0$, $m = 0$, the energy level is nondegenerate; the first-order perturbation energy is

$$\lambda H^{(1)}_{100,100} = eF \iiint r \cos \theta \, \frac{1}{\pi a^3} e^{-2(r/a)} r^2 \sin \theta \, dr \, d\theta \, d\varphi \tag{10–121}$$

$$= 0.$$

It is zero because the wave function is spherically symmetric. For the first excited state, $n = 2$, we have a fourfold degeneracy. ψ_{200}, ψ_{211}, ψ_{210}, ψ_{21-1} all have the same energy E_2. Equation (10–58) becomes

$$\begin{vmatrix} H^{(1)}_{200,200} - E^{(1)} & H^{(1)}_{200,211} & H^{(1)}_{200,210} & H^{(1)}_{200,21-1} \\ H^{(1)}_{211,200} & H^{(1)}_{211,211} - E^{(1)} & H^{(1)}_{211,210} & H^{(1)}_{211,21-1} \\ H^{(1)}_{210,200} & H^{(1)}_{210,211} & H^{(1)}_{210,210} - E^{(1)} & H^{(1)}_{210,21-1} \\ H^{(1)}_{21-1,200} & H^{(1)}_{21-1,211} & H^{(1)}_{21-1,210} & H^{(1)}_{21-1,21-1} - E^{(1)} \end{vmatrix} = 0.$$

$$\tag{10–122}$$

The matrix elements may be evaluated in a straightforward way. For example,

$$H^{(1)}_{210,200} = \frac{1}{32\pi} \frac{1}{a^3} \iiint \left(\frac{r}{a}\right)\left(2 - \frac{r}{a}\right) e^{-r/a} \cos \theta (r \cos \theta) r^2 \sin \theta \, dr \, d\theta \, d\varphi$$

$$= \frac{a}{32\pi} \int_0^\infty \sigma^4 (2 - \sigma) e^{-\sigma} \, d\sigma \int_0^\pi \cos^2 \theta \sin \theta \, d\theta \int_0^{2\pi} d\varphi$$

$$= \frac{a}{32\pi} (-5! + 2 \cdot 4!) \cdot \frac{2}{3} \cdot 2\pi$$

$$= -3 \frac{\hbar^2}{MZe^2} ;$$

$$H^{(1)}_{200,210} = H^{(1)}_{210,200} = -3 \frac{\hbar^2}{MZe^2}.$$

All other matrix elements turn out to be zero. Equation (10–122) thus becomes

$$
\begin{vmatrix}
-E^{(1)} & 0 & -\dfrac{3\hbar^2}{MZe^2} & 0 \\[2mm]
0 & -E^{(1)} & 0 & 0 \\[2mm]
-\dfrac{3\hbar^2}{MZe^2} & 0 & -E^{(1)} & 0 \\[2mm]
0 & 0 & 0 & -E^{(1)}
\end{vmatrix} = 0. \qquad (10\text{–}123)
$$

The four roots of this equation are

$$
E^{(1)} = 0, \quad 0, \quad +\frac{3\hbar^2}{MZe^2}, \quad -\frac{3\hbar^2}{MZe^2}, \qquad (10\text{–}124)
$$

and the corresponding perturbed energy values are

$$
\wedge E^{(1)} = 0, \quad 0, \quad +\frac{3F\hbar^2}{Mze}, \quad -\frac{3F\hbar^2}{MZe}. \qquad (10\text{–}125)
$$

We may compare the result with that obtained before, i.e., Eq. (10–117). For $n = 2$, the possible values of n_1 are 0, 1. Those of n_2 are also 0, 1. Thus there are four possible values of $n_1 - n_2$, i.e., 0, 0, 1, -1. Equation (10–117) gives the values of the perturbed energy:

$$
\lambda E_2^{(1)} = 0, \quad 0, \quad +\frac{3F\hbar^2}{MZe}, \quad -\frac{3F\hbar^2}{MZe}. \qquad (10\text{–}126)
$$

Thus the two methods lead to the same results.

The physical interpretation of the results is essentially the same as that in Section 10–1. The first-order perturbation of energy is a result of the asymmetry of the charge distribution (given by the wave functions) with respect to the $z = 0$ plane. If the wave function is symmetric with respect to the $z = 0$ plane, the first-order perturbation vanishes. However, the field distorts any wave function by an amount of the first order. The distorted wave function gives rise to an average value of the electrostatic potential eFz which in turn is a second-order small quantity. Thus the second-order perturbation of energy derives its origin from the polarization of the hydrogen atom. We shall treat the second-order perturbation of the ground state of the hydrogen atom by the general perturbation theory of Section 10–3.

(D) *The generalized perturbation method.* The perturbed energy of the ground state to the second order may be obtained from Eq. (10–75) where

n is to be identified with the ground state. We rewrite the Schrödinger equation (10–92) as follows:

$$\nabla^2\psi + \frac{2MZe^2}{\hbar^2}\frac{1}{r}\psi - \frac{2MeF}{\hbar^2}r\cos\theta\psi = -\frac{2ME}{\hbar^2}\psi. \quad (10\text{–}127)$$

Change the scale of the radius vector by the following transformation:

$$\xi = \frac{2r}{a}, \qquad a = \frac{\hbar^2}{ZMe^2}. \quad (10\text{–}128)$$

Equation (10–127) may thus be transformed to

$$\nabla^2\psi + \left(\frac{1}{\xi} - \frac{1}{4}\right)\psi - \Lambda\xi\cos\theta\psi = W\psi, \quad (10\text{–}129)$$

where ∇^2 is expressed in terms of ξ, and

$$\Lambda = \frac{2MeF}{\hbar^2}\frac{a}{2}\frac{a^2}{4} = \frac{a^2}{4Ze}F, \quad (10\text{–}130)$$

$$W = -\frac{2M}{\hbar^2}E\frac{a^2}{4} - \frac{1}{4} = -\frac{a}{2Ze^2}E - \frac{1}{4}. \quad (10\text{–}131)$$

The eigenvalue problem is thus specified by the Hamiltonian operator

$$H = \nabla^2 + \left(\frac{1}{\xi} - \frac{1}{4}\right) - \Lambda\xi\cos\theta. \quad (10\text{–}132)$$

The eigenvalue is denoted by W. The constant Λ is proportional to F and thus corresponds to the parameter λ in the perturbation theory of Sections 10–1 and 10–2. By the method of Section 10–3 we need not separate H into $H^{(0)}$ and $\lambda H^{(1)}$; on the other hand, we need a set of orthogonal functions $F_n(x)$. The following set of functions is introduced for this purpose:

$$F_{\nu\lambda\mu}(\xi, \theta, \varphi) = \Xi_{\nu\lambda}(\xi)Y_{\lambda\mu}(\theta, \varphi), \quad (10\text{–}133)$$

where $Y_{\lambda\mu}(\theta, \varphi)$ is the spherical harmonic and

$$\Xi_{\nu\lambda}(\xi) = \left[\frac{(\nu - \lambda - 1)!}{\{(\nu + \lambda)!\}^3}\right]^{1/2}\xi^\lambda L_{\nu+\lambda}^{2\lambda+1}(\xi)e^{-\xi/2}, \quad (10\text{–}134)$$

$L_{\nu+\lambda}^{2\lambda+1}$ being the associated Laguerre polynomial. The function $F_{\nu\lambda\mu}(\xi, \theta, \varphi)$ is similar to the hydrogen wave function except that the exponential factor is independent of the principal quantum number. This difference presents no difficulty in making mathematical computations involving $\Xi_{\nu\lambda}(\xi)$ if a proper change of scale of ξ is made in the associated Laguerre polynomials.

From the properties of the associated Laguerre polynomials, the following properties of $F_{\nu\lambda\mu}$ may be established. As the mathematical details are not essential for the understanding of the physical problem, we shall state the results without proof.

1. The functions $F_{\nu\lambda\mu}$ satisfy the differential equation

$$\nabla^2 F_{\nu\lambda\mu} + \left(\frac{1}{\xi} - \frac{1}{4}\right) F_{\nu\lambda\mu} = -\frac{(\nu - 1)}{\xi} F_{\nu\lambda\mu}. \qquad (10\text{–}135)$$

2. Their orthonormal property, as defined by Eq. (10–68), is represented by the following equation:

$$\int_0^{2\pi} \int_0^{\pi} \int_0^{\infty} F_{\nu\lambda\mu}^* F_{\nu'\lambda'\mu'} \xi\, d\xi \sin\theta\, d\theta\, d\varphi = \delta_{\nu\nu'}\delta_{\lambda\lambda'}\delta_{\mu\mu'}. \qquad (10\text{–}136)$$

The density function is

$$\rho(\xi, \theta, \varphi) = \xi \sin\theta. \qquad (10\text{–}137)$$

Since $\rho\, d\xi\, d\theta\, d\varphi$ is not the volume element $\xi^2 \sin\theta\, d\xi\, d\theta\, d\varphi$, the integral of $|F_{\nu\lambda\mu}|^2$ over the whole space may not equal unity. Actually, this integral has a value given by

$$\gamma_{\nu\lambda\mu,\nu'\lambda'\mu'} = \int_0^{2\pi} \int_0^{\pi} \int_0^{\infty} F_{\nu\lambda\mu}^* F_{\nu'\lambda'\mu'} \xi^2\, d\xi \sin\theta\, d\theta\, d\varphi = 2\delta_{\nu\nu'}\delta_{\lambda\lambda'}\delta_{\mu\mu'},$$
$$(10\text{–}138)$$

$\gamma_{\nu\lambda\mu,\nu'\lambda'\mu'}$ being defined according to Eq. (10–69).

3. They are related among themselves by the following equation:

$$\xi F_{\nu\lambda\mu} = -\{(\nu - \lambda)(\nu + \lambda + 1)\}^{1/2} F_{\nu+1,\lambda,\mu}$$
$$+ 2\nu F_{\nu\lambda\mu} - \{(\nu + \lambda)(\nu - \lambda - 1)\}^{1/2} F_{\nu-1,\lambda,\mu}. \qquad (10\text{–}139)$$

From this general relation we can prove, in particular, that

$$\xi^2 \cos\theta F_{100} = 4\sqrt{2}\, F_{210} - 2\sqrt{2}\, F_{310}. \qquad (10\text{–}140)$$

This set of functions $F_{\nu\lambda\mu}$ will now be used in the general perturbation theory of Section 10–3. Equation (10–75), in the more general form of Eq. (10–72), may now be written as follows:

$$\begin{vmatrix} H_{100,100}-2W & H_{100,210} & H_{100,310} & H_{100,200} & H_{100,211} & H_{100,21\text{-}1} & H_{100,300} & \cdots \\ H_{210,100} & H_{210,210}-2H_{100,100} & 0 & 0 & \vdots & \vdots & \vdots & \\ H_{310,100} & 0 & H_{310,310}-2H_{100,100} & 0 & \vdots & \vdots & \vdots & \\ H_{200,100} & 0 & 0 & H_{200,200}-2H_{100,100} & \vdots & \vdots & \vdots & \\ \vdots & \vdots & \vdots & \vdots & & & & \end{vmatrix} = 0,$$

$$(10\text{–}141)$$

where the values of $\gamma_{\nu\lambda\mu,\nu'\lambda'\mu'}$ given by Eq. (10–138) are used. The matrix elements are given explicitly by

$$H_{\nu\lambda\mu,\nu'\lambda'\mu'} = \int_0^{2\pi} \int_0^\pi \int_0^\infty F_{\nu\lambda\mu}^* \left\{ \nabla^2 + \left(\frac{1}{\xi} - \frac{1}{4} \right) - \Lambda\xi\cos\theta \right\}$$
$$\times F_{\nu'\lambda'\mu'}\xi^2 \, d\xi \sin\theta \, d\theta \, d\varphi. \quad (10\text{–}142)$$

They may be evaluated by the properties of $F_{\nu\lambda\mu}$ listed above. By Eqs. (10–135) and (10–136),

$$H_{\nu\lambda\mu,\nu'\lambda'\mu'} = -(\nu' - 1)\delta_{\nu\nu'}\delta_{\lambda\lambda'}\delta_{\mu\mu'}$$
$$- \Lambda \int_0^{2\pi} \int_0^\pi \int_0^\infty F_{\nu\lambda\mu}^* \xi^2 \cos\theta F_{\nu'\lambda'\mu'}\xi \, d\xi \sin\theta \, d\theta \, d\varphi. \quad (10\text{–}143)$$

The matrix elements in the first row and in the first column may be calculated by making use of Eq. (10–140). Because of the orthogonality relation expressed by Eq. (10–136), Eq. (10–140) requires that all of them vanish except those connecting F_{100} with F_{210} or F_{310}. Equation (10–141) becomes

$$\begin{vmatrix} -2W & H_{100,210} & H_{100,310} \\ H_{210,100} & H_{210,210} & 0 \\ H_{310,100} & 0 & H_{310,310} \end{vmatrix} = 0. \quad (10\text{–}144)$$

The numerical values of the nonvanishing matrix elements are readily evaluated by Eqs. (10–143), (10–139), and (10–136). Thus

$$\begin{vmatrix} -2W & -4\sqrt{2}\,\Lambda & 2\sqrt{2}\,\Lambda \\ -4\sqrt{2}\,\Lambda & -1 & 0 \\ 2\sqrt{2}\,\Lambda & 0 & -2 \end{vmatrix} = 0. \quad (10\text{–}145)$$

The solution of Eq. (10–145) is

$$W = 18\Lambda^2. \quad (10\text{–}146)$$

By Eqs. (10–130) and (10–131), the above equation becomes

$$-\frac{9}{2Ze^2} E_{100} - \frac{1}{4} = \frac{18a^4}{16Z^2e^2} F^2 \quad (10\text{–}147)$$

or

$$E_{100} = -\frac{MZ^2e^4}{2\hbar^2} - \frac{9}{4}\frac{a^3}{Z} F^2. \quad (10\text{–}148)$$

The first term is the unperturbed energy value of the ground state of the hydrogen atom. The second term, proportional to the square of the field, is the second-order perturbation energy. The *polarizability* α is defined as the proportionality constant of the induced electric dipole moment *vs.* the inducing electric field. The energy of the induced dipole is thus

$$\varepsilon = -\int_0^F \alpha F \, dF = -\tfrac{1}{2}\alpha F^2. \qquad (10\text{--}149)$$

Comparing the second term of Eq. (10–147) with Eq. (10–149), we obtain the polarizability of the hydrogen atom at the ground state:

$$\alpha = \frac{9}{2}\frac{a^3}{Z} = 0.68 \times 10^{-24} \text{ cm}^3. \qquad (10\text{--}150)$$

Problems

10–1. A linear harmonic oscillator is perturbed by an additional potential λx^3. Determine the eigenvalues and eigenfunctions. Repeat for the perturbation potential λx^4. Compare the results with those of Problem 7–2.

10–2. A simple pendulum may be regarded as a linear harmonic oscillator to the first approximation. Repeat Problem 4–1 in the second and third approximations.

10–3 Discuss the perturbation of the isotropic, three-dimensional harmonic oscillator by an additional potential λr. Repeat for λr^3 and λr^4.

*10–4. Discuss the perturbation of an isotropic, four-dimensional harmonic oscillator (Problem 9–2) by an additional potential $\lambda_1(x^2 + y^2) + \lambda_2(z^2 + w^2)^2$.

10–5. Consider the problem specified by the potential shown in Fig. 10–1. Solve the Schrödinger equation by the perturbation method, the unperturbed problem being that of Section 5–1.

10–6. Show that the function $\psi_n^{(0)}(1 \leq n \leq g)$ is an eigenfunction of $H^{(1)}$, the eigenvalue $E_n^{(1)}$ being a root of Eq. (10–58).

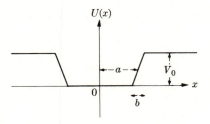

FIGURE 10–1

10–7. Discuss the Stark effect of the third energy level ($n = 3$) of the hydrogen atom by the method of Section 10–4(B).

10–8. Verify the result of the above problem by the method of Section 10–4(C).

*10–9. The μ-mesic x-ray spectra do not agree exactly with the Balmer formula; the deviation is attributed to the fact that the nucleus has a finite size, instead of being a point charge as tacitly assumed in the theory of the hydrogen atom. Calculate the shift of energy levels by the perturbation theory assuming the nucleus to be a charged spherical shell of radius a. Discuss how to use the μ-meson as a probe to investigate the charge distribution inside the nucleus.

*10–10. The three-dimensional spherical well potential is perturbed by a small potential λr where λ is a small negative constant and r is the radius vector. Does the perturbation theory lead to a correct solution in this case? [*Hint:* The perturbation changes all bound states to resonance states (or virtual states, see Problem 5–5). The perturbation theory gives rise to an energy shift but fails to account for the resonance character. When the width of the well approaches zero (δ-function potential), the exact solution by the method of complex eigenvalues is given by Titchmarsh, Proc. Roy. Soc. **207**, 321 (1951).]

*10–11.# *Molecular physics.* Sketch the study of the inter-molecular force (van der Waals force) between neutral atoms at a large distance, which is not zero, very small and always attractive, by the second order perturbation theory.

10–12.# *Molecular physics.* Molecules can be neutral or with a dipole. Explain the dipole-dipole force, dipole-induced-dipole force, and induced-dipole-induced-dipole force between two molecules. The third one, also known as the London force, is of quantum mechanical origin, calculable by the second order perturbation theory. See the previous problem. In general the three are of the same order of magnitude in chemistry.

10–13.# *Molecular biology.* Explain that the London force tends to pull two overlapping molecules to overlap in as much area as possible.

10–14.# *Molecular biology.* Explain why the DNA double helix is wound *tightly twisted*. Watson and Crick showed this as a fact but did not explain its dynamic origin.

10–15.# *Molecular biology.* Can a single strand DNA form a single helix? If so, how would its radius compare with that of the double helix?

---

* Indicates more difficult problems.

\# Indicates new problems added to the expanded edition.

CHAPTER 11

TIME-DEPENDENT PERTURBATION THEORY

The perturbation theory developed in the last chapter dealt with the time-independent Schrödinger equation. The time variable was not involved. Nevertheless, the general idea of the perturbation method may also be applied for the solution of the time-dependent Schrödinger equation. In this chapter we shall develop such a theory. Since the time-dependent Schrödinger equation is the equation of motion in quantum mechanics and its exact solution is in general difficult to obtain, this approximate method is important for the solution of many dynamical problems. In atomic physics, particularly in spectral analysis, the main concern is the quantization of energy; the time-independent perturbation theory of the last chapter finds wide application. In nuclear physics, particularly in nuclear reactions and disintegrations, the main concern is the determination of the outcome of a specific process; the time-dependent perturbation method is frequently used. In the following, we first develop the general theory; this will be followed by a discussion of applications in nuclear physics and in radiation theory. Incidentally, this method is also known as the method of variation of constants, the reason for which will be made clear later in the course of development.

11-1 General theory of time-dependent perturbation.* The purpose of the time-independent perturbation theory is to solve a time-independent partial differential equation with the help of a set of known solutions of an approximate equation, i.e., the unperturbed equation. Similarly, the purpose of the time-dependent perturbation theory to be developed here is to solve a time-dependent partial differential equation with the help of a set of known solutions of an approximate equation which will again be called the unperturbed equation.

In the last chapter we wrote the time-independent Schrödinger equation in the operator form

$$H\psi = E\psi. \tag{11-1}$$

Since the time-dependent Schrödinger equation (2–41) is obtained from the time-independent Schrödinger equation by replacing $(2M/\hbar^2)E\Psi$ with

---

* P. A. M. Dirac, *Proc. Roy. Soc.* **A112**, 661 (1926); **A114**, 243 (1927).

$(2iM/\hbar)(\partial\Psi/\partial t)$, it may be written in operator form as follows:

$$H\Psi = -\frac{\hbar}{i}\frac{\partial\Psi}{\partial t}. \tag{11-2}$$

We may consider Eq. (11-2) to be obtained from Eq. (11-1) by the following operator substitution:

$$E \rightarrow -\frac{\hbar}{i}\frac{\partial}{\partial t}. \tag{11-3}$$

The reader may recognize the similarity and difference between Eq. (11-3) and Eq. (10-3).

We consider those problems the Hamiltonians of which may be written as the sum of two terms, the second term being much smaller than the first:

$$H = H^{(0)} + \lambda H^{(1)}. \tag{11-4}$$

As before, λ is an infinitesimal parameter which determines the order of perturbation. For generality $H^{(1)}$ may contain (but does not necessarily) the time variable t. This happens when the potential contains a small time-dependent part. In classical mechanics such a force field is not conservative, and conservation of energy does not hold. We shall see similar situations in quantum mechanics. Our purpose is to solve Eq. (11-2) with the help of the known solutions of the unperturbed equation

$$H^{(0)}\Psi^{(0)} = -\frac{\hbar}{i}\frac{\partial\Psi^{(0)}}{\partial t}. \tag{11-5}$$

The general solution of Eq. (11-5) is given by

$$\Psi^{(0)} = \sum_n a_n(0)e^{-(i/\hbar)E_n^{(0)}t}\psi_n^{(0)}, \tag{11-6}$$

where the coefficients $a_n(0)$ are constants and the functions $\psi_n^{(0)}$ are eigenfunctions of the following time-independent Schrödinger equation:

$$H^{(0)}\psi_n^{(0)} = E_n^{(0)}\psi_n^{(0)}, \tag{11-7}$$

$E_n^{(0)}$ being the corresponding eigenvalue. The $\psi_n^{(0)}$'s and $E_n^{(0)}$'s are assumed to be known.

Since any well-behaved function of x, y, z may be expanded in a series of orthonormal functions, the general solution of Eq. (11-2) at a particular time t_0, being a function of x, y, z, may be expressed by a series of orthonormal functions $F_n(x, y, z)$. Thus

$$\Psi(x, y, z, t_0) = \sum_n a_n(t_0)F_n(x, y, z). \tag{11-8}$$

The coefficients of expansion, of course, depend on t_0. Similar expansions may be made at other times, and all of these expressions may be summarized by the following equation:

$$\Psi(x, y, z, t) = \sum_n a_n(t) F_n(x, y, z), \tag{11-9}$$

where the coefficients of expansion $a_n(t)$ are considered as functions of time. As the functions $F_n(x, y, z)$ are known quantities, the unknown function $\Psi(x, y, z, t)$ is thus replaced by a set of unknown quantities $a_n(t)$, $n = 1, 2, \ldots$ (This is a mathematical transformation similar to the Fourier transformation.) The equation of motion governing $\Psi(x, y, z, t)$ is to be transformed into a set of equations determining the coefficients $a_n(t)$. This may be carried out by substituting Eq. (11-9) in Eq. (11-2), the result being

$$\sum_n a_n(t) H F_n = -\frac{\hbar}{i} \sum_n \dot{a}_n(t) F_n. \tag{11-10}$$

Multiplying this equation by F_m^* and integrating over the space variables, we obtain the following equation:

$$\dot{a}_m(t) = -\frac{i}{\hbar} \sum_n a_n(t) H_{mn}, \qquad m = 1, 2, \ldots, \tag{11-11}$$

where

$$H_{mn} \equiv \int F_m^* H F_n \, dt. \tag{11-12}$$

The H_{mn}'s are known quantities. Equations (11-11), being of the first order in time, are a set of simultaneous ordinary differential equations which completely determine the set of functions $a_n(t)$ once their initial values $a_n(t_0)$ are given. Since the set $a_n(t)$ determines the function $\Psi(x, y, z, t)$ completely, Eqs. (11-11), determining the set $a_n(t)$, are equivalent to Eq. (11-2), which determines $\Psi(x, y, z, t)$.. A *partial* differential equation (11-2) is thus replaced by a set of simultaneous *ordinary* differential equations (11-11) which may be solved more conveniently. The initial condition specified by $\Psi(x, y, z, t_0)$ is now replaced by the initial condition specified by the set $a_n(t_0)$. An equivalent mathematical problem is thus formulated and Eqs. (11-11) will now be considered as the quantum-mechanical equations of motion in terms of the functions $a_n(t)$.

A trivial case may be mentioned first. If the set of functions $F_n(x, y, z)$ are the energy eigenfunctions, i.e.,

$$F_n = \psi_n, \qquad H\psi_n = E_n\psi_n, \tag{11-13}$$

then the equations of motion, Eqs. (11–11) become

$$\dot{a}_m(t) = -\frac{i}{\hbar} E_m a_m(t), \qquad m = 1, 2, \ldots . \qquad (11\text{--}14)$$

Their solutions are

$$a_m = a_m(0)e^{-(i/\hbar)E_m t}, \qquad m = 1, 2, \ldots, \qquad (11\text{--}15)$$

where $a_m(0)$ is the initial value of a_m at $t = 0$. It follows that

$$\Psi(x, y, z, t) = \sum_n a_n(0)e^{-(i/\hbar)E_n t}\psi_n(x, y, z), \qquad (11\text{--}16)$$

and the wave function at any time t is expressed in terms of its initial value at $t = 0$ represented by $a_n(0)$. Thus the present method reproduces the well-known result that the general solution of the time-dependent Schrödinger equation may be expressed by a series of energy eigenfunctions each with a proper time factor attached.

When the above result is applied to the unperturbed equation, Eq. (11–6) is reproduced. When the perturbation is small the perturbed wave function $\Psi(x, y, z, t)$ should not differ greatly from the unperturbed wave function $\Psi^{(0)}(x, y, z, t)$. Since $\Psi^{(0)}(x, y, z, t)$ is given by Eq. (11–6), where the coefficients $a_n(0)$ are constants, $\Psi(x, y, z, t)$ may be assumed to take the form

$$\Psi(x, y, z, t) = \sum_n \alpha_n(t)e^{-(i/\hbar)E_n^{(0)}t}\psi_n^{(0)}, \qquad (11\text{--}17)^*$$

where the functions $\alpha_n(t)$ approach the constants $a_n(0)$ when the perturbation is reduced to zero. Comparing this equation with Eq. (11–9), we note that the functions $\psi_n^{(0)}$ are used merely as a set of orthonormal functions, the coefficients of expansion being

$$a_n(t) = \alpha_n(t)e^{-(i/\hbar)E_n^{(0)}t}.$$

For a small perturbation we may expect from the above consideration that $a_n(t)$'s are slowly varying functions of time. Under this condition Eqs. (11–11) may be solved by appropriate approximations. It is here the idea of the perturbation method enters. Let us assume, by anology to Eqs. (10–9) and (10–10), that

$$\alpha_n(t) = \alpha_n^{(0)}(t) + \lambda\alpha_n^{(1)}(t) + \lambda^2\alpha_n^{(2)}(t) + \cdots, \quad n = 1, 2, \ldots . \qquad (11\text{--}18)$$

---

* The differences between Eqs. (11–16) and (11–17) should be clearly recognized.

Multiplying $\alpha_n(t)$ by $e^{-(i/\hbar)E_n^{(0)}t}$, we obtain $a_n(t)$. Substitution of $a_n(t)$ in Eq. (11–11) results in the following:

$$[\dot{\alpha}_m^{(0)}(t) + \lambda\dot{\alpha}_m^{(1)}(t) + \lambda^2\dot{\alpha}_m^{(2)}(t) + \cdots]e^{-(i/\hbar)E_m^{(0)}t}$$

$$- \frac{i}{\hbar} E_m^{(0)}[\alpha_m^{(0)}(t) + \lambda\alpha_m^{(1)}(t) + \lambda^2\alpha_m^{(2)}(t) + \cdots]e^{-(i/\hbar)E_m^{(0)}t}$$

$$= -\frac{i}{\hbar}\sum_n[\alpha_n^{(0)}(t) + \lambda\alpha_n^{(1)}(t) + \lambda^2\alpha_n^{(2)}(t) + \cdots]e^{-(i/\hbar)E_n^{(0)}t}(H_{mn}^{(0)} + \lambda H_{mn}^{(1)}).$$

$$(11\text{--}19)$$

Since we now use the unperturbed wave functions $\psi_n^{(0)}$ as the orthonormal functions F_n, the matrix elements are given by

$$H_{mn}^{(0)} = E_m^{(0)}\delta_{mn}. \qquad (11\text{--}20)$$

Equation (11–19) thus becomes

$$[\dot{\alpha}_m^{(0)}(t) + \lambda\dot{\alpha}_m^{(1)}(t) + \lambda^2\dot{\alpha}_m^{(2)}(t) + \cdots]e^{-(i/\hbar)E_m^{(0)}t}$$

$$= -\frac{i}{\hbar}\sum_n[\lambda H_{mn}^{(1)}\alpha_n^{(0)}(t) + \lambda^2 H_{mn}^{(1)}\alpha_n^{(1)}(t) + \lambda^3 H_{mn}^{(1)}\alpha_n^{(2)}(t) + \cdots]e^{-(i/\hbar)E_n^{(0)}t}.$$

$$(11\text{--}21)$$

As before, coefficients of the same powers of λ on both sides of the equation must be equal. These conditions give rise to the following equations from which the values of the various orders of perturbation of $\alpha_n(t)$ may be determined in succession:

$$\dot{\alpha}_m^{(0)}(t) = 0, \qquad (11\text{--}22)$$

$$\dot{\alpha}_m^{(1)}(t) = -\frac{i}{\hbar}\sum_n H_{mn}^{(1)}\alpha_n^{(0)}(t)e^{(i/\hbar)(E_m^{(0)}-E_n^{(0)})t}, \qquad (11\text{--}23)$$

$$\dot{\alpha}_m^{(2)}(t) = -\frac{i}{\hbar}\sum_n H_{mn}^{(1)}\alpha_n^{(1)}(t)e^{(i/\hbar)(E_m^{(0)}-E_n^{(0)})t}, \qquad (11\text{--}24)$$

$$\cdots \cdots \cdots \cdots \cdots \cdots \cdots ,$$

$$\dot{\alpha}_m^{(i+1)}(t) = -\frac{i}{\hbar}\sum_n H_{mn}^{(1)}\alpha_n^{(i)}(t)e^{(i/\hbar)(E_m^{(0)}-E_n^{(0)})t}. \qquad (11\text{--}25)$$

Equations (11–22) determine the zero-th order $\alpha_m^{(0)}(t)$; the solutions are all constants. Thus the zero-th order approximation reduces to the trivial case we have already discussed. Before we proceed to solve the next equation, we have to consider the initial conditions of Eqs. (11–22) to (11–25). The initial condition of Eqs. (11–11) is that at time $t = 0$ the $a_n(t)$ take the initial values $a_n(0)$. Since in Eq. (11–17) we require $\alpha_n(t)$ to approach $a_n(0)$ when the perturbation is reduced to zero, Eq. (11–18) requires that

$$\alpha_n^{(0)}(t) = a_n(0). \tag{11–26}$$

This is consistent with Eq. (11–22), according to what we have just discussed. From Eqs. (11–26) and (11–18) we may obtain the initial conditions of Eqs. (11–22) to (11–25). The result is that the initial values of $\alpha_n^{(1)}$, $\alpha_n^{(2)}$, ... $\alpha_n^{(i)}$, ..., at $t = 0$ must all be zero, i.e.,

$$\alpha_n^{(1)}(0) = 0, \quad \alpha_n^{(2)}(0) = 0, \ldots, \quad \alpha_n^{(i)}(0) = 0, \ldots, \quad n = 1, 2, \ldots.$$

Equations (11–26) may now be substituted into Eqs. (11–23) to determine $\alpha_m^{(1)}$. With the initial condition that $\alpha_m^{(1)}(0) = 0$, Eqs. (11–23) completely determine $\alpha_m^{(1)}$. The results may now be put into Eqs. (11–24) to determine $\alpha_m^{(2)}$. The procedure may be carried out to as many orders as we wish. The initial conditions just stated enable us to determine uniquely the solutions of Eqs. (11–22) to (11–25).

We now proceed to solve Eqs. (11–23) for the first-order perturbation. Introduce ω_{mn} by the definition

$$\omega_{mn} \equiv \frac{E_m^{(0)} - E_n^{(0)}}{\hbar}. \tag{11–27}$$

Equations (11–23) may be rewritten as follows:

$$\dot{\alpha}_m^{(1)}(t) = -\frac{i}{\hbar} \sum_n H_{mn}^{(1)} a_n(0) e^{i\omega_{mn}t}. \tag{11–28}$$

When $H^{(1)}$ is independent of time, Eqs. (11–28) may be solved easily. (For a time-dependent $H^{(1)}$ the solution is somewhat different; a special case will be discussed in Section 11–5.) The constants of integration are determined by the initial conditions, i.e., $\alpha_m^{(1)}(0) = 0$. The solutions thus obtained are

$$\alpha_m^{(1)}(t) = -\frac{1}{\hbar} \sum_n H_{mn}^{(1)} a_n(0) \frac{e^{i\omega_{mn}t} - 1}{\omega_{mn}}, \qquad m = 1, 2, \ldots. \tag{11–29}$$

In many practical applications the initial conditions are such that at $t = 0$ the system is described by one of the unperturbed eigenfunctions, say, $\psi_k^{(0)}(x, y, z)$. The mathematical expression for this condition is

$$a_n(0) = \delta_{nk}, \qquad n = 1, 2, \ldots. \tag{11–30}$$

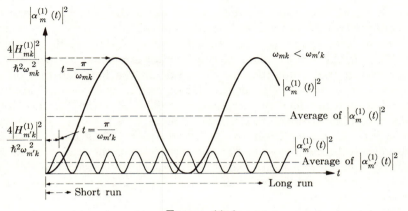

FIGURE 11–1

The physical situation corresponding to such an initial condition will be discussed in the next section. From now on we shall always assume the initial conditions to be given by Eq. (11–30) unless otherwise specified. Substitution of Eq. (11–30) in Eq. (11–29) results in the equation

$$\alpha_m^{(1)}(t) = -\frac{1}{\hbar} H_{mk}^{(1)} \frac{e^{i\omega_{mk}t} - 1}{\omega_{mk}}, \qquad m = 1, 2, \dots \qquad (11\text{–}31)$$

For the special case $m = k$ the fraction $(e^{i\omega_{mk}t} - 1)/\omega_{mk}$ is to be replaced by its limiting value it (see Problem 11–9 for further discussion). Making use of the identical equation

$$e^{ix} - 1 = 2i\, e^{ix/2} \sin \frac{x}{2}, \qquad (11\text{–}32)$$

we derive the following result:

$$|\alpha_m^{(1)}(t)|^2 = \frac{4|H_{mk}^{(1)}|^2 \sin^2(\omega_{mk}t/2)}{\hbar^2 \omega_{mk}^2}, \qquad m = 1, 2, \dots \qquad (11\text{–}33)$$

The time variation of $|\alpha_m^{(1)}(t)|^2$ is plotted in Fig. 11–1 where two curves for two values of m (m and m' are such that $\omega_{mk} < \omega_{m'k}$) are given. The curves are all of the $\sin^2 x$ type. For large values of ω_{mk}, the amplitudes are small because they are proportional to $1/\omega_{mk}^2$; the periods are also small because they are proportional to $1/\omega_{mk}$. Since the average value of $\sin^2 x$ over a period is $\frac{1}{2}$, the amplitude multiplied by $\frac{1}{2}$ may be regarded as the average value of $|\alpha_m^{(1)}(t)|^2$ for a time period t long compared with $2\pi/\omega_{mk}$. The results may be summarized thus: If a system at time 0 is described by a single term $\psi_k^{(0)}$ in the expansion of Eq. (11–17), then at time t long compared with the periods $2\pi/\omega_{mk}$, the system will be represented by an expansion in which $\psi_k^{(0)}$ is again the dominant term; but other terms $\psi_m^{(0)}$ also appear with coefficients $\lambda \alpha_m^{(1)}(t) e^{-(i/\hbar) E_m^{(0)} t}$, the average value of the square of the absolute value of the coefficient being $2|\lambda H_{mk}^{(1)}|^2 / \hbar^2 \omega_{mk}^2$. Since these average values are time independent we may regard the composition of the series, in the long run and on the average, to be time independent. (However, this result cannot apply in the short run in which the amplitudes of different terms $\psi_m^{(0)}$ increase in time according to different rates.) Since $\hbar \omega_{mk} = E_m^{(0)} - E_k^{(0)}$, the average of $|\alpha_m^{(1)}(t)|^2$ is proportioned to $1/(E_m^{(0)} - E_k^{(0)})^2$. Thus those terms $\psi_m^{(0)}$ with eigenvalues $E_m^{(0)}$ close to $E_k^{(0)}$, the eigenvalue of the original term $\psi_k^{(0)}$, will be built up to a larger extent. Other $\psi_m^{(0)}$'s with eigenvalues $E_m^{(0)}$ far away from $E_k^{(0)}$ will appear in the series Eq. (11–17) only in very small amounts.

Having obtained the first-order solution we proceed to solve the second-order equations (11–24). The initial conditions Eq. (11–30) are again assumed. Equations (11–24) become

$$\dot{\alpha}_m^{(2)}(t) = + \frac{i}{\hbar^2} \sum_n H_{mn}^{(1)} H_{nk}^{(1)} \frac{e^{i\omega_{nk} t} - 1}{\omega_{nk}} e^{i\omega_{mn} t}, \quad m = 1, 2, \ldots. \quad (11\text{–}34)$$

As the matrix elements are assumed to be time independent, Eqs. (11–34) may be integrated. The integration constants are to be determined by the initial conditions $\alpha_m^{(2)}(0) = 0$. Making use of the relation

$$\omega_{mn} + \omega_{nk} = \omega_{mk}, \quad (11\text{–}35)$$

we write the solutions thus obtained as follows:

$$\alpha_m^{(2)}(t) = \frac{1}{\hbar^2} \sum_n \frac{H_{mn}^{(1)} H_{nk}^{(1)}}{\omega_{nk}} \left(\frac{e^{i\omega_{mk} t} - 1}{\omega_{mk}} - \frac{e^{i\omega_{mn} t} - 1}{\omega_{mn}} \right). \quad (11\text{–}36)$$

Higher order perturbations $\alpha_m^{(i)}(t)$ may be obtained in a similar manner.

We now discuss a simple example for illustration. Consider a charged linear harmonic oscillator placed in a weak electric field. The eigenvalues and eigenfunctions of this system have been obtained by the time-independent perturbation method of Section 10–1. Knowing these solutions, we may write down the general solution of the time-dependent Schrödinger equation. However, we shall solve the time-dependent Schrödinger equation by the time-dependent perturbation method and show that the results of the two methods are the same. The equation to be solved is

$$\left(-\frac{\hbar^2}{2M} \frac{d^2}{dx^2} + \frac{1}{2} kx^2 - qFx \right) \Psi = -\frac{\hbar}{i} \frac{\partial \Psi}{\partial t}. \quad (11\text{–}37)$$

The unperturbed equation

$$\left(-\frac{\hbar^2}{2M}\frac{d^2}{dx^2} + \frac{1}{2}kx^2\right)\Psi^{(0)} = -\frac{\hbar}{i}\frac{\partial\Psi^{(0)}}{\partial t} \tag{11-38}$$

has already been solved in Chapter 4. Its general solution is a superposition of eigenfunctions with time factors attached:

$$\Psi^{(0)}(x, t) = \sum_{n=0}^{\infty} a_n e^{-(i/\hbar)E_n^{(0)}t}\psi_n^{(0)}(x), \tag{11-39}$$

where $E_n^{(0)}$ and $\psi_n^{(0)}$ are given by Eqs. (4–25) and (4–34). The term $-qFx$ is to be considered as perturbation. We let

$$\lambda = -qF, \qquad H^{(1)} = x. \tag{11-40}$$

The perturbation theory developed here is based on the initial condition of Eqs. (11–30). Let us assume that the initial state of the system at $t = 0$ is represented by one of the $\psi_n^{(0)}$'s, say, $\psi_k^{(0)}$,

$$\Psi(x, 0) = \psi_k^{(0)}(x).$$

As time goes on, the wave function $\Psi(x, t)$ changes according to the dictation of the time-dependent Schrödinger equation. We may expand $\Psi(x, t)$, as a function of x, in terms of a set of orthonormal functions. Let this set be the $\psi_n^{(0)}(x)$. Thus

$$\Psi(x, t) = \sum_n a_n(t)\psi_n^{(0)}(x).$$

At $t = 0$, we have

$$a_n(0) = \begin{cases} 1, & \text{for } n = k, \\ 0, & \text{for } n \neq k. \end{cases} \tag{11-41}$$

As time goes on, $a_k(t)$ will remain near unity while other $a_n(t)$'s increase gradually from zero. Consider the first-order perturbation. The matrix elements of $H^{(1)} = x$ with respect to the eigenfunctions of the oscillator are already obtained in Eq. (10–42). Since the matrix elements are non-vanishing only when the two quantum numbers differ by one, Eqs. (11–31) give rise to two nonvanishing equations

$$\alpha_{k+1}^{(1)}(t) = -\frac{1}{\hbar}\sqrt{\frac{\hbar}{M\omega}}\frac{k+1}{2}\frac{e^{i\omega t}-1}{\omega},$$

$$\alpha_{k-1}^{(1)}(t) = -\frac{1}{\hbar}\sqrt{\frac{\hbar}{M\omega}}\frac{k}{2}\frac{e^{-i\omega t}-1}{-\omega},$$

$$\alpha_m^{(1)}(t) = 0, \qquad m \neq k \pm 1.$$

The equation $E_n = \hbar\omega(n + \frac{1}{2})$ is used in writing the above equations. The first-order perturbed wave function is thus

$$\Psi(x, t) = e^{-(i/\hbar)E_k^{(0)}t}\psi_k^{(0)} + qF\sqrt{\frac{\hbar}{M\omega}} \frac{k + 1}{2} \frac{e^{i\omega t} - 1}{\hbar\omega} e^{-(i/\hbar)E_{k+1}^{(0)}t} \psi_{k+1}^{(0)}$$

$$+ qF\sqrt{\frac{\hbar}{M\omega}} \frac{k}{2} \frac{1 - e^{-i\omega t}}{\hbar\omega} e^{-(i/\hbar)E_{k-1}^{(0)}t}\psi_{k-1}^{(0)}. \qquad (11\text{--}42)$$

Since the only nonvanishing $\alpha_m^{(1)}(t)$'s are $\alpha_{k+1}^{(1)}(t)$ and $\alpha_{k-1}^{(1)}(t)$, Eqs. (11–24) tell us that the only nonvanishing $\alpha_m^{(2)}(t)$'s are those for which $m = k + 2$, k, $k - 2$. Thus the second-order perturbed wave function contains terms $\lambda^2\psi_{k+2}^{(0)}$, $\lambda^2\psi_k^{(0)}$, $\lambda^2\psi_{k-2}^{(0)}$; in other words,

$$\Psi(x, t) = [1 + 0(\lambda^2)]\psi_k^{(0)} + 0(\lambda)\psi_{k+1}^{(0)} + 0(\lambda)\psi_{k-1}^{(0)} + 0(\lambda^2)\psi_{k+2}^{(0)} + 0(\lambda^2)\psi_{k-2}^{(0)},$$

$$(11\text{--}43)$$

where the notation $0(\lambda^n)$ designates a quantity of the order of λ^n. Higher order perturbations may be obtained accordingly. The ith-order perturbation is to introduce in $\Psi(x, t)$ two new terms $0(\lambda^i)\psi_{k+i}^{(0)}$ and $0(\lambda^i)\psi_{k-i}^{(0)}$.

We now consider the physical meaning of this solution. This particular solution is such that at time $t = 0$, its spatial distribution is given by the function $\psi_k^{(0)}$. After $t = 0$, the system evolves in such a manner that, in the short run, other terms $\psi_n^{(0)}$ in the series expansion Eq. (11–17) appear and increase gradually. After the coefficients of the $\psi_n^{(0)}$'s have reached their maximum values, they will decrease and then become oscillatory. Hence in the long run, we may consider their average values (shown in Fig. 11–1); and the composition of $\Psi(x, t)$ in terms of $\psi_n^{(0)}$'s may be regarded, on the average, as time independent. The $\psi_n^{(0)}$'s closer to $\psi_k^{(0)}$ are built up to a greater extent than those far away from $\psi_k^{(0)}$.

This particular solution may be considered from a different point of view, i.e., that of the time-independent perturbation theory. Since we already know the perturbed eigenvalues E_n and eigenfunctions ψ_n of Eq. (11–37) in Chapter 10, the general solution of Eq. (11–37) may be expressed in terms of these quantities. Thus

$$\Psi(x, t) = \sum_n C_n e^{-(i/\hbar)E_n t}\psi_n(x). \qquad (11\text{--}44)$$

As a special case, the particular solution we obtained in this section Eq. (11–42) may be expressed in the form of Eq. (11–44). This may be accomplished by first making a series expansion of our initial wave function $\psi_k^{(0)}(x)$ in terms of the perturbed eigenfunctions $\psi_n(x)$. Let the result be expressed by

$$\psi_k^{(0)}(x) = \sum_n d_n \psi_n(x). \tag{11-45}$$

Then we have

$$\Psi(x, t_0) = \sum_n d_n \psi_n(x).$$

From this expression we conclude

$$\Psi(x, t) = \sum_n d_n e^{-(i/\hbar) E_n t} \psi_n(x). \tag{11-46}$$

We shall show that Eq. (11–46) is equivalent to Eq. (11–42) (extended to higher orders), and therefore the solutions of this initial value problem obtained by both methods, time-dependent and time-independent perturbation, are identical. To do this we have to determine explicitly the coefficients of expansion d_n. From Eq. (10–48) we know the perturbed eigenfunctions $\psi_k(x)$ to be

$$\psi_k(x) = \psi_k^{(0)}(x) + qF\sqrt{\frac{k+1}{2\hbar\omega K}}\, \psi_{k+1}^{(0)} - qF\sqrt{\frac{k}{2\hbar\omega K}}\, \psi_{k-1}^{(0)}$$

$$+ q^2F^2\frac{\sqrt{(k+1)(k+2)}}{4\hbar\omega K}\, \psi_{k+2}^{(0)} + q^2F^2\frac{\sqrt{k(k-1)}}{4\hbar\omega K}\, \psi_{k-2}^{(0)} + \cdots. \tag{11-47}$$

Since the dominant term of the above equation is $\psi_k^{(0)}(x)$, the dominant term in Eq. (11–45) must be $d_k \psi_k(x)$ and d_k must be unity to the zero-th approximation. Transferring the term $d_k \psi_k(x)$ to the left-hand side of Eq. (11–45), we obtain

$$\psi_k^{(0)}(x) - \psi_k(x) = \sum_{n=0}^{k-1} d_n \psi_n(x) + \sum_{n=k+1}^{\infty} d_n \psi_n(x). \tag{11-48}$$

The left-hand side of this equation, according to Eq. (11–47), is a small quantity of the first order; therefore, d_n's for $n \neq k$ must be at least first-order infinitesimal. Actually, comparing Eqs. (11–47) with Eq. (11–48) we can determine d_{k+1} and d_{k-1}. The results are

$$d_{k+1} = -qF\sqrt{(k+1)/2\hbar\omega K},$$
$$d_{k-1} = +qF\sqrt{k/2\hbar\omega K}. \tag{11-49}$$

The other d_n's are small quantities of higher orders. The energy eigenvalues E_n are the same as $E_n^{(0)}$ to the first order. Thus

$$E_n = E_n^{(0)} + 0(\lambda^2). \tag{11-50}$$

Equations (11–47), (11–49), and (11–50) may now be substituted in Eq. (11–46) to obtain the first-order approximation of $\Psi(x, t)$. Thus

$$
\begin{aligned}
\Psi(x, t) \\
&= \psi_k e^{-(i/\hbar)E_k t} - qF\sqrt{\frac{k+1}{2\hbar\omega K}}\,\psi_{k+1}\,e^{-(i/\hbar)E_{k+1}t} \\
&\qquad + qF\sqrt{\frac{k}{2\hbar\omega K}}\,\psi_{k-1}\,e^{-(i/\hbar)E_{k-1}t} + \cdots \\
&= e^{-(i/\hbar)E_k^{(0)}t}\left[\psi_k^{(0)}(x) + qF\sqrt{\frac{k+1}{2\hbar\omega K}}\,\psi_{k+1}^{(0)} - qF\sqrt{\frac{k}{2\hbar\omega K}}\,\psi_{k-1}^{(0)} + \cdots\right] \\
&\quad - qF\sqrt{\frac{k+1}{2\hbar\omega K}}\,e^{-(i/\hbar)E_{k+1}^{(0)}t}[\psi_{k+1}^{(0)}(x) + \cdots] \\
&\quad + qF\sqrt{\frac{k}{2\hbar\omega K}}\,e^{-(i/\hbar)E_{k-1}^{(0)}t}[\psi_{k-1}^{(0)}(x) + \cdots] \\
&\quad + \cdots,
\end{aligned}
\tag{11–51}
$$

or

$$
\begin{aligned}
\Psi(x, t) \\
&= e^{-(i/\hbar)E_k^{(0)}t}\psi_k^{(0)}(x) + qF\sqrt{\frac{k+1}{2\hbar\omega K}}\,(1 - e^{-i\omega t})e^{-(i/\hbar)E_k^{(0)}t}\psi_{k+1}^{(0)}(x) \\
&\qquad - qF\sqrt{\frac{k}{2\hbar\omega K}}\,(1 - e^{i\omega t})e^{-(i/\hbar)E_k^{(0)}t}\psi_{k-1}^{(0)}(x) + \cdots.
\end{aligned}
\tag{11–52}
$$

Remembering the force constant $K = M\omega^2$, we rewrite the above equation as follows:

$$
\begin{aligned}
\Psi(x, t) &= e^{-(i/\hbar)E_k^{(0)}t}\psi_k^{(0)}(x) \\
&\quad + qF\sqrt{\frac{\hbar}{M\omega}\frac{k+1}{2}}\,\frac{e^{i\omega t} - 1}{\hbar\omega}\,e^{-(i/\hbar)E_{k+1}^{(0)}t}\psi_{k+1}^{(0)}(x) \\
&\quad + qF\sqrt{\frac{\hbar}{M\omega}\frac{k}{2}}\,\frac{1 - e^{-i\omega t}}{\hbar\omega}\,e^{-(i/\hbar)E_{k-1}^{(0)}t}\psi_{k-1}^{(0)}(x) + \cdots.
\end{aligned}
\tag{11–53}
$$

Equation (11–53) is identical with Eq. (11–42). Thus the two solutions obtained by the two perturbation methods, to their first order, are identical. This discussion illustrates the relation between these two methods. The time-independent perturbation theory places its emphasis on the eigenvalues and eigenfunctions. Once these are obtained, the evolution of a system in time may be ascertained by Eq. (11–46) after making a series expansion of the initial wave function in terms of the perturbed energy eigenfunctions. The time-dependent perturbation theory places its emphasis on the evolution of the system in time, which is determined by making use of a set of orthonormal functions—the unperturbed eigenfunctions. The two methods lead to the same result.

It may be remarked that in the time-dependent perturbation, *the unperturbed eigenfunctions are not energy eigenfunctions of the perturbed system.* They are energy eigenfunctions of a different problem—the unperturbed problem. For the perturbed problem, the energy eigenfunctions are those given by Eq. (10–48). Now let us consider the physical meaning of Eq. (11–42). Since Eq. (11–42) is equivalent to Eq. (11–51) through Eq. (11–53), the physical meaning of Eq. (11–42) may be obtained from Eq. (11–51). According to Born's second assumption, Eq. (11–51) represents a state which has probabilities $|d_n|^2$ of finding energy equal to E_n, the perturbed energy eigenvalues given by Eq. (10–47). Since

$$d_k = 1 + 0(\lambda^2), \quad d_{k\pm 1} = 0(\lambda), \quad d_{k\pm 2} = 0(\lambda^2), \ldots d_{k\pm i} = 0(\lambda^i), \ldots,$$

$$(11\text{–}54)$$

the energy value is most likely to be E_k, with small chance of it being E_{k+1} or E_{k-1}, and even smaller chance of it being E_{k+2} and E_{k-2}, etc. These probabilities are time independent. Therefore, at any time, when an observation is performed to ascertain the energy, the same information of probability distribution of the energy value is to be obtained. This fact, as we have mentioned in Chapter 6, exemplifies conservation of energy in the quantum-mechanical sense. This is not obvious from Eq. (11–42). The mathematical expression of Eq. (11–42) is such that at time $t = 0$ the wave function reduces to $\psi_k^{(0)}$ and after $t = 0$ the other terms $\psi_{k\pm 1}^{(0)}, \psi_{k\pm 2}^{(0)}, \ldots$, increase at rates proportional to $\lambda, \lambda^2, \ldots$. The reader is warned not to be tempted to interpret this mathematical result to mean that at time $t = 0$ the energy value is definitely $E_k^{(0)}$ and as time goes on the probabilities of finding the energy equal to $E_{k\pm 1}^{(0)}, E_{k\pm 2}^{(0)}, \ldots$, increase according to rates proportional to $\lambda, \lambda^2, \ldots$. The reason is simply that $E_k^{(0)}, E_{k\pm 1}^{(0)}, E_{k\pm 2}^{(0)}, \ldots$, are not energy eigenvalues of the perturbed problem. Upon an observation, the energy value obtained must be an energy eigenvalue, i.e., one of the following: $E_k, E_{k\pm 1}, E_{k\pm 2}, \ldots$. In spite of the fact that at $t = 0$ the wave function is $\psi_k^{(0)}$, its energy value is not definite and the probability distribution over the E_n spectrum (not $E_n^{(0)}$ spectrum) is given by $|d_n|^2$. In spite of the fact that $\psi_{k\pm 1}^{(0)}, \psi_{k\pm 2}^{(0)}, \ldots$ terms increase in time, the probability distribution of energy remains the same, as the values of $|d_n|^2$ are independent of time. The functions $\psi_n^{(0)}$ are merely a set of orthonormal functions. The relations among the $E_n^{(0)}$

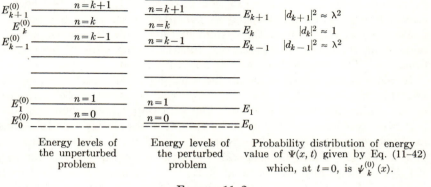

Energy levels of the unperturbed problem

Energy levels of the perturbed problem

Probability distribution of energy value of $\Psi(x, t)$ given by Eq. (11–42) which, at $t = 0$, is $\psi_k^{(0)}(x)$.

FIGURE 11–2

spectrum, the E_n spectrum, and the probability distribution of energy are shown in Fig. 11–2.

11–2 The sudden and the adiabatic approximations. In steady state problems the solution developed in the last section is hardly applicable because a system seldom starts at $t = 0$ from an unperturbed eigenfunction $\psi_k^{(0)}$. The initial condition Eq. (11–30) seldom represents an actual situation. The reason for developing the theory with such an initial condition is that in nonsteady state problems the theory thus developed may be applicable. Let us consider a linear harmonic oscillator, its motion being described by the kth energy eigenfunction. The oscillator is assumed to be charged, but there is no electric field before $t = 0$. At $t = 0$, a weak electric field is switched on *suddenly*. By this we mean that the field reaches its full value in a time duration much smaller than the period of oscillation of the oscillator. After a long period of time T has elaspsed (long compared with the period of oscillation) the field is switched off *suddenly*. The time variation of the electric field is shown in Fig. 11–3. We ask what is the state of motion of the oscillator after time T. This is not a steady state problem. Actually, this consists of three problems. The first corresponds to the time interval $-\infty < t \leq 0$; the oscillator moves under the influence of the elastic force only and the solution is

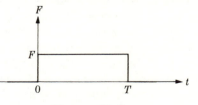

FIGURE 11–3

already given in Chapter 4. The solution enables us to write down the wave function at the last moment of this period, i.e., $t = 0$. This wave function $\Psi(x, 0)$ is to be taken as the initial condition of the second problem, which corresponds to the time interval $0 \leq t \leq T$ when the oscillator is perturbed by the weak electric field. The perturbation methods of the last or the present chapter may be used to solve this problem. The wave function $\Psi(x, T)$ so obtained is to be used as the initial condition for the third problem which corresponds to the time interval $T \leq t < \infty$ when the perturbation is removed and the equation of motion returns to that of the first problem, $-\infty < t \leq 0$. The solution of this problem may be written according to Chapter 4. Using the notations of the last section, the equations of motion in the first and the third time intervals may be expressed by Eq. (11–38) and that in the second time interval may be expressed by Eq. (11–37). The notations of the solutions expressed by Eq. (11–38) and (11–37) in the last section may be adopted for the present problem without change. However, it must be remembered that these expressions have different physical meanings in the present problem.

The solution of the whole problem becomes simple if in the first time interval the oscillator is assumed to be in one particular energy eigenstate, say the kth. The wave function describing the oscillator in this time interval is thus

$$\Psi(x, t) = e^{-(i/\hbar)E_k^{(0)}t}\psi_k^{(0)}(x), \qquad -\infty < t \leq 0. \qquad (11\text{–}55)$$

The initial condition for the second problem is thus

$$\Psi(x, 0) = \psi_k^{(0)}(x). \qquad (11\text{–}56)$$

Equation (11–56) is exactly the initial condition we adopted in the last section, i.e., Eq. (11–30). Therefore to solve the second problem is equivalent to solving Eq. (11–37) with the initial condition Eq. (11–30), which is already given in the previous section. The result to the first order is expressed by Eq. (11–42) and its physical meaning is given by Eq. (11–51). It is for problems of this kind that the theory of the last section is developed. Equation (11–42) enables us to write down the wave function at $t = T$, which is to be the initial condition of the third problem,

$$\Psi(x, T) = e^{-(i/\hbar)E_k^{(0)}T}\psi_k^{(0)} + qF\sqrt{\frac{\hbar}{M\omega}}\frac{k+1}{2}\frac{e^{i\omega T} - 1}{\hbar\omega}e^{-(i/\hbar)E_{k+1}^{(0)}T}\psi_{k+1}^{(0)}$$

$$+ qF\sqrt{\frac{\hbar}{M\omega}}\frac{k}{2}\frac{1 - e^{-i\omega T}}{\hbar\omega}e^{-(i/\hbar)E_{k-1}^{(0)}T}\psi_{k-1}^{(0)}.$$

$$(11\text{–}57)$$

Let us abbreviate Eq. (11–57) by

$$\Psi(x, T) = C_k(T)\psi_k^{(0)} + qFC_{k+1}(T)\psi_{k+1}^{(0)} + qFC_{k-1}(T)\psi_{k-1}^{(0)}. \quad (11\text{–}58)$$

The solution of the third problem may be obtained by solving Eq. (11–38) with Eq. (11–58) as the initial condition. Since $\psi_k^{(0)}$, $\psi_{k+1}^{(0)}$, and $\psi_{k-1}^{(0)}$ are already eigenfunctions of Eq. (11–38), the solution follows immediately:

$$\Psi(x, t) = C_k(T)e^{-(i/\hbar)E_k^{(0)}t}\psi_k^{(0)} + qFC_{k+1}(T)e^{-(i/\hbar)E_{k+1}^{(0)}t}\psi_{k+1}^{(0)}$$

$$+ qFC_{k-1}(T)e^{-(i/\hbar)E_{k-1}^{(0)}t}\psi_{k-1}^{(0)}, \quad \text{for } T < t < \infty. \quad (11\text{–}59)$$

As $\psi_n^{(0)}$'s are the energy eigenfunctions in the third time interval, the physical meaning of this equation is that for $T \leq t < \infty$, the oscillator is to have probabilities $|C_k(T)|^2$, $|qFC_{k+1}(T)|^2$, $|qFC_{k-1}(T)|^2$ for its energy value to be $E_k^{(0)}$, $E_{k+1}^{(0)}$, $E_{k-1}^{(0)}$ respectively. Before $t = 0$, the oscillator has a definite energy value $E_k^{(0)}$. After the field has been switched on for a duration T and then off, the energy information of the oscillator is changed and there is now a small probability for the energy to be $E_{k\pm1}^{(0)}$. Considering the second-order perturbation, there is an even smaller probability for the energy to be $E_{k\pm2}^{(0)}$, etc. These probabilities are dependent on the time duration T. However, Fig. 11–1 tells us that after a long period of time the probabilities oscillate according to $\sin^2 x$ and we may take their time averages. The average values are

$$\left.\begin{array}{l} \overline{|C_k(T)|^2} = 1, \\[2mm] \overline{|qFC_{k+1}(T)|^2} = q^2F^2\,\dfrac{k+1}{\hbar\omega K}, \\[2mm] \overline{|qFC_{k-1}(T)|^2} = q^2F^2\,\dfrac{k}{\hbar\omega K}, \end{array}\right\} \quad \begin{array}{l} \text{to the first order of } \lambda \\ (\lambda = -qF). \end{array} \quad (11\text{–}60)$$

Thus the switch-on of a field F for a finite duration T induces transitions of the oscillator from the original state $\psi_k^{(0)}$ to neighboring states $\psi_{k\pm1}^{(0)}$ (first-order effect), $\psi_{k\pm2}^{(0)}$ (second-order effect), etc., with probabilities given by Eq. (11–60).

Many problems of a similar nature occur in physics, particularly in nuclear physics. A familiar one concerns the ionization of atoms or molecules of a gas when a high-energy charged particle passes through it. This phenomenon is responsible for the appearance of the tracks in the Wilson cloud chamber. The effect of the temporary presence of the fast-moving charged particle in the immediate neighborhood of an atom is equivalent to the switch-on of a Coulomb field for a finite duration and may be treated as a perturbation. The electrons in the atom, under the influence of this perturbation, may undergo transitions to excited states of which ionization is a special case. Another similar problem concerns the Coulomb excitation of the nucleus. A high-energy charged particle passing by a nucleus may induce transitions in the nuclear states. Problems of this kind may be treated by the above method, which is known as the *sudden approximation*.

The transition of states involves a change in energy. This fact does not contradict conservation of energy, since the system is not in a steady state. The energy required for the transition is supplied by the electric field. In the cases of ionization of atoms and Coulomb excitation of nuclei, the necessary amount of energy supplied by the high-energy charged particle that produces the ionization and excitation. The transfer of energy to the oscillator from an electric field turned on and off may be understood from a classical analogy. A classical oscillator is represented by a point moving back and forth between $x = A$ and $x = -A$, where A is the amplitude. If the field is switched on suddenly when the classical oscillator is at $x = -A$, the energy of the oscillator will gain an amount $+qFA$. If the field is switched off when the oscillator is at $x = A$, then the oscillator will lose an amount of energy $-qFA$. After the field is thus switched on and off the oscillator gains a net amount of energy $2qFA$. Switching on the field at any position other than $x = -A$ will result in a smaller amount of energy gain and switching off the field at any position other than $x = A$ will result in a larger loss. Thus the maximum energy transfer to the oscillator is $2qFA$. In a similar manner we may show that the minimum energy transfer is $-2qFA$. The amount of energy transfer as a function of the time duration T is given in Fig. 11–4. If the field is switched off immediately after it is switched on, i.e., $T \doteq 0$, then the energy transfer will be zero because the oscillator does not have time to move and the work done by the field on the oscillator is zero. This is indicated in Fig. 11–4. If a long enough time has elapsed between switching the field on and off, i.e., T is large, then the amount of energy transfer will be a quantity oscillating with an amplitude qFA as indicated in Fig. 11–4. These classical results have their analogy in quantum mechanics. If the field is switched off shortly after it is switched on, then the coefficients of $\psi_{k\pm1}^{(0)}$ in Eq. (11–42) will be nearly zero ($e^{i\omega t} - 1 \doteq 0$), and the probability of finding the energy equal to $E_{k\pm1}^{(0)}$ will be nearly zero. The probabilities for $E_{k\pm2}^{(0)}$, $E_{k\pm3}^{(0)} \ldots$, will be even smaller; the energy is thus essentially unchanged from $E_k^{(0)}$. Therefore for a short duration T, no energy transfer to the oscillator may take place. If a long enough time T has elapsed, then ($e^{i\omega t} - 1$) will have oscillated between 0 and 2 many times, and the absolute value of the coefficient of $\psi_{k+1}^{(0)}$ will have

Energy transfer

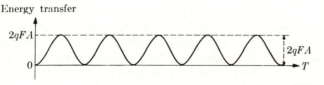

Field switched on at $t = 0$, when $x = -A$; off at $t = T$.

Energy transfer

Field switched on at $t = 0$, when $x = 0$; off at $t = T$.

Energy transfer

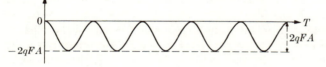

Field switched on at $t = 0$, when $x = A$; off at $t = T$.

FIGURE 11–4

oscillated many times between 0 and $(qF/\hbar\omega)\sqrt{2\hbar(k+1)/M\omega}$. Let us digress for a moment to consider the meaning of this amplitude. When $k \gg 1$, we may find a classical analogy of $\sqrt{2\hbar(k+1)/M\omega}$, i.e.,

$$\sqrt{2\hbar(k+1)/M\omega} \approx \sqrt{2\hbar\omega(k+\tfrac{1}{2})/M\omega^2} = \sqrt{2E_k^{(0)}/K} = A_k, \quad (11\text{–}61)$$

where A_k is the classical amplitude corresponding to the energy $E_k^{(0)}$. Thus the amplitude of the coefficient oscillates between 0 and $qFA_k/\hbar\omega$. The numerator qFA_k is the amplitude of the classical energy transfer. The denominator is the spacing between two adjacent energy levels. The ratio

$$Q \equiv \frac{qFA_k}{\hbar\omega} \qquad (11\text{–}62)$$

thus indicates the size of the classical energy transfer relative to the energy quantum $\hbar\omega$. Since the perturbation theory is based on the requirement that the series expansion, such as Eq. (11–42), converges sufficiently rapidly, $qFA_k/\hbar\omega$ must be much smaller than unity ($Q \ll 1$) in order that the perturbation theory may be valid. Thus the field must be so weak that the classical energy transfer qFA_k is a small fraction of the energy quantum $\hbar\omega$. In classical mechanics, energy may be changed con-

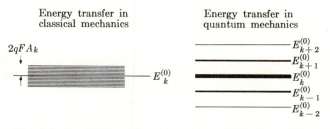

After transfer of energy, the energy value will be represented by one of the lines (levels). In the right-hand figure, the darkness of the line represents the relative probability of its occurrence.

FIGURE 11–5

tinuously, so that the energy of the oscillator after a field is on and off may be changed to any amount within the range from $2qFA_k$ to $-2qFA_k$ (see Fig. 11–4). In quantum mechanics, energy is quantized and can change only by discrete steps. For an oscillator, energy can change only by units of the quantum $\hbar\omega$, which is much greater than the classical quantity qFA_k. Equation (11–42) asserts that the energy transfer due to the perturbation, if it occurs at all, is at least one unit of the quantum (transition to state $\psi_{k+1}^{(0)}$ or $\psi_{k-1}^{(0)}$). Thus the amount of energy change is much greater (by a factor of $1/Q$) in quantum mechanics than in classical mechanics. However, this increase in amount of energy transfer is offset by the fact that such transition takes place only with a very small probability (of the order of Q^2). Hence on the average, the quantization of energy actually hinders the transfer of energy (by a factor of Q). (This hindrance effect has a bearing on the Planck theory of blackbody radiation discussed before in Section 1–1.) Returning to the analogy between the classical and quantum-mechanical energy transfer, we note that a *definite* amount of energy transfer in classical mechanics *of the order of* qFA_k is now replaced in quantum mechanics by a *probable* energy transfer of a *much larger but fixed amount* $\hbar\omega$, which takes place with a *small but varying* probability. The *variation* of the amount of energy transfer in classical mechanics (with an amplitude of qFA_k as indicated in Fig. 11–4) is now replaced by the *variation* of the absolute value of the coefficient of $\psi_{k+1}^{(0)}$ in Eq. (11–42) (between 0 and $qFA_k/\hbar\omega$). This comparison is graphically represented in Fig. 11–5. Another difference may be mentioned here. In classical mechanics the amount of energy transfer is limited between $\pm 2qFA$. In quantum mechanics the energy transfer in first-order perturbation is one quantum, $\hbar\omega$; in second-order perturbation is two quanta $2\hbar\omega$; and in nth-order perturbation is n quanta $n\hbar\omega$. Thus there is a very small chance for a very large amount of energy transfer to take place. This fact has no classical analogy. It arises as a result of the fact that the wave function in quantum mechanics has a "tail" extending to in-

infinity, whereas the classical oscillator is strictly confined in the region $-A \leq x \leq A$.

The energy transfer in quantum mechanics may be given a physical interpretation as follows. Before the field is turned on, the wave function $\psi_k^{(0)}$ is an energy eigenfunction. The distribution of probability elements* is given by $|\Psi(x, t)|^2$ or $|\psi_k^{(0)}(x)|^2$ and is independent of time. Such a state is a *stationary state*. After the field is switched on, $\psi_k^{(0)}(x)e^{-(i/\hbar)E_k^{(0)}t}$ is no longer an energy eigenfunction, $|\Psi(x, t)|^2$ is no longer time independent, and the state is no longer stationary. The probability elements will continuously redistribute themselves in space without reaching equilibrium. When the probability elements move in the electric field, the latter does work on them with amounts depending on the locality and velocity of the probability elements. The small fraction of them at infinity are to receive very large amounts of energy. Upon an observation, the energy value of the oscillator may be found equal to many possible values, including very large ones. This qualitative argument cannot be carried out quantitatively, since it does not give rise to quantized energy values. (It is the failure in energy interpretation that led de Broglie to abandon the pilot wave theory in favor of Born's second assumption.)

The perturbation represented by Fig. 11–3 is an ideal case. In actual cases it always takes some time to turn the field on or off. The sharp rise and fall of the perturbing field shown in Fig. 11–3 is an idealization. However, if the time required to turn the field on or off is short compared with the period of oscillation of the system, Fig. 11–3 may be shown presently to be a valid approximation (the sudden approximation). On the other hand, we have another approximation called the *adiabatic approximation*,† for the other extreme case where the field is switched on or off very slowly. This approximation will now be considered.

Let us simplify the problem by considering a perturbing field which rises linearly in time (Fig. 11–6). The linear rise may be approximated by a series of step rises in intervals of time ΔT, at the beginning of which

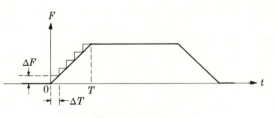

FIGURE 11–6

---

* See Section 2–7.

† M. Born and V. Fock, *Zeits, f. Phys.* **51**, 165 (1928); P. Güttinger, *Zeits. f. Phys.* **73**, 169 (1931).

the field rises suddenly by an amount ΔF (see Fig. 11–6). When ΔF and ΔT approach zero, the steps approach a straight line. Let us assume the initial condition to be such that before $t = 0$, there is no perturbing field and the oscillator is in a stationary state $\psi_k^{(0)}$. At $t = 0$, a field ΔF is switched on suddenly for a period of ΔT. The evolution of the wave function in the period $0 \leq t \leq \Delta T$ is a problem we have already discussed. Therefore the wave function at $t = \Delta T$ may be written down by Eq. (11–42). In adiabatic approximation we require the perturbation to be switched on so slowly that after the period T of the linear rise has passed, the coefficients of Eq. (11–42) have completed a large number of oscillations. Thus, after the interval ΔT, the phases of these coefficients may be finite fractions of 2π. At $t = \Delta T$, an additional amount of field ΔF is switched on. The evolution of the wave function after $t = \Delta T$ may be treated similarly. But we cannot apply Eq. (11–42) directly, because the initial condition, Eq. (11–30), is not satisfied. The initial wave function at $t = \Delta T$ is not a stationary state of the system in the time interval $0 \leq t \leq \Delta T$; instead, it is a superposition of states given by Eq. (11–51). The new initial condition, Eq. (11–51) taken at $t = \Delta T$, may be written as follows:

$$\Psi(x, \Delta T) = e^{-(i/\hbar)(E_k)^{\mathrm{I}} \Delta T} (\psi_k)^{\mathrm{I}} - q\,\Delta F \sqrt{(k+1)/2\hbar\omega K}\; e^{-(i/\hbar)(E_{k+1})^{\mathrm{I}} \Delta T}$$

$$\times (\psi_{k+1})^{\mathrm{I}} + q\,\Delta F \sqrt{k/2\hbar\omega K}\; e^{-(i/\hbar)(E_{k-1})^{\mathrm{I}} \Delta T} (\psi_{k-1})^{\mathrm{I}}, \qquad (11\text{–}63)$$

where $(\psi_n)^{\mathrm{I}}$'s are energy eigenfunctions and $(E_n)^{\mathrm{I}}$'s are energy eigenvalues in the period $0 \leq t \leq \Delta T$, i.e., in the first step of the series of rises. The solution in the second step, i.e., in the period $\Delta T \leq t \leq 2\,\Delta T$, may be found by the time-independent perturbation method. First, $(\psi_k)^{\mathrm{I}}$ and $(\psi_{k+1})^{\mathrm{I}}$ are to be expanded in terms of the energy eigenfunctions $(\psi_n)^{\mathrm{II}}$ of the period $\Delta T \leq t \leq 2\,\Delta T$. To do this we consider a more general problem of expanding $(\psi_k)^R$, eigenfunctions in the Rth step, in terms of $(\psi_k)^{R+1}$, eigenfunctions in the $(R+1)$th step. Taking variation of Eq. (11–47) with respect to F, we obtain

$$(\psi_k)^{R+1} - (\psi_k)^R = q\,\Delta F \sqrt{(k+1)/2\hbar\omega K}\; \psi_{k+1}^{(0)}$$

$$- q\,\Delta F \sqrt{k/2\hbar\omega K}\; \psi_{k-1}^{(0)} + \text{higher order terms.}$$

Therefore

$$(\psi_k)^R = (\psi_k)^{R+1} - q\,\Delta F \sqrt{(k+1)/2\hbar\omega K}\; (\psi_{k+1})^{R+1}$$

$$+ q\,\Delta F \sqrt{k/2\hbar\omega K}\; (\psi_{k-1})^{R+1} + \text{higher order terms.}$$

It is found that Eq. (11–49) represents a special case of this general result. Applying it to $(\psi_k)^{\mathrm{I}}$ we have

$$(\psi_k)^{\mathrm{I}} = (\psi_k)^{\mathrm{II}} - q\,\Delta F\sqrt{\frac{k+1}{2\hbar\omega K}}\,(\psi_{k+1})^{\mathrm{II}} + q\,\Delta F\sqrt{\frac{k}{2\hbar\omega K}}\,(\psi_{k-1})^{\mathrm{II}} + \cdots,$$

$$(\psi_{k+1})^{\mathrm{I}} = (\psi_{k+1})^{\mathrm{II}} + \cdots,$$

$$(\psi_{k-1})^{\mathrm{I}} = (\psi_{k-1})^{\mathrm{II}} + \cdots. \tag{11–64}$$

Substituting Eqs. (11–64) in Eq. (11–63) we have

$$\Psi(x, \Delta T) = e^{-(i/\hbar)(E_k)^{\mathrm{I}}\,\Delta T}(\psi_k)^{\mathrm{II}}$$

$$- q\,\Delta F\sqrt{\frac{k+1}{2\hbar\omega K}}\,(e^{-(i/\hbar)(E_k)^{\mathrm{I}}\,\Delta T} + e^{-(i/\hbar)(E_{k+1})^{\mathrm{I}}\,\Delta T})(\psi_{k+1})^{\mathrm{II}}$$

$$+ q\,\Delta F\sqrt{\frac{k}{2\hbar\omega K}}\,(e^{-(i/\hbar)(E_k)^{\mathrm{I}}\,\Delta T} + e^{-(i/\hbar)(E_{k-1})^{\mathrm{I}}\,\Delta T})(\psi_{k-1})^{\mathrm{II}} + \cdots. \tag{11–65}$$

Thus the time-dependent wave function in the period $\Delta T \le t \le 2\,\Delta T$ is

$$\Psi(x, t) = e^{-(i/\hbar)(E_k)^{\mathrm{I}}\,\Delta T}(\psi_k)^{\mathrm{II}}e^{-(i/\hbar)(E_k)^{\mathrm{II}}(t-\Delta T)}$$

$$- q\,\Delta F\sqrt{\frac{k+1}{2\hbar\omega K}}\,(e^{-(i/\hbar)(E_k)^{\mathrm{I}}\,\Delta T} + e^{-(i/\hbar)(E_{k+1})^{\mathrm{I}}\,\Delta T})$$

$$\times\,(\psi_{k+1})^{\mathrm{II}}e^{-(i/\hbar)(E_{k+1})^{\mathrm{II}}(t-\Delta t)}$$

$$+ q\,\Delta F\sqrt{\frac{k}{2\hbar\omega K}}\,(e^{-(i/\hbar)(E_k)^{\mathrm{I}}\,\Delta T} + e^{-(i/\hbar)(E_{k-1})^{\mathrm{I}}\,\Delta T})$$

$$\times\,(\psi_{k-1})^{\mathrm{II}}e^{-(i/\hbar)(E_{k-1})^{\mathrm{II}}(t-\Delta T)},$$

$$\Delta T \le t \le 2\,\Delta T. \tag{11–66}$$

At $t = 2\,\Delta T$, the wave function is given by

$$\Psi(x, 2\,\Delta T) = e^{-(i/\hbar)[(E_k)^{\mathrm{I}}+(E_k)^{\mathrm{II}}]\,\Delta T}(\psi_k)^{\mathrm{II}}$$

$$- q\,\Delta F\sqrt{\frac{k+1}{2\hbar\omega K}}\,(e^{-(i/\hbar)(E_k)^{\mathrm{I}}\,\Delta T} + e^{-(i/\hbar)(E_{k-1})^{\mathrm{I}}\,\Delta T})e^{-(i/\hbar)(E_{k+1})^{\mathrm{II}}\,\Delta T}(\psi_{k+1})^{\mathrm{II}}$$

$$+ q\,\Delta F\sqrt{\frac{k}{2\hbar\omega K}}\,(e^{-(i/\hbar)(E_k)^{\mathrm{I}}\,\Delta T} + e^{-(i/\hbar)(E_{k-1})^{\mathrm{I}}\,\Delta T})e^{-(i/\hbar)(E_{k-1})^{\mathrm{II}}\,\Delta T}(\psi_{k-1})^{\mathrm{II}}. \tag{11–67}$$

This procedure may be repeated to obtain $\Psi(x, 3\,\Delta T)$, $\Psi(x, 4\,\Delta T)$, etc.

Finally

$$\Psi(x, N\,\Delta T) = e^{-(i/\hbar)[(E_k)^{\mathrm{I}}+(E_k)^{\mathrm{II}}+\cdots+(E_k)^N]\Delta T}(\psi_k)^N$$

$$-\,q\,\Delta F\,\sqrt{\frac{k+1}{2\hbar\omega K}}\left(\sum_{M=1}^{N}e^{-(i/\hbar)[\sum_{S=1}^{M-1}(E_k)^S+\sum_{S=M}^{N}(E_{k+1})^S]\Delta T}\right)(\psi_{k+1})^N$$

$$+\,q\,\Delta F\,\sqrt{\frac{k}{2\hbar\omega K}}\left(\sum_{M=1}^{N}e^{-(i/\hbar)[\sum_{S=1}^{M-1}(E_k)^S+\sum_{S=M}^{N}(E_{k-1})^S]\Delta T}\right)(\psi_{k-1})^N.$$

$$(11\text{–}68)$$

Assume that after N steps the field reaches its full value F, i.e.,

$$N\,\Delta F = F,$$
$$N\,\Delta T = T.$$

$$(11\text{–}69)$$

Equation (11–68) thus gives us the probabilities of finding the energy value equal to eigenvalues $(E_k)^N$, $(E_{k+1})^N$, and $(E_{k-1})^N$. Let us examine the coefficients of $(\psi_{k+1})^N$ and $(\psi_{k-1})^N$. Inside the parentheses of each coefficient there is a summation of complex numbers all of modulus unity. Under the assumption of adiabatic approximation, ΔT is large enough so that the phases of these complex numbers are completely at random. Consequently, the sum of these complex numbers is zero and Eq. (11–68) reduces to one term $(\psi_k)^N$,

$$\Psi(x, T) = e^{-(i/\hbar)\bar{E}_k T}(\psi_k)^N,$$

$$(11\text{–}70)$$

where $\bar{E}_k$ is the average of $E_k^{\mathrm{I}}, E_k^{\mathrm{II}}, \ldots, E_k^N$. As a result,

$$\Psi(x, t) = e^{-(i/\hbar)\bar{E}_k T}e^{-(i/\hbar)(E_k)^N(t-T)}(\psi_k)^N, \quad \text{for } t \geq T. \quad (11\text{–}71)$$

Thus the quantum state for $t \geq T$ after the field has reached its full value is an eigenstate $(\psi_k)^N$ and has a definite energy value $(E_k)^N$. This conclusion may be stated differently as follows: in adiabatic approximation a quantum state, if started as an energy eigenstate $\psi_k^{(0)}$, remains an energy eigenstate instantaneously. This is different from the sudden approximation in which the quantum state, after the field is switched on suddenly, is no longer an energy eigenstate. In the adiabatic approximation, if the field is switched off gradually the quantum state $(\psi_k)^N$ will gradually change back to $\psi_k^{(0)}$, remaining an eigenstate all the time. Thus when a field is switched on and then off very slowly, the quantum state, after some transitory changes in response to the perturbation, returns to the initial state. As a result, there is to be no energy exchange between the oscillator and the field (hence the term adiabatic approximation). In the sudden approximation, energy exchange does take place. That there is no energy exchange in adiabatic approximation may also be understood by a classical analogy.   If the field is nearly constant throughout one period of vibration of the classical oscillator, then the work done on the oscillator when it moves in the $+x$ direction is compensated for by that in the $-x$ direction.   Therefore the field adds no energy to the system. From the viewpoint of the motion of the probability elements, we may also reach the same conclusion. Since the system remains in an eigenstate instantaneously, it is stationary and so does not exchange energy with the field.

Equation (11–68) reduces to the result of the sudden approximation if we set $T = 0$ for a quick switch-on of the field.   From Eq. (11–69) we have $\Delta T = 0$. The quantities in each of the two parentheses in Eq. (11–68) thus reduce to a sum of complex numbers all having the same phase, i.e., zero.   Instead of destructive interference among the complex numbers, as in the case of the adiabatic approximation, we now have reinforcement. The sum is simply N. Since $N \Delta F$ equals F, Eq. (11–68) reduces to Eq. (11–51), as it should.

It will be left to the reader to consider the case in which the field is switched on gradually but is turned off suddenly; also the case in which the field is switched on quickly but is turned off slowly (Problem 11–10).

We have considered as an example of the sudden approximation the ionization of gas atoms or molecules by a fast-moving charged particle. If the charged particle moves so slowly that the electrons complete a large number of revolutions before the charged particle moves over an appreciable distance in their immediate neighborhood, then the adiabatic approximation will be valid. The electron wave functions (or orbits) will undergo transitory distortions in response to the presence of the charged particle in the neighborhood of the atom but will return to the original state when the charged particle has moved far away.   The slow-moving charged particle will produce no permanent effect on the atoms or molecules, unlike the fast-moving one which causes permanent change by inducing excitation and ionization.

The adiabatic approximation has an application in molecular physics. Consider a diatomic molecule made of atoms A and B. Electrons in atoms A and B are arranged in "orbits" described by wave functions. These wave functions are dependent on the interatomic distance of the molecule.   As the molecule vibrates and rotates, the electronic wave functions change accordingly. Because the nuclear mass is much greater than the electron mass, the period of the nuclear motion (i.e., the vibration and rotation of the molecule) is much longer than that of the electronic motion. Thus the disturbance of electronic motion due to nuclear motion may be treated by adiabatic approximation. We may speak of the electrons being in certain eigenstates with definite energy eigenvalues while the vibrational and rotational motion of the molecule is going on. Actually the electronic orbits suffer transitory but reversible distortions in response to the vibration and rotation, and the energy eigenvalues vary accordingly. These variations are periodic and are in phase with the vibration and rotation. They are so small that in the long run we may consider them by their time averages. The molecular problem may thus be treated rather simply. The adiabatic approximation is also applied in nuclear physics to treat the collective motion of the nucleons. The nucleonic states and rotational states of a nucleus are analogous to the electronic states and rotational states of a molecule. According to the adiabatic approximation the nucleonic states adjust themselves to follow the nuclear rotation without losing their identity as nucleonic states. The nuclear energy levels may thus be analyzed just as the analysis of molecular energy levels.

The sudden and the adiabatic approximations represent two extreme cases. Practical problems may be solved approximately, more or less, by one of the two methods. The conditions for the validity of the sudden and adiabatic approximations are actually the conditions of reinforcement and interference of the wave functions summed over in Eq. (11–68). The phase difference between successive terms in Eq. (11–68) is $(i/\hbar) \times (\overline{E}_k - \overline{E}_{k-1}) \Delta T$. If the change of phase after N terms is much greater than 2π, then the N phases will be completely at random and the destructive interference will be nearly complete. Thus the condition for the adiabatic approximation to be valid is

$$\frac{i}{\hbar}(\overline{E}_k - \overline{E}_{k-1})N \Delta T \gg 2\pi i \qquad (11\text{–}72)$$

or

$$T \gg \frac{h}{\overline{E}_k - \overline{E}_{k-1}}. \qquad (11\text{–}73)$$

Similarly, the condition for the sudden approximation is

$$\frac{i}{\hbar}(\overline{E}_k - \overline{E}_{k-1})N \Delta T \ll 2\pi i \qquad (11\text{–}74)$$

or

$$T \ll \frac{h}{\overline{E}_k - \overline{E}_{k-1}}. \qquad (11\text{–}75)$$

The right-hand side of Eqs. (11–73) and (11–75) is the period of oscillation of the oscillator. Thus we justify the previous statements concerning the condition of validity of these two approximations.

Let us apply this condition to the ionization of an atom by a charged particle. The field experienced by the atom is schematically represented in Fig. 11-7. The field is appreciable when the charged particle is within

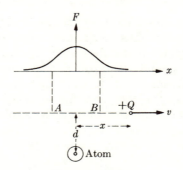

FIGURE 11–7

the line segment AB, the length of which is comparable to the impact parameter d. Let the velocity of the charged particle be v; we have

$$T \approx \frac{d}{v}.$$

Now, the denominator in Eq. (11–73) is actually the ionization potential E_i. The condition for the adiabatic approximation to be valid is thus

$$\frac{d}{v} \gg \frac{h}{E_i} \tag{11–76}$$

or

$$d \gg \frac{h}{E_i} v.$$

Consequently, for impact parameters d much greater than $(h/E_i)v$, there will be no ionization. For this reason we may regard the quantity $(h/E_i)v$ as a measure of the radius of influence of the ionizing particle. For slow-moving particles, v is small and the sphere of influence is small. Thus few ions will be produced. On the other hand, within the sphere of influence where the sudden approximation is valid, the probability of inducing transition depends on the length of time the perturbing field is present, which is also of the order of d/v. When v is so large that d/v is much smaller than (h/E_i), there is little time for the transition probability to build up to an appreciable amount, as may be seen in Fig. 11–1. Thus very fast charged particles again produce little ionization. Between these two extreme cases there exists a velocity at which maximum ionization is produced. This happens when the velocity of the charged particle is comparable with the velocity of the electron inside the atom.

11–3 Transition to continuous states. The perturbation theory developed in the previous sections applies to systems with discrete energy levels. Yet we have seen in Chapter 3 (the free particle) and Section 8–3 physical systems with continuous energy levels. Problems involving continuous energy levels may also be treated by the perturbation method. The *Rutherford scattering* process, i.e., the scattering of an α-particle by the Coulomb field of an atomic nucleus, may be treated as a perturbation problem; the Coulomb potential is considered as a perturbation which induces transitions of the unperturbed states—the free particle states. For problems of this kind, we need a perturbation theory dealing with continuous levels. The physical principles of this theory are the same as those in the previous sections. Nevertheless, special mathematical techniques are required.

Since in a large number of cases the unperturbed wave equation is the equation of the free particle, the results of Chapter 3 will be reviewed here for later use. The eigenfunctions, given by Eq. (3–8), are plane waves,

$$\psi_E(x, y, z) = e^{(i/\hbar)(p_x x + p_y y + p_z z)}, \tag{11–77}$$

where p_x, p_y, p_z are three integration constants identified with the components of momentum. The energy eigenvalues E are related to p_x, p_y, p_z by Eq. (3–9):

$$E = \frac{1}{2M} (p_x^2 + p_y^2 + p_z^2). \tag{11–78}$$

E may take on any positive value and the energy is not quantized. In Chapter 3 we discussed the normalization of these eigenfunctions. It is accomplished by confining the free particle in a large but finite volume Ω. The normalized wave functions are given by

$$\psi_E(x, y, z) = \frac{1}{\sqrt{\Omega}} e^{(i/\hbar)(p_x x + p_y y + p_z z)}. \tag{11–79}$$

We now consider this normalization procedure from a different angle. Let us start with a different problem, i.e., a particle in a square box, which was discussed at the end of Section 8–2. Assume the lengths of the three sides of the box to be the same, say a. Equation (8–69) gives the normalized eigenfunctions as follows:

$$\psi_{n_x n_y n_z}(x, y, z) = \frac{1}{(a/2)^3} \sin \frac{n_x \pi}{a} x \sin \frac{n_y \pi}{a} y \sin \frac{n_z \pi}{a} z, \tag{11–80}$$

and Eq. (8–70) gives the eigenvalues of energy

$$E = \frac{\hbar^2 \pi^2}{2Ma^2} (n_x^2 + n_y^2 + n_z^2), \qquad \left.\begin{matrix} n_x \\ n_y \\ n_z \end{matrix}\right\} = 1, 2, 3, \ldots. \tag{11–81}$$

In this problem the energy is quantized. Now let the value of a increase.

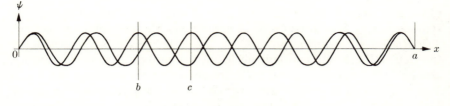

FIGURE 11-8

As a approaches infinity, this problem approaches that of a free particle and the eigenvalues and eigenfunctions of this problem approach those of the free particle. Thus we may regard the solution of the free particle as a limiting case of that of a particle in a box. Consider the limit of the energy eigenvalues. Since the *level spacing*, i.e., the energy difference between two successive levels, is inversely proportional to a^2, it decreases rapidly as a increases. When a approaches infinity the levels become so closely spaced that they may be regarded as forming a continuous band. This result agrees with the fact that the energy of a free particle is not quantized. Consider next the limit of the eigenfunctions. As the size of the box approaches infinity, Eq. (11-80) may be considered to be equivalent to Eq. (11-79) in the following sense. It is always possible to choose from the numerous solutions of Eq. (11-80) two waves of nearly the same wavelengths such that, within a finite range, say, $b < x < c$ (consider the one-dimensional case for simplicity), these two waves seem to have the same wavelength but differ in phase by a constant amount $\pi/2$. A graphical illustration is provided in Fig. 11-8. These two waves may be regarded as a sine wave and a cosine wave having equal wavelengths so far as the region $b < x < c$ is concerned. A linear combination of them gives rise to the exponential wave function of Eq. (11-79). In practical applications, wave functions are used to evaluate matrix elements of the perturbation potential. Since the perturbation potential usually is appreciable only in a limited region in space, say, $b < x < c$, the exponential wave functions are equivalent to the trigonometric wave functions so far as the matrix element is concerned. The change from the sine functions to exponential functions may be regarded as a change of the base functions of a manifold of degenerate eigenfunctions, the physical results being the same whichever set of base is used for calculation. Having shown the identity of the two sets of functions we may use Eqs. (11-80) and (11-81) (with $a \to \infty$) to represent the eigenfunctions and eigenvalues of the free particle. Since the energy of a particle in a box is quantized, we may apply the theory of the previous sections, which is developed for quantized energy levels. After having applied the perturbation theory, we let the size of the box approach infinity. The limiting result thus obtained is the

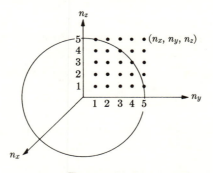

FIGURE 11–9

result for continuous energy levels. This is the general procedure of the perturbation theory for continuous energy levels.

When the size of the box is large but not infinite, the level spacings are small but finite. Thus within a small energy interval ΔE there exist a large number of energy levels. Let the number of levels be ΔN. When the level spacing is much smaller than ΔE, we may define a level density function $\rho(E)$ as follows:

$$\rho(E) = \lim_{\Delta E \to 0} \frac{\Delta N}{\Delta E}. \tag{11–82}$$

$\rho(E)$ is considered as a continuous function. For a particle confined in a square box of length a, the level density may be calculated as follows. Let us draw a set of coordinate axes representing the three quantum numbers n_x, n_y, n_z (Fig. 11–9). Each combination of the three quantum numbers will be represented by a point in this space. Since n_x, n_y, n_z are all positive integers, the representative points form a cubic lattice in the first octant with unit cell $(1, 1, 1)$. Let a sphere of radius R be drawn. A lattice point on this sphere satisfies the equation

$$R^2 = n_x^2 + n_y^2 + n_z^2. \tag{11–83}$$

This point represents an energy level the energy value of which is, according to Eq. (11–81),

$$E = \frac{\hbar^2 \pi^2 R^2}{2Ma^2}. \tag{11–84}$$

The total number of lattice points within this sphere is simply the volume of the sphere in the first octant divided by the volume of the unit cell, the result being

$$\frac{1}{8} \frac{4\pi}{3} R^3.$$

Since the value of $(n_x^2 + n_y^2 + n_z^2)$ of all these points is less than R^2, the corresponding states have energy values less than that given by Eq. (11–84). Thus the total number of states with energy value less than E is

$$\frac{1}{8}\,\frac{4\pi}{3}\left(\frac{2MEa^2}{\hbar^2\pi^2}\right)^{3/2}.$$

This represents the integral of the level density function $\rho(E)$. Therefore $\rho(E)$ may be obtained by differentiation:

$$\rho(E) = \frac{d}{dE}\left[\frac{1}{8}\,\frac{4\pi}{3}\left(\frac{2MEa^2}{\hbar^2\pi^2}\right)^{3/2}\right] = \frac{1}{4}\,\frac{a^3(2M)^{3/2}}{\hbar^3\pi^2}\,E^{1/2} = \frac{1}{4}\,\frac{\Omega(2M)^{3/2}}{\hbar^3\pi^2}\,E^{1/2},$$

(11–85)

where Ω, being the volume of the box, is given by

$$\Omega = a^3. \tag{11–86}$$

Another useful expression may be written here:

$$\rho(E) = \frac{d}{dE}\left(\frac{\pi}{6}\,\frac{p^3 a^3}{\hbar^3\pi^3}\right) = \frac{\pi a^3}{2\hbar^3\pi^3}\,p^2\,\frac{dp}{dE} = \frac{\Omega}{2\hbar^3\pi^2}\,\frac{p^2}{v}, \quad (11–87)$$

where p and v are the momentum and velocity of the particle respectively. From this equation, we derive the following:

$$\rho(E)\,dE = \frac{1}{h^3}\,\Omega 4\pi p^2\,dp. \tag{11–88}$$

Since Ω is the volume of the *configuration space** and $4\pi p^2\,dp$ is the volume of the *momentum space*† corresponding to the momentum range dp, the product of the two volumes $\Omega 4\pi p^2\,dp$, by definition, is the volume of the corresponding *phase space*.‡ The above equation may thus be interpreted as meaning that each volume h^3 in phase space is represented by one energy level. We digress for a moment to mention that the validity of this result is not limited to the free particle. In fact this result is taken

---

* This term is used for the physical space extended by the coordinates x, y, z, to distinguish it from the mathematical spaces such as the momentum space and phase space.

† The momentum space is defined as a three-dimensional space the three coordinates of which are p_x, p_y, p_z, the components of momentum.

‡ The phase space is defined as a six-dimensional space, the coordinates being x, y, z, p_x, p_y, p_z. This definition is a generalization of that of the phase space of the one-dimensional motion discussed in Chapter 1.

as a fundamental assumption in quantum statistical mechanics. The level density expressions derived here will be used in later applications.

Since the volume Ω in which we confine the particle is eventually to be made to approach infinity, it may be useful to summarize here the dependence on Ω of many important quantities. The exponential wave functions given by Eq. (11–79) are proportional to $\Omega^{-1/2}$, as the normalization constants are $\Omega^{-1/2}$. The level density varies as Ω. The matrix elements, containing two wave functions, vary as Ω^{-1}. Thus a single matrix element is infinitesimal in magnitude. However, the summation of the matrix elements of all levels in a finite energy range is independent of Ω because the level density varies as Ω; such a summation is thus finite.

Now we are ready to develop the perturbation theory for continuous levels. Our purpose is to solve the perturbed equation

$$(H^{(0)} + \lambda H^{(1)})\Psi = -\frac{\hbar}{i}\frac{\partial \Psi}{\partial t} \tag{11–89}$$

by the knowledge of the solutions of the unperturbed equation

$$H^{(0)}\psi^{(0)} = E^{(0)}\psi^{(0)}, \tag{11–90}$$

which has continuous energy eigenvalues. Consider the case in which the unperturbed equation is the free particle equation. For this case we first confine the particle in a large volume Ω. The energy is then quantized, although the levels are very close together. The perturbation theory of quantized energy levels of Section 11–1 may thus be applied. The solution of Eq. (11–89) may be assumed to take the form

$$\Psi(x, y, z, t) = \sum_n \alpha_n(t)e^{-(i/\hbar)E_n^{(0)}t}\psi_n^{(0)}, \tag{11–91}$$

where the undetermined parameters $\alpha_n(t)$ are functions of time and may be broken down in different orders of the perturbation parameter λ. Thus

$$\alpha_n(t) = \alpha_n^{(0)}(t) + \lambda\alpha_n^{(1)}(t) + \lambda^2\alpha_n^{(2)}(t) + \cdots. \tag{11–92}$$

The initial condition is assumed to be that at $t = 0$, $\Psi(x, y, z, 0) = \psi_k^{(0)}$, i.e.,

$$\alpha_n(0) = \delta_{nk}, \qquad n = 1, 2, \ldots, k, \ldots. \tag{11–93}$$

The first-order solution is given by Eq. (11–31),

$$\alpha_m^{(1)}(t) = -\frac{1}{\hbar} H_{mk}^{(1)} \frac{e^{i\omega_{mk}t} - 1}{\omega_{mk}}. \tag{11–94}$$

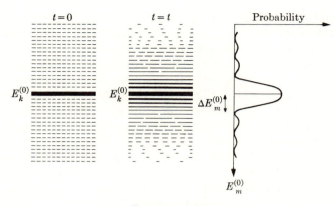

FIGURE 11-10

Equation (11–33) gives us the relative probability of finding the particle in the state $\psi_m^{(0)}$ at time t, which is

$$|\lambda\alpha_m^{(1)}(t)|^2 = \frac{4\lambda^2|H_{mk}^{(1)}|^2 \sin^2\left(\frac{1}{2}\omega_{mk}t\right)}{\hbar^2\omega_{mk}^2}.\tag{11-95}$$

Because of the factor $1/\omega_{mk}^2$, the probability is largest for those states $\psi_m^{(0)}$ in which the unperturbed energies $E_m^{(0)}$ are closest to $E_k^{(0)}$. If $E_m^{(0)}$ is far away from $E_k^{(0)}$, the probability is small. Since the levels are very close to one another, there are a large number of levels having an appreciable probability of occurring at time t (see Fig. 11–10). These levels form a cluster around $E_m^{(0)} = E_k^{(0)}$ with a range of $\pm\Delta E_m^{(0)}$. As the levels almost form a continuous band, the physical properties of a level may be regarded as continuously varying from one level to the other. Since all levels in the cluster represent nearly the same physical properties, we may add up the probabilities of all levels in the cluster and consider them as a whole. The total probability thus obtained which depends on time t is called the *transition probability* at time t from the initial state $\psi_k^{(0)}$ to the cluster of levels. The summation may be evaluated by the following integral in which $\rho(E)\,dE$ is the number of levels in the infinitesimal energy range dE within which the matrix elements of all levels may be regarded as the same:

$$\sum_{m\neq k}|\lambda\alpha_m^{(1)}(t)|^2 = \int_{E_k^{(0)}-\Delta E_m^{(0)}}^{E_k^{(0)}+\Delta E_m^{(0)}} \frac{4\lambda^2|H_{mk}^{(1)}|^2}{\hbar^2}\frac{\sin^2\left(\frac{1}{2}\omega_{mk}t\right)}{\omega_{mk}^2}\rho(E)\,dE.\tag{11-96}$$

Since the largest contribution to the integral comes from the levels for which $E_m^{(0)} \cong E_k^{(0)}$, we may replace $H_{mk}^{(1)}$ and $\rho(E)$ by their values at

$E = E_k^{(0)}$ and take them out of the integral without affecting to a large extent the value of the integral. Also we may extend the limits of integration to $-\infty$ and $+\infty$, since only negligibly small amounts will be added by this extension. Let $\psi_{k'}^{(0)}$ be a typical level in the cluster and $E_{k'}^{(0)} = E_k^{(0)}$. We have

$$\sum_{m \neq k} |\lambda \alpha_m^{(1)}(t)|^2 = \frac{4\lambda^2 |H_{k'k}^{(1)}|^2}{\hbar^2} \rho(E_k^{(0)}) \int_{-\infty}^{\infty} \frac{\sin^2 \left(\frac{1}{2}\omega_{mk}t\right)}{\omega_{mk}^2} \, dE$$

$$= \frac{\lambda^2}{\hbar} |H_{k'k}^{(1)}|^2 \rho(E_k^{(0)}) \int_{-\infty}^{\infty} \frac{\sin^2 \left(\frac{1}{2}\omega_{mk}t\right)}{(\omega_{mk}/2)^2} \, d\omega_{mk}$$

$$= \frac{2\pi}{\hbar} |\lambda H_{k'k}^{(1)}|^2 \rho(E_k^{(0)})t. \tag{11–97}$$

The last step is rendered by using the following definite integral:*

$$\int_{-\infty}^{\infty} \frac{\sin^2 x}{x^2} \, dx = \pi. \tag{11–98}$$

The important result is that the transition probability to the cluster (designated by k') is proportional to time t. This is similar to the law of radioactive decay, and we may define the transition probability per unit time $\Lambda_{k \to k'}$ and the mean lifetime $\tau_{k \to k'}$ analogously. Thus

$$\Lambda_{k \to k'} = \frac{\sum_{m \neq k} |\lambda \alpha_m^{(1)}(t)|^2}{t} = \frac{2\pi}{\hbar} |\lambda H_{k'k}^{(1)}|^2 \rho(E_k^{(0)}), \quad E_{k'}^{(0)} = E_k^{(0)}, \tag{11–99}$$

and

$$\frac{1}{\tau_{k \to k'}} = \frac{2\pi}{\hbar} |\lambda H_{k'k}^{(1)}|^2 \rho(E_k^{(0)}), \quad E_{k'}^{(0)} = E_k^{(0)} \quad \text{(first order).} \tag{11–100}$$

The last equation or its equivalent is one of the most frequently used formulas in quantum mechanics. It gives the lifetime of a state $\psi_k^{(0)}$ against transition into a cluster of levels designated by k'. It may be remarked that the above summation method needs modification for physical systems where the matrix element depends not only on the unperturbed energies $E_k^{(0)}$ and $E_m^{(0)}$ but also on some other variables such as θ, φ representing the angular orientation in space. In such cases we first sort out in the cluster a subcluster of levels with fixed values of θ, φ, etc., in which $E_m^{(0)}$ is the only variable. The above summation method may thus be applied to the subcluster. The lifetime $\tau_{k \to k''}$ obtained by summation over the subcluster is the lifetime against transition into this particular subcluster

---

* Peirce's *Table of Integrals*, p. 63.

of levels designated by k'' (k'' specifies the physical properties of the sub-cluster of levels including θ, φ, etc.). The total lifetime τ_k of a state $\psi_k^{(0)}$ may be calculated by combining these partial lifetimes $\tau_{k \to k''}$. Since the transition probabilities are additive, we obtain the total transition probability of a state $\psi_k^{(0)}$ as follows:

$$\Lambda_k = \sum_{k''} \Lambda_{k \to k''}. \tag{11-101}$$

Hence the total lifetime is

$$\frac{1}{\tau_k} = \sum_{k''} \frac{1}{\tau_{k \to k''}}. \tag{11-102}$$

These results are analogous to those in some radioactive decay processes in which a number of decay modes are possible. The application of the results obtained here will be illustrated by a discussion of the Rutherford scattering process in the next section.

We note that the transition probability depends on the normalization volume Ω. As the matrix element is proportional to $1/\Omega$ and the level density is proportional to Ω, the transition probability per unit time is proportional to $1/\Omega$. However, as we shall see in the next section, the physically significant quantity, e.g., the scattering cross section, calculated from the transition probability, is independent of Ω and thus remains the same when we let Ω approach infinity. It is this result that justifies the normalization procedure of confining the particle in a finite volume Ω for the unperturbed eigenfunctions with continuous energy eigenvalues. For those cases in which the unperturbed energy is quantized but the levels are very closely spaced, the perturbation theory developed in this section is readily applicable, there being no need of introducing the normalization volume Ω.

A remark may be added here concerning the cluster of levels designated by k'. The $E_m^{(0)}$ values of these levels are close to $E_k^{(0)}$ but are spreading out about $E_k^{(0)}$ with a width $\Delta E_m^{(0)}$ (see Fig. 11–10). The width may be calculated from Eq. (11–95) as follows. The first minimum in Fig. 11–10 occurs when in Eq. (11–95)

$$\tfrac{1}{2}\omega_{mk}t = \pi \tag{11-103}$$

or

$$\left(\frac{\Delta E_m^{(0)}}{\hbar}\right) t \cong 2\pi \tag{11-104}$$

or

$$\Delta E_m^{(0)} t \cong h. \tag{11-105}$$

Thus the energy width of the cluster is inversely proportional to time t. This result may be interpreted in a different manner as follows: Suppose we switch on a perturbation potential at $t = 0$ and later switch it off at $t = t$. Then a physical system originally represented by a pure state $\psi_k^{(0)}$ before $t = 0$ is to be represented by a mixed state $\sum_m \psi_m^{(0)}$ after $t = t$. Since before $t = 0$ and after $t = t$ the functions $\psi_n^{(0)}$ represent energy eigenstates, the perturbation changes the energy information of the system from a definite value $E_k^{(0)}$ before $t = 0$ to a probability distribution of $E_m^{(0)}$ after $t = t$. According to Eq. (11–105) the width of this probability distribution $\Delta E_m^{(0)}$ and the time duration of the perturbation t are inversely proportional to each other and their product is the Planck constant h. The longer the duration of the perturbation, the more well-defined is the energy of the system. In experiments we used to prepare a system with definite properties by a set of physical operations. For example, slits, diaphragms, electric fields, magnetic fields, etc., may be used to prepare a beam with definite kinematical and dynamical properties. Each operation is actually a perturbation on the beam being prepared. The above result thus means that *to prepare a system to possess a well-defined energy value must take an infinitely long time.* This statement, as well as Eq. (11–105), resembles the statement of Eq. (6–10) concerning the uncertainty relation between energy and time. Actually this result may be shown to be a manifestation of the uncertainty relation as follows: Let us consider a specific experimental set-up, the Dempster mass spectrograph. A stream of particles (positive ions) produced by a hot filament pass through an accelerating field in a chamber and enter another chamber, through a slit in a diaphragm, where a magnetic field is applied to deflect the particles. We are interested in the properties of the beam in the second chamber which are determined by the source and the slit. The slit is assumed to be closed before $t = 0$. After an exposure of t' seconds, the slit is closed again. If the slit remained closed all the time, there would be no positive ions in the deflection chamber. But the opening of the slit from $t = 0$ to $t = t'$ has the effect that after $t = t'$ there will be particles present in the deflection chamber. The wave function describing these particles in the deflection chamber is generated by the perturbation, i.e., the opening of the slit from $t = 0$ to $t = t'$, according to Eq. (11–95). That the opening of a slit may be considered as the switching-on of a perturbation potential may be explained as follows: When the slit is closed, the whole diaphragm may be considered as a surface over which the potential is infinitely large so that it acts as an impenetrable barrier. The opening of the slit may be regarded as altering the potential in such a way that its value over the slit is reduced to zero. The closing of the slit restores the potential to the original, infinite value. Therefore the opening and closing of the slit is equivalent to switching a perturbing potential on and off. Now let us assume that the stream of particles in the first chamber forms a uniform beam with definite momentum and energy described by a monochromatic plane wave. If the exposure time t is short, the slit allows a finite wave train or a wave packet to pass. For a wave packet the uncertainty relations between p_x and x, and between E and t have been established in Chapters 2 and 6. It was shown that a short wave packet contains a large number of Fourier components corresponding to a wide spread in energy and momentum values. In order that the energy and momentum have well-defined values, the wave train, containing only one Fourier component, must be infinitely long. This requires an infinitely long exposure time. This problem may also be discussed by using a different mathematical method, namely, the time-dependent perturbation method. The conclusion according to Eq. (11–105) is that the wave in the deflection chamber, given by Eq. (11–95), is to have a less spread in energy when the time of exposure t is long. Since the two lines of approach deal with the same phenomenon, their underlying physical principles must be the same. Therefore Eq. (11–105) is a manifestation of the uncertainty relation.

As a special case let us consider Eq. (11–105) for the time $t = \tau_{k \to k'}$. At this time the transition probability is unity and a transition to the cluster k' is likely to have happened. The spread of $E_m^{(0)}$ of the final states in the cluster, i.e., $\Delta E_m^{(0)}$, is given by

$$\Delta E_m^{(0)} \cong \frac{h}{\tau_{k \to k'}}. \tag{11–106}$$

This is a frequently used formula relating the "energy" spread of the final states to the lifetime of the initial state. In case there are several partial lifetimes, the "energy" spread of any subcluster k'' has to be measured at the time $t = \tau_k$, where τ_k is the total lifetime. Since $\tau_k \leq \tau_{k \to k''}$ the energy spread becomes larger. Thus the opening-up of additional decay channels widens the "energy" spread of the final states.

Having discussed the first-order perturbation, we now proceed to the second-order perturbation. It is not unusual that the matrix element in Eq. (11–100) connecting states of nearly the same unperturbed energy may vanish. Under such circumstances no transition corresponding to the first-order perturbation may take place; the second-order perturbation becomes important. Equation (11–36) states the result of the second-order perturbation for discrete energy levels. It is reproduced here:

$$\alpha_m^{(2)}(t) = \frac{1}{\hbar^2} \sum_n \frac{H_{mn}^{(1)} H_{nk}^{(1)}}{\omega_{nk}} \left(\frac{e^{i\omega_{mk}t} - 1}{\omega_{mk}} - \frac{e^{i\omega_{mn}t} - 1}{\omega_{mn}} \right). \tag{11–107}$$

The parentheses may be regarded as a weight function for the summation. When the normalization volume approaches infinity, the levels become more closely spaced, $H_{mn}^{(1)} H_{nk}^{(1)}/\omega_{nk}$ approaches a continuous function of n, and $(e^{i\omega_{mn}t} - 1)/\omega_{mn}$ approaches a rapidly oscillating function of n. The weighted summation

$$\sum_n \frac{H_{mn}^{(1)}H_{nk}^{(1)}}{\omega_{nk}} \left(\frac{e^{i\omega_{mn}t} - 1}{\omega_{mn}} \right)$$

then approaches zero. Note that $H_{mn}^{(1)}H_{nk}^{(1)}/\omega_{nk}$ remains finite in spite of the denominator, since we assume that matrix elements connecting $\psi_k^{(0)}$ with states of nearly the same unperturbed energy all vanish. Therefore we may drop the second term in the parentheses (see also Problem 11–7). As a result,

$$\alpha_m^{(2)}(t) = \frac{1}{\hbar^2} \left(\sum_n \frac{H_{mn}^{(1)}H_{nk}^{(1)}}{\omega_{nk}} \right) \frac{e^{i\omega_{mk}t} - 1}{\omega_{mk}}$$

$$= -\frac{1}{\hbar} \left(\sum_n \frac{H_{mn}^{(1)}H_{nk}^{(1)}}{E_k^{(0)} - E_n^{(0)}} \right) \frac{e^{i\omega_{mk}t} - 1}{\omega_{mk}}. \qquad (11\text{--}108)$$

The above equation becomes identical with Eq. (11–94) if in the latter we make the substitution

$$H_{mk}^{(1)} \rightarrow \sum_n \frac{H_{mn}^{(1)}H_{nk}^{(1)}}{E_k^{(0)} - E_n^{(0)}}. \qquad (11\text{--}109)$$

As a result all equations and conclusions following Eq. (11–94), except for the order of λ, may be taken over for the second-order perturbation provided the above substitution is assumed. In particular, the second-order perturbation causes a transition to take place leading from $\psi_k^{(0)}$ to a cluster of levels with $E_m^{(0)}$ values close to $E_k^{(0)}$. The transition probability per unit time to a cluster k' is

$$\frac{1}{\tau_{k \to k'}} = \frac{2\pi}{\hbar} \left| \sum_n \frac{\lambda H_{k'n}^{(1)} \lambda H_{nk}^{(1)}}{E_k^{(0)} - E_n^{(0)}} \right|^2 \rho(E_k^{(0)}), \quad E_{k'}^{(0)} = E_k^{(0)} \text{ (second order)}.$$

$$(11\text{--}110)$$

Equation (11–109) is a convenient mnemonic aid as it applies not only to the time-dependent perturbation theory but also to the time-independent perturbation theory; the latter may be verified by comparing Eqs. (10–32) and (10–33) with Eqs. (10–30) and (10–31). [*Note:* In Eq. (10–33) $H_{nn}^{(1)}$ is assumed to be zero.]

11–4 The Rutherford scattering. Rutherford's experiments on the scattering of α-particles by various materials are the foundation of the modern atomic theory. They established the concept of the nuclear atom and determined the size and charge of the nucleus. We shall first derive the Rutherford scattering equation according to classical mechanics as

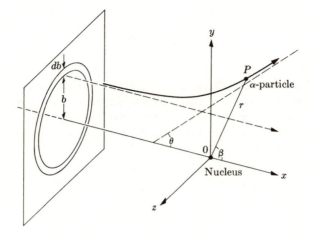

FIGURE 11-11

Rutherford did. Then the same equation will be derived according to quantum mechanics by using the time-dependent perturbation method. Finally the time-independent perturbation method will be used for the same purpose as a mathematical alternative.

(A) *The classical theory.* In classical mechanics both the α-particle, with charge $Z_1 e$ and mass M, and the nucleus, with charge $Z_2 e$ and mass M', are considered as dimensionless points. The force between them is the Coulomb force $Z_1 Z_2 e^2 / r^2$. Classical mechanics gives the result that the orbit of the α-particle relative to the nucleus is a hyperbola, a general result of all repulsive, inverse-square-law forces. In Rutherford's experiments a uniform beam of α-particles with given velocity v along the x-axis impinges on a thin foil of metal, for example, a gold foil. Consider in particular one gold nucleus, assumed to be fixed at the origin of the coordinates.* This nucleus sees the α-particles coming in at different impact parameters b (see Fig. 11-11). The angle of deflection θ, i.e., the angle between the two asymptotes, depends on the impact parameter b. This relation may be obtained as follows. By conservation of angular momentum we have

$$Mvb = -Mr^2 \frac{d\beta}{dt}. \tag{11-111}$$

---

* Actually it is the center of mass of M and M' that is fixed in space. However, according to a result in classical mechanics, we may consider M' as fixed provided the value of M is replaced by the *reduced mass* $MM'/(M + M')$. As the nucleus is usually much heavier than the α-particle, $MM'/(M + M')$ is not much different from M. We shall ignore this complication by assuming $M' \to \infty$.

Applying Newton's second law in the y-direction, we obtain

$$M \frac{dv_y}{dt} = \frac{Z_1 Z_2 e^2}{r^2} \sin \beta. \tag{11–112}$$

Combining the two equations we have

$$M \frac{dv_y}{dt} = - \frac{Z_1 Z_2 e^2}{vb} \sin \beta \frac{d\beta}{dt}. \tag{11–113}$$

Integration of the above equation for $-\infty < t < \infty$, which corresponds to $\pi > \beta > \theta$ and $0 < v_y < v \sin \theta$, results in

$$Mv \sin \theta = \frac{Z_1 Z_2 e^2}{vb} (\cos \theta + 1). \tag{11–114}$$

The impact parameter b is thus a function of the deflection angle θ, which may be expressed as follows:

$$b = \frac{Z_1 Z_2 e^2}{Mv^2} \cot \frac{\theta}{2}. \tag{11–115}$$

From the above equation we calculate a quantity $2\pi b \, db$,

$$2\pi b \, db = \pi \, d(b^2)$$

$$= \frac{\pi Z_1^2 Z_2^2 e^4}{M^2 v^4} \frac{d}{d\theta} \left(\cot^2 \frac{\theta}{2} \right) d\theta. \tag{11–116}$$

This quantity is the area of a ring (Fig. 11–11) on a plane far from the origin, perpendicular to the x-axis. If an α-particle hits this area, it will be scattered in an angle between θ and $\theta + d\theta$. This area is defined as the *differential scattering cross section* $\sigma(\theta) \, d\theta$. Thus,

$$\sigma(\theta) \, d\theta = \frac{\pi Z_1^2 Z_2^2 e^4}{M^2 v^4} \frac{d}{d\theta} \left(\cot^2 \frac{\theta}{2} \right) d\theta. \tag{11–117}$$

The physical significance of this quantity is this: When a uniform beam of particles with density D and velocity v impinges on the nucleus, the number of particles scattered in an angle between θ and $\theta + d\theta$ per unit time is $D \cdot v \cdot \sigma(\theta) \, d\theta$. In other words,

scattering current within $d\theta$ = incident current density $\times \sigma(\theta) \, d\theta$.

$$\tag{11–118}$$

In most scattering processes the experimental results may be stated in terms of the differential scattering cross section, and the purpose of a scattering theory is to derive an expression for it. Equation (11–117) is the *Rutherford formula* for the differential scattering cross section. It was verified by experiments. There is another way of expressing the scattering cross section. The solid angle $d\omega$ between θ and $\theta + d\theta$ is $-2\pi \sin \theta\, d\theta$, the sign being negative as $d\theta$ is a negative quantity for a positive db. The scattering cross section per solid angle is thus

$$\frac{\sigma(\theta)\, d\theta}{d\omega} = -\frac{Z_1^2 Z_2^2 e^4}{2M^2 v^4 \sin \theta} \frac{d}{d\theta}\left(\cot^2 \frac{\theta}{2}\right)$$

$$= \frac{Z_1^2 Z_2^2 e^4}{4M^2 v^4} \frac{1}{\sin^4 (\theta/2)}. \tag{11-119}$$

Hence the differential scattering cross section in the direction (θ, φ) within the solid angle $d\omega$ is

$$\sigma(\theta, \varphi)\, d\omega = \frac{Z_1^2 Z_2^2 e^4}{4M^2 v^4} \frac{1}{\sin^4 (\theta/2)}\, d\omega. \tag{11-120}$$

(B) *The time-dependent perturbation theory.* We now consider the same problem from the quantum-mechanical point of view. The equation of motion for this problem is the time-dependent Schrödinger equation in which the potential is the Coulomb potential of the nucleus:

$$-\frac{\hbar^2}{2M} \nabla^2 \Psi + \frac{Z_1 Z_2 e^2}{r} \Psi = -\frac{\hbar}{i} \frac{\partial \Psi}{\partial t}.$$

This equation may be solved by treating the entire potential as a perturbation. The unperturbed equation is thus the free particle equation and the perturbation Hamiltonian is

$$\lambda H^{(1)} = \frac{Z_1 Z_2 e^2}{r}. \tag{11-121}$$

This method of treating the scattering potential as a perturbation is called *the Born approximation.* As the perturbation theory is based on the assumption that the perturbation term is small, this method is suited for those cases in which the scattering potential is small compared with the kinetic energy of the incident particle (the unperturbed energy).

Let us assume that the scattering foil is inserted in position at $t = 0$. Before $t = 0$ the α-particle is described by a plane wave along the x-axis,

$$\psi_k^{(0)} = \frac{1}{\sqrt{\Omega}} e^{(i/\hbar)px} \equiv \frac{1}{\sqrt{\Omega}} e^{(i/\hbar)\vec{p_0} \cdot \vec{r}},$$

where p is the magnitude of the momentum vector $\vec{p_0}$ of the α-particle. Therefore the initial condition is that at $t = 0$ the particle is described by a single unperturbed eigenfunction. The perturbation theory developed in the last section is based on such an initial condition and thus is readily applicable. At $t = 0$ the perturbation is switched on. According to the theory in the last section, transitions may occur to clusters of levels the unperturbed energy values $E_m^{(0)}$ of which are close to $E_k^{(0)}$. Let us examine these levels. They are all plane waves (unperturbed eigenfunctions), the momentum vectors of which have nearly the same magnitudes but different directions, i.e.,

$$\psi_m^{(0)} = \frac{1}{\sqrt{\Omega}} e^{(i/\hbar)\vec{p}' \cdot \vec{r}}, \qquad |\vec{p}'| \cong p. \qquad (11\text{–}122)$$

These levels may be grouped into subclusters, each containing waves with momentum vectors in a fixed direction (θ, φ). The transition matrix elements of all levels in such a subcluster may be shown presently to depend on the unperturbed energy only. Therefore the total transition probability to such a subcluster may be evaluated by the summation method of the last section. Take a typical level $\psi_{k'}^{(0)}$ in a given subcluster for which $|\vec{p}'| = p$. For this level the momentum vector may be denoted by $\vec{p}$. Thus

$$\psi_{k'}^{(0)} = \frac{1}{\sqrt{\Omega}} e^{(i/\hbar)\vec{p} \cdot \vec{r}}. \qquad (11\text{–}123)$$

The matrix element for the transition is given by

$$\lambda H_{k'k}^{(1)} = \frac{Z_1 Z_2 e^2}{\Omega} \iiint \frac{e^{(i/\hbar)(\vec{p_0} - \vec{p}) \cdot \vec{r}}}{r} \, d\tau. \qquad (11\text{–}124)$$

Once we have evaluated this matrix element and found the density of levels of this subcluster, Eq. (11–100) gives us the transition probability per unit time from the initial state $\psi_k^{(0)}$ to this subcluster. As waves in this subcluster represent particles moving in the direction (θ, φ) and having energy values near $E_k^{(0)}$ (with a spread $\Delta E_m^{(0)}$ specified before), this transition probability represents the probability of elastic scattering of the α-particle in the direction (θ, φ). The same procedure may be repeated for other subclusters corresponding to different directions of scattering. The differential scattering cross section may thus be derived.

We now proceed to evaluate the matrix element of Eq. (11–124). The direction of the vector $\vec{p_0} - \vec{p}$ is graphically shown in Fig. 11–12 and the magnitude of this vector is found to be $2p \sin(\theta/2)$ where θ is the scattering angle. Thus we have

$$(\vec{p_0} - \vec{p}) \cdot \vec{r} = 2pr \sin \frac{\theta}{2} \cos \theta'. \qquad (11\text{–}125)$$

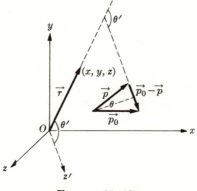

FIGURE 11–12

In carrying out the volume integration we shall use a set of spherical coordinates (r, θ', φ) the origin being at 0 but the polar axis being parallel to the vector $\overrightarrow{p_0} - \overrightarrow{p}$. It follows that

$$
\begin{aligned}
\lambda H_{k'k}^{(1)} &= \frac{Z_1 Z_2 e^2}{\Omega} \int_0^\infty \int_0^\pi \frac{e^{(i/\hbar) 2pr \sin (\theta/2) \cos \theta'}}{r} 2\pi r^2 \, dr \sin \theta' \, d\theta' \\
&= \frac{2\pi Z_1 Z_2 e^2}{\Omega} \int_0^\infty \left[\frac{e^{(i/\hbar) 2pr \sin (\theta/2) \cos \theta'}}{-(i/\hbar) 2pr \sin (\theta/2)} \right]_{\theta'=0}^{\theta'=\pi} r \, dr \\
&= \frac{2\pi Z_1 Z_2 e^2 \hbar}{\Omega p \sin (\theta/2)} \int_0^\infty \sin \left(\frac{2pr}{\hbar} \sin \frac{\theta}{2} \right) dr \\
&= \frac{2\pi Z_1 Z_2 e^2 \hbar^2}{2\Omega p^2 \sin^2 (\theta/2)} \int_0^\infty \sin x \, dx.
\end{aligned}
\tag{11--126}
$$

When the upper limit approaches infinity the value of the last integral fluctuates between 0 and 2; thus we have no definite value for the matrix element. This situation may be examined in the light of the following consideration. In Fig. 11–13 we show the graphs of the function $y_1(x) = \sin x$ and its integral $y_2(x) = \int_0^x \sin X \, dX$. Let us now consider a scattering problem in which the force is nearly, but not exactly, an inverse-square-law force, expressed by

$$
F = \frac{Z_1 Z_2 e^2}{r^{2+q}}, \qquad q \ll 1.
\tag{11--127}
$$

The corresponding potential is

$$
U = \frac{Z_1 Z_2 e^2}{(1 + q) r^{1+q}}, \qquad q \ll 1.
\tag{11--128}
$$

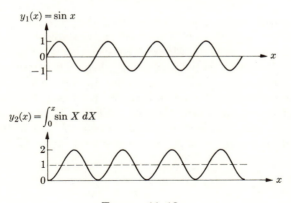

$$y_1(x) = \sin x$$

$$y_2(x) = \int_0^x \sin X \, dX$$

FIGURE 11–13

The last integral in Eq. (11–126) is to be replaced by

$$y_4(\infty) \equiv \int_0^\infty \frac{\sin x}{x^q} \, dx. \qquad (11\text{–}129)$$

The functions

$$y_3(x) = \frac{\sin x}{x^q} \qquad \text{and} \qquad y_4(x) = \int_0^x \frac{\sin X}{X^q} \, dX$$

are plotted in Fig. 11–14. The integral $y_4(x)$ converges to unity as $x \to \infty$. When q is comparatively large, the convergence takes place rapidly; when q is small, it takes place slowly. For any finite value q, no matter how small, convergence may eventually be realized. Bear in mind that Coulomb's law is an experimental law which is subject to experimental error. When Cavendish investigated the law of force between electric

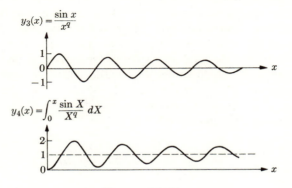

$$y_3(x) = \frac{\sin x}{x^q}$$

$$y_4(x) = \int_0^x \frac{\sin X}{X^q} \, dX$$

FIGURE 11–14

charges (before Coulomb did) he believed the expression for the force to be an equation like Eq. (11–127) and he was able to narrow down the value of q to within $\pm 1/50$. Maxwell later claimed that q cannot be greater than $1/21600$. Still it is not vanishing. From the experimental point of view it is perfectly legitimate to use a law of force with a nonvanishing q smaller than the experimental error, say $1/21600$. Such a law of force will reproduce, for practical purposes, all results of the inverse square law and will have the additional advantage of making the integral in Eq. (11–126) converge. Using the limiting value 1 for $\int_0^\infty \sin x \, dx$ we have

$$\lambda H_{k'k}^{(1)} = \frac{2\pi Z_1 Z_2 e^2 \hbar^2}{2\Omega p^2 \sin^2 (\theta/2)}. \tag{11–130}$$

We see that the matrix element for scattering in a given direction (θ, φ) depends on the unperturbed energy $E_m^{(0)} = p^2/2M$ only, as mentioned previously. It may be remarked here that the original method to make the integral converge was due to Wentzel who replaced the Coulomb potential by

$$\lim_{k \to 0} \frac{Z_1 Z_2 e^2}{r} e^{-kr}.$$

The last integral in Eq. (11–126) thus becomes

$$\int_0^\infty e^{-kr} \sin r \, dr = \frac{1}{\left(\dfrac{\hbar}{2p \sin (\theta/2)} k\right)^2 + 1} \to 1 \quad \text{when } k \to 0,$$

and the result obtained is the same as before. De Broglie interpreted Wentzel's method in the following manner. The plane wave used for calculation represents an infinite wave, whereas the actual beam is finite and limited in dimension laterally. By using an infinite wave we introduce a part which has no real existence. The effect of this error may be annulled by the cut-off factor e^{-kr} introduced in the potential.

We next calculate the level density of the subcluster of levels corresponding to scattering in the direction (θ, φ). The total level density consisting of all levels near $E_k^{(0)}$ is given by Eq. (11–87). As the space is isotropic, the density of levels with momentum vectors pointing in the direction (θ, φ) within a solid angle $d\omega$ is

$$\rho(E, \theta, \varphi) \, d\omega = \frac{\rho(E)}{4\pi} \, d\omega = \frac{\Omega}{(2\pi\hbar)^3} \frac{p^2}{v} \, d\omega. \tag{11–131}$$

We are now ready to write down the transition probability per unit time for scattering in the direction (θ, φ) within a solid angle $d\omega$, according to Eq. (11–100);

$$\frac{1}{\tau_{k \to k'}} = \frac{2\pi}{\hbar} \left(\frac{2\pi Z_1 Z_2 e^2 \hbar^2}{2\Omega p^2 \sin^2 (\theta/2)}\right)^2 \frac{\Omega}{(2\pi\hbar)^3} \frac{p^2}{v} d\omega$$

$$= \frac{Z_1^2 Z_2^2 e^4}{4M^2 v^4} \frac{1}{\sin^4 (\theta/2)} \frac{v}{\Omega} d\omega. \qquad (11–132)$$

This result may be expressed in terms of the scattering cross section. The initial wave function represents a beam of α-particles the velocity of which is v and the density of which is $1/\Omega$. Let the scattering cross section be denoted by $\sigma(\theta, \varphi) \, d\omega$. Since the transition probability per unit time for scattering in the direction (θ, φ) represents the scattered current in the direction (θ, φ), Eq. (11–118) yields

$$\frac{1}{\tau_{k \to k'}} = \frac{1}{\Omega} v \, \sigma(\theta, \varphi) \, d\omega \qquad (11–133)$$

or

$$\sigma(\theta, \varphi) \, d\omega = \frac{Z_1^2 Z_2^2 e^4}{4M^2 v^4} \frac{1}{\sin^4 (\theta/2)} d\omega. \qquad (11–134)$$

Equation (11–134) is identical with Eq. (11–120); thus quantum mechanics in the Born approximation gives the same result as classical mechanics.

How can classical mechanics and quantum mechanics, the mathematical apparatuses of which are drastically different, render exactly the same result? That this happens by chance is extremely improbable. The answer may be found in Section 2–7 where we discuss the relation between quantum mechanics and the Hamiltonian theory of classical mechanics. We have shown that the quantity $\Psi^*\Psi$, in the limit $\hbar \to 0$, may be regarded as a density function of a "fluid" made of a large number of classical particles moving independently of one another; these particles obey the laws of classical mechanics and are described in the Hamiltonian theory by a given principal function ϕ. Now a uniform beam of α-particles actually represents an ensemble of particles befitting this description. The future distribution of the particles, which is determined by the classical trajectories of the particles, therefore may be represented by $\Psi^*\Psi$, the wave function being obtained by solving the Schrödinger equation in quantum mechanics. Consequently the classical result expressed in Eq. (11–120) may be reproduced automatically in quantum mechanics in the approximation $\hbar \to 0$. The quantum-mechanical result is given in Eq. (11–134), in which $\hbar$ does not appear.

(C) *The time-independent perturbation theory.* The above relation between classical and quantum mechanics is perhaps obscured by the complicated mathematics of the time-dependent perturbation theory. Since a beam of α-particles undergoing scattering is an ensemble of particles describable by the motion of a fluid, it should be possible to find a solution $\Psi(x, y, z, t)$ of the time-dependent Schrödinger equation such that $\Psi^*\Psi$ describes the density of this fluid. Furthermore, for a steady beam (a beam turned on for an infinitely long time) undergoing scattering, the density distribution of the representative fluid must be stationary (independent of time). This requires that $\Psi^*\Psi$ be independent of time or $\Psi(x, y, z, t)$ be a stationary solution of the form $\psi(x, y, z)e^{-(i/\hbar)E_n t}$. It follows that such a wave function $\Psi(x, y, z, t)$ must be an energy eigenfunction and is obtainable by solving the time-independent Schrödinger equation. To find such a wave function $\psi(x, y, z)$ we shall now proceed to solve the time-independent Schrödinger equation by a method which is essentially a time-independent perturbation method. We shall see that this wave function $\psi(x, y, z)$ also leads to the Rutherford equation, thus demonstrating the fact that the two perturbation theories are equivalent.

The time-independent Schrödinger equation for the scattering problem is as follows:

$$\nabla^2\psi + \frac{2M}{\hbar^2}\left(E - \frac{Z_1 Z_2 e^2}{r}\right)\psi = 0. \tag{11-135}$$

It may be noted that this equation is identical with that of the hydrogen atom except for the values of M, Z_1, and Z_2. However, for scattering problems we are interested in solutions of positive energy values which satisfy certain boundary conditions (the eigenfunctions of the hydrogen atom correspond to negative energy values). As before, we consider the Coulomb potential as a perturbation, the unperturbed equation being again the free particle equation. The beam of α-particles before being scattered may be represented by an unperturbed eigenfunction denoted by $\psi^{(0)}$, which is a plane wave in the x-direction represented by $(1/\sqrt{\Omega})e^{(i/\hbar)px}$. The perturbation changes the wave function slightly and we may write

$$\psi = \psi^{(0)} + \psi^{(1)}, \tag{11-136}$$

where $\psi^{(1)}$ is small compared with $\psi^{(0)}$. $\psi^{(1)}$ represents the scattered particles. To obtain $\psi^{(1)}$ we put Eq. (11-136) in Eq. (11-135) and make use of the unperturbed equation, $\nabla^2\psi^{(0)} + (2ME/\hbar^2)\psi^{(0)} = 0$. The result is

$$\nabla^2\psi^{(1)} + \frac{2M}{\hbar^2}E\psi^{(1)} = \frac{2M}{\hbar^2}\frac{Z_1 Z_2 e^2}{r}(\psi^{(0)} + \psi^{(1)}). \tag{11-137}$$

So far the derivation is exact. In the spirit of the perturbation theory we

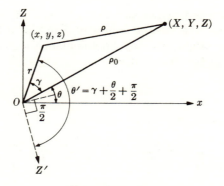

FIGURE 11–15

now make an approximation by neglecting $\psi^{(1)}$ in comparison with $\psi^{(0)}$ in the right-hand side. Thus

$$\nabla^2\psi^{(1)} + \frac{2M}{\hbar^2}E\psi^{(1)} = \frac{2M}{\hbar^2}\frac{Z_1Z_2e^2}{r}\frac{1}{\sqrt{\Omega}}e^{(i/\hbar)px}. \qquad (11\text{–}138)$$

To solve this inhomogeneous differential equation for $\psi^{(1)}$, we deviate from the standard perturbation method of Chapter 10 and take a different approach. Multiplying this equation by the time factor $e^{-(i/\hbar)Et}$, we obtain

$$\nabla^2\psi^{(1)}e^{-(i/\hbar)Et} + \frac{2M}{\hbar^2}E\psi^{(1)}e^{-(i/\hbar)Et} = \frac{2M}{\hbar^2}\frac{Z_1Z_2e^2}{r}\frac{1}{\sqrt{\Omega}}e^{(i/\hbar)(px-Et)}.$$

$$(11\text{–}139)$$

Then

$$\nabla^2(\psi^{(1)}e^{-(i/\hbar)Et}) - \frac{1}{(E/p)^2}\frac{\partial^2}{\partial t^2}(\psi^{(1)}e^{-(i/\hbar)Et}) = \frac{2M}{\hbar^2}\frac{Z_1Z_2e^2}{\sqrt{\Omega}}\frac{e^{(i/\hbar)(px-Et)}}{r}.$$

$$(11\text{–}140)$$

This equation takes the form of the equation for the retarded potential in electrodynamics and thus may be solved by Kirchhoff's formula (see Fig. 11–15). The result is

$$(\psi^{(1)}(X,\,Y,\,Z)e^{-(i/\hbar)Et})$$

$$= -\frac{1}{4\pi}\iiint\left[\frac{\dfrac{2M}{\hbar^2}\dfrac{Z_1Z_2e^2}{\sqrt{\Omega}}\dfrac{e^{(i/\hbar)(px-Et)}}{r}}{\rho}\right]_{t-[\rho/(E/p)]} dx\,dy\,dz \qquad (11\text{–}141)$$

$$= -\frac{1}{4\pi}\frac{2M}{\hbar^2}\frac{Z_1Z_1e^2}{\sqrt{\Omega}}e^{-(i/\hbar)Et}\iiint\frac{e^{(i/\hbar)p(x+\rho)}}{r\rho}dx\,dy\,dz. \qquad (11\text{–}142)$$

We are only interested in the value of $\psi^{(1)}(X, Y, Z)$ at a large distance from the origin. As the integral is largely determined by the integrand in a region near the origin, we may approximate the ρ in the denominator by ρ_0 and the ρ in the exponent by

$$\rho = \rho_0 - r \cos \gamma.$$

Hence

$$\psi^{(1)}(X, Y, Z)e^{-(i/\hbar)Et}$$

$$= -\frac{1}{4\pi} \frac{2M}{\hbar^2} \frac{Z_1 Z_2 e^2}{\sqrt{\Omega}} \frac{e^{(i/\hbar)(p\rho_0 - Et)}}{\rho_0} \iiint \frac{e^{(i/\hbar)p(x - r\cos\gamma)}}{r} r^2 \, dr \, d\omega.$$

$$(11\text{-}143)$$

The exponent may be reduced as follows:

$$p(x - r\cos\gamma) = pr\left[\cos(\gamma + \theta) - \cos\gamma\right]$$

$$= pr\left[\cos\left(\gamma + \frac{\theta}{2} + \frac{\theta}{2}\right) - \cos\left(\gamma + \frac{\theta}{2} - \frac{\theta}{2}\right)\right]$$

$$= -2pr\sin\left(\gamma + \frac{\theta}{2}\right)\sin\frac{\theta}{2}$$

$$= 2pr\cos\left(\gamma + \frac{\theta}{2} + \frac{\pi}{2}\right)\sin\frac{\theta}{2}$$

$$= 2pr\cos\theta'\sin\frac{\theta}{2}. \qquad (11\text{-}144)$$

The integral in Eq. (11–143) is thus exactly the same as that in Eq. (11–126), as the solid angle $d\omega$ may be written as $2\pi \sin \theta' \, d\theta'$, referring to the direction OZ' as the polar axis. The result of Eq. (11–126) may thus be taken over and we have

$$\psi^{(1)}(X, Y, Z)e^{-(i/\hbar)Et} = -\frac{1}{4\pi}\frac{2M}{\hbar^2}\frac{1}{\sqrt{\Omega}}\frac{e^{(i/\hbar)(p\rho_0 - Et)}}{\rho_0}\frac{2\pi Z_1 Z_2 e^2 \hbar^2}{2p^2 \sin^2(\theta/2)}$$

$$= -\frac{1}{\sqrt{\Omega}}\frac{e^{(i/\hbar)(p\rho_0 - Et)}}{\rho_0}\frac{Z_1 Z_2 e^2}{2Mv^2 \sin^2(\theta/2)}.$$

It may be remarked that Eq. (11–144) is physically equivalent to the equation (11–125) obtained by geometrical method, and the direction OZ' is actually the direction of $\vec{p}_0 - \vec{p}$ in Fig. 11–12. The point (X, Y, Z) may be expressed in the polar coordinates $(\rho_0, \theta, \varphi)$. Thus

$$\psi^{(1)}(\rho_0, \theta, \varphi)e^{-(i/\hbar)Et} = -\frac{1}{\sqrt{\Omega}}\frac{Z_1 Z_2 e^2}{2Mv^2 \sin^2(\theta/2)}\frac{e^{(i/\hbar)(p\rho_0 - Et)}}{\rho_0}. \qquad (11\text{-}145)$$

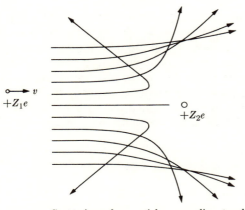

Scattering of α-particles according to classical mechanics

(a)

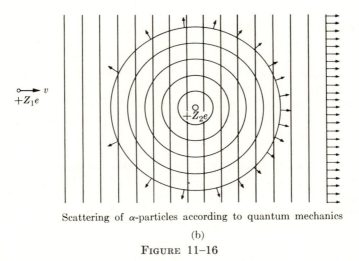

Scattering of α-particles according to quantum mechanics

(b)

FIGURE 11–16

The last factor in the above expression represents a radial spherical wave. Thus Eq. (11–145) represents a wave the amplitude of which depends on the azimuth angle θ according to the second factor, but is independent of the longitude angle φ, i.e., the wave exhibits axial symmetry. The first factor $1/\sqrt{\Omega}$ fixes the absolute intensity of this wave. This result may be interpreted to mean that the effect of the perturbation of the Coulomb potential on $\psi^{(0)}$, the plane wave along the x-axis, is to generate a weak spherical wave from the center of the force field. The student may visualize this situation by considering a plane wave in water impinging on a fixed object from which a spherical wave is generated. Since the scattered wave $\psi^{(1)}$ is to represent the scattered α-particles, Eq. (11–145) tells us that the α-particles will be scattered with different intensities in different directions. Again the result may be expressed in terms of the scattering cross section. By definition the incident current density multiplied by $\sigma(\theta, \varphi)\, d\omega$ equals the scattered current in the solid angle $d\omega$. The latter equals the intensity of the scattered wave at $(\rho_0, \theta, \varphi)$ multiplied by the surface area $\rho_0^2\, d\omega$ covering the solid angle $d\omega$, and by the velocity of the scattered particle, which equals v for elastic scattering:

$$\left(\frac{1}{\sqrt{\Omega}}\right)^2 v\, \sigma(\theta, \varphi)\, d\omega = \left(\frac{1}{\sqrt{\Omega}} \frac{Z_1 Z_2 e^2}{2Mv^2 \sin^2(\theta/2)} \frac{1}{\rho_0}\right)^2 (\rho_0^2\, d\omega) v. \quad (11\text{–}146)$$

Thus the differential scattering cross section is

$$\sigma(\theta, \varphi)\, d\omega = \frac{Z_1^2 Z_2^2 e^4}{4M^2 v^4} \frac{1}{\sin^4(\theta/2)}\, d\omega, \quad (11\text{–}147)$$

which is identical with Eq. (11–120) and Eq. (11–134).

It may be remarked that the wave functions used in quantum mechanics actually represent the motion of one particle. For many particles, the results are additive. However, the cross sections remain the same since they are geometrical quantities independent of the beam intensity. In Fig. 11–16 we sketch the scattering process symbolically according to both classical and quantum mechanics. The density distributions of the α-particles are the same in both cases.

11–5 The radiation processes. Bohr's theory is based on two assumptions: the quantum condition and the frequency rule. He restricted classical mechanics by the quantum condition and replaced classical electrodynamics by the frequency rule. In spite of the remarkable successes achieved, it remains an incoherent theory. Yet there is no doubt that the two assumptions are the master keys to atomic physics and a more refined theory will put these assumptions on a firm basis within a coherent theoretical structure. So far quantum mechanics has succeeded in incorporating the quantum condition into mechanics, but the frequency rule remains unexplained. In fact, quantum mechanics as we have discussed it so far does not allow any radiative transition to take place as specified by the Bohr frequency rule. Suppose that at $t = 0$ the initial wave function of the electron of a hydrogen atom is ψ_{210}. The equation of motion, i.e., the time-dependent Schrödinger equation, tells us that the wave function at any later time t is $\psi_{210} e^{-(i/\hbar)E_2 t}$. This means that after $t = 0$ the electron always stays at the $2p$ state, unable to come down to the $1s$ state accompanied by the emission of the first line in the Lyman series. Clearly *quantum mechanics as we have discussed it so far is not* *complete when a radiation process is involved. It is complete for purely mechanical problems,* as may be seen in its application to the Rutherford scattering. On the other hand, the classical theory of electrodynamics is inadequate in many applications, such as the photoelectric effect and the Compton effect. It has to be modified just as classical mechanics has been modified (by passing to a quantized theory). A radiation process such as the emission of the first Lyman line by a hydrogen atom, involving both the mechanics of the atom and the electrodynamics, can be successfully treated only on the basis of a quantized theory of electrodynamics. Much work has been done in developing a quantum theory of radiation, which is an advanced topic in quantum mechanics and will not be included in this volume. In this theory the quantum nature of the radiation process is accounted for, and not only the frequency rule but also the selection, intensity, and polarization rules may be derived from the theory. However, before such a systematic treatment of the radiation process can be made, many of the results may be obtained within the present scheme of quantum mechanics by treating the radiation field classically and considering its influence on the atom as a perturbation. Such a theory, considering the atom quantum-mechanically while considering the radiation field classically, may be called a *semiclassical theory.* Because it overlooks the quantum nature of radiation, it is approximate from the beginning and is not expected to be wholly satisfactory. However, the success of the semiclassical theory is an important step in the development of the theory of radiation; actually the quantum theory of radiation is more or less a refinement of it. We discuss the semiclassical theory of radiation here not only for our interest in radiation, but also as an illustration of the perturbation theory.

(A) *The perturbation treatment of the radiative transitions.* Consider an atom exposed to a radiation field. We know as experimental fact that the atom may absorb radiation and thereby raises itself to an excited state. From the point of view of the semiclassical theory the atom, treated quantum-mechanically, may undergo a transition from one energy level to another under the perturbation of the radiation field. Along this line of thought we may formulate a theory of *absorption.* Also, a theory of *induced emission*—emission of light induced by an external field—may be worked out along a line parallel to the theory of absorption. However, the semiclassical theory was unable to account for the *spontaneous emission,* such as the emission of the Lyman lines without the presence of an external field. This is one of the limitations of the semiclassical theory. As the radiation field is treated classically, it may be expressed by a time-dependent potential function. We have treated the perturbation by a static electric field, the perturbation potential being a function of the coordinates x, y, z only. For the perturbation by a radiation field the per-

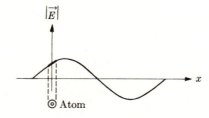

FIGURE 11–17

turbation potential is not only a function of the position x, y, z but also is dependent on time t. Except for this point the perturbation theory to be developed is the same as that developed before.

Consider a monochromatic electromagnetic wave propagating in the x-direction with its electric field polarized along the z-axis:

$$E_z = A \cos\left(\omega t - \frac{x}{\lambda}\right), \tag{11–148}$$

where A is the amplitude of the wave, ω is the angular frequency and λ is the wave length divided by 2π. The magnetic field H of the wave is equal in magnitude to the electric field E, but points in the $-y$ direction. When an atom is exposed to this wave, both the electric and magnetic fields act on the electron, the Lorentz force being $q\vec{E} + (q/c)\vec{v} \times \vec{H}$. However, the ratio of the magnetic force (the second term) to the electric force (the first term) is of the order of v/c, since $\vec{E}$ and $\vec{H}$ are equal in magnitude. For electrons moving slowly compared with the velocity of light, the effect of the magnetic field may be neglected. As the electron in the first Bohr orbit of the hydrogen atom has a velocity $c/137$, the effect of the magnetic field may be neglected in atomic physics. The dimension of the atom is about 10^{-8} cm while the wavelength of light in the visible region is of the order of 10^{-5} cm, much greater than the dimension of the atom. Thus the variation of the electric field over the space of the atom is small. (See Fig. 11–17.) Consequently, the field may be regarded as constant with respect to the space variables. With these two approximations the radiation field as seen by the atom may be represented by the equation

$$E_z = A \cos \omega t. \tag{11–149}$$

It may be remarked that these two approximations are valid only within certain limitations. When higher accuracy is required, both the magnetic field and the space variation of the electric field must also be considered.

The potential corresponding to the field of Eq. (11–149) is $-E_z \cdot z$ and the potential energy of an electron with charge $-e$ in this field is ezE_z.

Thus the perturbation potential is

$$\lambda H^{(1)} = ezA \cos \omega t \qquad \text{(for linearly polarized light)}. \qquad (11\text{–}150)$$

With the perturbation potential known, the machinery of the time-dependent perturbation theory may be applied to calculate the probability of transition. However, the present case differs from the previous ones in that the perturbation potential is time-dependent, a possibility we did not exclude in formulating the time-dependent perturbation theory. As a time-dependent Hamiltonian represents a nonconservative force, the law of conservation of energy may not hold. This point will be discussed in more detail later.

Before we apply the perturbation theory we first express the wave amplitude A in terms of more familiar quantities. The energy density of an electromagnetic wave is $(E^2 + H^2)/8\pi$, the time-average of which is

$$\overline{\left(\frac{E_z^2 + H_y^2}{8\pi}\right)} = \overline{\left(\frac{E_z^2}{4\pi}\right)} = \frac{A^2}{8\pi}. \qquad (11\text{–}151)$$

In actual cases a light beam of a continuous spectrum is often used, for which the intensity distribution with respect to frequency may be specified by a *spectral distribution function* $U(\omega)$. Within an infinitesimal range $d\omega$, all waves may be lumped together as a single monochromatic wave for which Eq. (11–151) applies. Thus

$$U(\omega)\, d\omega = \frac{A^2}{8\pi} \qquad (11\text{–}152)$$

or

$$A^2 = 8\pi U(\omega)\, d\omega. \qquad (11\text{–}153)$$

Consider an atomic system exposed to a radiation field expressed by Eq. (11–149). For simplicity we consider the hydrogen atom because its wave functions ψ_{nlm} are well known. The results obtained may easily be generalized to other atoms. The electron in the hydrogen atom is assumed, for generality, to be in a state ψ_{NLM} before the light is turned on at $t = 0$. Therefore the initial condition is

$$\Psi(x, y, z, 0) = \psi_{NLM}(r, \theta, \varphi). \qquad (11\text{–}154)$$

After $t = 0$ the perturbation Hamiltonian of Eq. (11–150) is switched on, and the time-dependent Schrödinger equation becomes

$$\left[\frac{1}{2M}\left(\frac{\hbar}{i}\right)^2 \nabla^2 - \frac{Ze^2}{r} + ezA \cos \omega t\right] \Psi = -\frac{\hbar}{i}\frac{\partial \Psi}{\partial t}. \qquad (11\text{–}155)$$

Since we consider the term $ezA \cos \omega t$ as a perturbation, the unperturbed equation is simply that of the hydrogen atom, and the unperturbed solutions are the ψ_{nlm}'s of Section 8–3. As the initial condition, Eq. (11–154), is such that the system is described by one of its unperturbed states, the perturbation theory developed in Section 11–1 is applicable and leads to the conclusion that other unperturbed eigenfunctions ψ_{nlm} may gradually appear in the wave function. The amplitude of ψ_{nlm} at time t, $\lambda \alpha_{nlm}^{(1)}$, according to Eq. (11–28), is determined by

$$\lambda \dot{\alpha}_{nlm}^{(1)}(t) = -\frac{i}{\hbar} \lambda H_{nlm;NLM}^{(1)} e^{i\omega_{nN} t}, \qquad (11\text{--}156)$$

where

$$\omega_{nN} = \frac{E_n - E_N}{\hbar}. \qquad (11\text{--}157)$$

The matrix element may be written as follows:

$$\lambda H_{nlm;NLM}^{(1)} = eA \cos \omega t\, z_{nlm;NLM}, \qquad (11\text{--}158)$$

where

$$z_{nlm;NLM} \equiv \iiint \psi_{nlm}^{*} z\, \psi_{NLM}\, d\tau. \qquad (11\text{--}159)$$

In integrating Eq. (11–156) we have to consider the time-dependence of the matrix element which we did not have to consider in Section 11–1. The function $\cos \omega t$ may be written in the exponential form

$$\cos \omega t = \tfrac{1}{2}(e^{i\omega t} + e^{-i\omega t}). \qquad (11\text{--}160)$$

Equation (11–156) may thus be written as follows:

$$\lambda \dot{\alpha}_{nlm}^{(1)}(t) = -\frac{i}{2\hbar}\, eA z_{nlm;NLM}(e^{i(\omega_{nN}+\omega)t} + e^{i(\omega_{nN}-\omega)t}). \qquad (11\text{--}161)$$

Integrating Eq. (11–161) with the initial condition that $\alpha_{nlm}^{(1)}(0) = 0$, we obtain

$$\lambda \alpha_{nlm}^{(1)}(t) = -\frac{eA}{2\hbar}\, z_{nlm;NLM} \left(\frac{e^{i(\omega_{nN}+\omega)t} - 1}{\omega_{nN} + \omega} + \frac{e^{i(\omega_{nN}-\omega)t} - 1}{\omega_{nN} - \omega} \right). \qquad (11\text{--}162)$$

When ω_{nN} is close to ω, the second term of the above equation becomes very large and the amplitude of the corresponding ψ_{nlm} becomes large. Thus at time t, there is greater probability of finding the electron in this state ψ_{nlm} than in others. This means that the radiation field may induce a transition from the initial state ψ_{NLM} to a final state ψ_{nlm} for which the resonance condition $\omega_{nN} \cong \omega$ is satisfied. This transition raises the elec-

tron from a lower energy level E_N to a higher one E_n, and the energy
necessary for the transition is supplied by the radiation field. (The situation is similar to the ionization or excitation of atoms by fast-moving
charged particles, discussed in Section 11–2.) Thus the transition of
atomic states is accompanied by the absorption of light. The result of
the semiclassical theory may be stated thus: strong absorption of light is
to take place when the light frequency is close to one of the Bohr frequencies; as a result the electron is raised to the corresponding excited
level. This explains the resonance absorption. *This statement was an assumption in the Bohr theory with regard to the absorption process. Here we
have derived it from quantum mechanics.* One difference may be noted. In
Bohr's theory only light waves of frequency $\omega = \omega_{nN}$ may be absorbed.
The quantum-mechanical result is that all waves with frequencies ω near
ω_{nN} may be absorbed, but the probability becomes small when ω is far
away from ω_{nN}. [In quantum mechanics a wave of *any* frequency may
induce transitions to *any* states according to Eq. (11–162); the probability is in general extremely small except when the resonance condition
is satisfied.] It may be added that even when $\omega = \omega_{nN}$, the second term
in Eq. (11–162) is not infinite, since the numerator also vanishes and the
ratio is finite.

We now consider the first term in Eq. (11–162). This term becomes
large when $\omega_{nN} = -\omega$. The value of ω_{nN} becomes negative when the
final energy E_n is lower than the initial energy E_N. Thus there is a large
probability that the electron may come down to a lower energy level ψ_{nlm}
when the resonance condition $\omega_{Nn} = \omega$ is satisfied. The semiclassical
theory thus concludes that under the perturbation of a light wave, an
electron may jump down to a lower energy level. We have mentioned before
that without any perturbation an electron cannot jump down. The semiclassical theory makes this possible with the help of an external field. In
doing so the electron loses energy. According to the law of conservation
of energy we may conclude that the loss of energy of the electron $E_N - E_n$
becomes the gain of the radiation field. Thus light emission accompanies
the transition of states. Assuming the light emitted to be a photon, we
find the angular frequency to be

$$\omega = \frac{E_N - E_n}{\hbar}. \tag{11–163}$$

Thus we have arrived at the Bohr frequency rule in induced emission.
It must be pointed out that the last step invoking conservation of energy
is physically convincing but logically unsound. Also the photon is introduced by an *ad hoc* assumption. By the quantum theory of the atom we
can only predict about the atom, but not about the radiation field. As

mentioned before, the semiclassical theory is not expected to be perfect. Furthermore the semiclassical theory is unable to account for the spontaneous emission—a most unsatisfactory feature of this theory. Again the difficulty lies in the fact that the quantum nature of the radiation field is ignored; the solution is to be found in the quantum theory of radiation.

(B) *Transition probabilities; selection, intensity, and polarization rules.* After the above qualitative discussions we return to Eq. (11–162) to derive some quantitative results. Consider first the absorption of light. For a state ψ_{nlm} satisfying the condition of resonance absorption, we may omit the first term in Eq. (11–162), which is small compared with the second. The square of the amplitude of this state is thus

$$|\lambda a_{nlm}^{(1)}(t)|^2 = \frac{e^2 A^2}{\hbar^2} |z_{nlm;NLM}|^2 \frac{\sin^2 \frac{1}{2}(\omega_{nN} - \omega)t}{(\omega_{nN} - \omega)^2}. \qquad (11\text{–}164)$$

In absorption experiments, light of a continuous spectrum is often used which may be specified by a spectral distribution function $U(\omega)$. Even a so-called monochromatic wave may be subject to a small frequency spread, since only an infinite wave train is strictly monochromatic. Such a wave may also be represented by a spectral distribution function $U(\omega)$, though this function is narrowly peaked. A light beam of a continuous spectrum may be considered as being made up of a large number of monochromatic waves each having an intensity $U(\omega)\, d\omega$. The quantity A^2 corresponding to each monochromatic wave is given by Eq. (11–153). Applying Eq. (11–164) to each wave, we have

$$|\lambda a_{nlm}^{(1)}(t)|^2 = \frac{8\pi e^2}{\hbar^2} U(\omega)\, d\omega |z_{nlm;NLM}|^2 \frac{\sin^2 \frac{1}{2}(\omega_{nN} - \omega)t}{(\omega_{nN} - \omega)^2}.$$

The total probability of finding the electron in the state ψ_{nlm} at time t due to all these waves may be obtained by integrating the above expression over ω. As the integral is largely determined by the value of the integrand near the resonance frequency, we may approximate it by replacing $U(\omega)$ with $U(\omega_{nN})$ and then taking it out of the integral. Thus the total transition probability to state ψ_{nlm} is

$$\sum_{\omega} |\lambda a_{nlm}^{(1)}(t)|^2 = \frac{8\pi e^2}{\hbar^2} U(\omega_{nN}) |z_{nlm;NLM}|^2 \int_{-\infty}^{\infty} \frac{\sin^2 \frac{1}{2}(\omega_{nN} - \omega)t}{(\omega_{nN} - \omega)^2}\, d\omega$$

$$= \frac{4\pi^2 e^2}{\hbar^2} U(\omega_{nN}) |z_{nlm;NLM}|^2 t. \qquad (11\text{–}165)$$

The result of Eq. (11–98) is used for the last step. As in Section 11–3, the total transition probability is proportional to the time t and therefore we may speak of a transition probability per unit time Λ or a mean lifetime τ,

$$\Lambda_{NLM \to nlm} = \frac{4\pi^2 e^2}{\hbar^2} U(\omega_{nN})|z_{nlm;NLM}|^2, \qquad (11\text{–}166)$$

$$\frac{1}{\tau_{NLM \to nlm}} = \frac{4\pi^2 e^2}{\hbar^2} U(\omega_{nN})|z_{nlm;NLM}|^2. \qquad (11\text{–}167)$$

Equation (11–167) or Eq. (11–166) is the basic equation for the absorption process. For the induced emission we retain the first term in Eq. (11–162) but neglect the second term. The result is the same as Eq. (11–167) if we remember that ω_{nN} is now a negative number. Thus Eq. (11–167) applies to induced emission as well as to absorption. For an electron in a given initial state ψ_{NLM}, exposed to a radiation field described by $U(\omega)$, we can calculate its lifetimes against transitions to all other states ψ_{nlm}; and the total lifetime of the state ψ_{NLM} is given by

$$\frac{1}{\tau_{NLM}} = \sum_{nlm} \frac{1}{\tau_{NLM \to nlm}}. \qquad (11\text{–}168)$$

In order to evaluate the lifetimes we must first know the matrix elements of z with respect to the final and initial states. Using the hydrogen wave functions of Section 8–3, we have

$z_{nlm;NLM}$

$$= \iiint \psi_{nlm}^* \, z \, \psi_{NLM} \, d\tau$$

$$= \iiint R_{nl}^*(r) P_l^m (\cos\theta) e^{-im\varphi} (r\cos\theta) R_{NL}(r) P_L^M (\cos\theta) e^{iM\varphi} r^2 \sin\theta \, dr \, d\theta \, d\varphi$$

$$= \int R_{nl}^*(r) R_{NL}(r) r^3 \, dr \int P_l^m (\cos\theta) P_L^M (\cos\theta) \cos\theta \sin\theta \, d\theta \int e^{i(M-m)\varphi} \, d\varphi.$$

$$(11\text{–}169)$$

The last integral vanishes unless $m = M$. It may be shown that the second integral vanishes unless $l = L \pm 1$. (For a mathematical proof, see, for example, Sommerfeld's *Wave Mechanics*.) This may be visualized as follows: The multiplication of $P_L^M (\cos\theta)$ by $\cos\theta$ raises the order of the polynomial with respect to the variable $\cos\theta$ by one so that the product, expanded in a series of spherical harmonics, contains at least a term $P_{L+1}^M (\cos\theta)$. By the orthonormal relation, the integral will not vanish when $l = L + 1$. Since l and L are in symmetric positions in the integral, we can interchange them and prove the same when $l = L - 1$. The values of the second and third integrals combined are listed below without proof:

$$\iint Y^*_{l+1,m} \cos\theta\, Y_{l,m}\, d\omega = \sqrt{\frac{(l+1)^2 - m^2}{(2l+3)(2l+1)}},$$

$$\iint Y^*_{l-1,m} \cos\theta\, Y_{l,m}\, d\omega = \sqrt{\frac{l^2 - m^2}{(2l+1)(2l-1)}}, \qquad (11\text{--}170)$$

$$\text{others} = 0.$$

From these results, Eq. (11–166) tells us that no transition will take place between two states unless

$$\Delta l = \pm 1, \quad \Delta m = 0 \text{ (for linearly polarized light)}, \qquad (11\text{--}171)$$

where Δl and Δm are the changes of the two quantum numbers l and m between the initial and final states. These equations represent the *selection rules* that specify the pairs of states between which transition is possible. Selection rules were first obtained empirically in spectral analysis. In the old quantum theory, selection rules may be derived from the correspondence principle which makes a correlation between classical and quantum-mechanical quantities. In the semiclassical theory of radiation, selection rules are derived on a more satisfactory basis. The selection rules of Eq. (11–171) are based on the perturbation potential of Eq. (11–150) which represents light polarized in the z-direction; therefore they are valid only for the linearly polarized light. Also their validity is based on the two assumptions that the magnetic field of the light wave and the space variation of the electric field over the volume of the atom may be ignored. Actually, the small effects of the magnetic field and the space variation of the electric field may induce transitions violating the selection rules of Eq. (11–171) (these are called *forbidden transitions*), but the transition probabilities are very small. In spectroscopy such transitions are actually observed with very weak intensity. The corresponding spectral lines are called *forbidden lines*. The effect of the variation of the electric field over the atom may be investigated by writing Eq. (11–148) in the following form:

$$E_z = A\cos\omega t + A\,\frac{x}{\lambda}\sin\omega t + \cdots.$$

In addition to the perturbation of Eq. (11–150) there is a new term $(eA/\lambda)zx\sin\omega t$, the matrix element of which is essentially determined by $(xz)_{nlm;NLM}$. The corresponding selection rule is $\Delta l = 0, \pm 2$. Higher order terms in the expansion introduce new matrix elements and thereby new selection rules. The transitions induced by the potential $ezA\cos\omega t$ are called *electric dipole transitions*. Those induced by $(eA/\lambda)zx\sin\omega t$ are called *electric quadrupole transitions*. The ratio of the two potentials is of

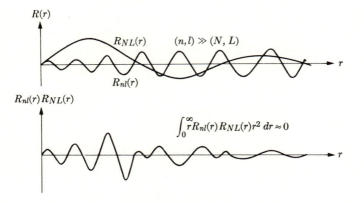

FIGURE 11–18

the order of $(x/\lambda) \approx (10^{-8}/10^{-5}) = 10^{-3}$. The ratio of the transition probabilities is the square of 10^{-3}, i.e., 10^{-6}. Thus the quadrupole transition is about a million times weaker than the dipole transition. Higher order terms result in higher order multipole transitions which decrease rapidly in intensity. The magnetic field of the light wave also induces multipole transitions with different selection rules. The intensity of the lines due to *magnetic dipole transition* is of the same order of magnitude as that of the electric quadrupole transition.

While the second and third integrals of Eq. (11–169) determine the selection rules, the first integral determines the magnitude of the matrix element and thus the intensity of the spectral lines. When (n, l) are identical with (N, L), the first integral is simply the average value of r over the state ψ_{NLM} and will have an order of magnitude comparable with the radius of the atom. For (n, l) different from (N, L), we may think of the integral as a sort of "mixed average" of r over the two states ψ_{NLM} and ψ_{nlm}. If the two sets of quantum numbers are not close to each other, the two radial wave functions R_{nl} and R_{NL} may "interfere" strongly in the radial integral, and the value of the integral may be much smaller than the radius of the atom (see Fig. 11–18). For the hydrogen atom, the wave functions are all known and the matrix elements of Eq. (11–169) may be calculated exactly. The transition probabilities may thus be calculated by Eq. (11–166). For several other kinds of atoms the wave functions have a structure similar to those of hydrogen atom in that they consist of a radial part multiplied by a spherical harmonic. Since the spherical harmonic determines the selection rules, these atoms have the same selection rules as the hydrogen atom. We need only change the radial integral in Eq. (11–169) in order to calculate the transition probabilities.

Having considered the linearly polarized light, we now consider the perturbation by a circularly polarized light. If the electric field vector rotates in the xy-plane in the counterclockwise sense, its components may be expressed as follows:

$$E_x = A \cos \omega t, \qquad E_y = A \sin \omega t. \qquad (11\text{--}172)$$

Equation (11–149) is to be replaced by Eqs. (11–172) and analogously Eq. (11–150) is to be replaced by

$$\lambda H^{(1)} = exA \cos \omega t + eyA \sin \omega t. \qquad (11\text{--}173)$$

Again, replace the trigonometric functions by exponential functions. Thus

$$\lambda H^{(1)} = \frac{1}{2} exA(e^{i\omega t} + e^{-i\omega t}) + \frac{1}{2i} eyA(e^{i\omega t} - e^{-i\omega t})$$

$$= \tfrac{1}{2} eA(x - iy)e^{i\omega t} + \tfrac{1}{2} eA(x + iy)e^{-i\omega t}. \qquad (11\text{--}174)$$

In evaluating the matrix elements, the role played by $z_{nlm;NLM}$ is to be taken by $(x \pm iy)_{nlm;NLM}$, defined in a similar way. Equation (11–161) is thus replaced by

$$\lambda \dot{a}_{nlm}^{(1)}(t) = - \frac{i}{2\hbar} eA(x - iy)_{nlm;NLM} \, e^{i(\omega_{nN} + \omega)t}$$

$$- \frac{i}{2\hbar} eA(x + iy)_{nlm;NLM} \, e^{i(\omega_{nN} - \omega)t}. \qquad (11\text{--}175)$$

Equations (11–166) and (11–167) remain valid for circularly polarized light if $z_{nlm;NLM}$ is replaced by $(x + iy)_{nlm;NLM}$ for the case of absorption and by $(x - iy)_{nlm;NLM}$ for induced emission. If the circularly polarized light is such that its electric vector rotates in the xy-plane in a clockwise sense, then in the above substitution, $(x \pm iy)$ is to be replaced by $(x \mp iy)$. Therefore the matrix elements determining the absorption and emission of circularly polarized light in the xy-plane are $(x \pm iy)_{nlm;NLM}$. In spherical coordinates $(x \pm iy) = r \sin \theta e^{\pm i\varphi}$. Their matrix elements may be evaluated accordingly. Again they may be written as a product of three integrals of r, θ, and φ. The radial integral is identical with the radial integral of Eq. (11–169) while the angular integrals may be evaluated by a straightforward calculation, the results of which are listed below without proof,

$$\int Y^*_{l+1,m} \sin \theta e^{\pm i\varphi} Y_{l,m\mp 1} \, d\omega = \mp \sqrt{\frac{(l \pm m)(l + 1 \pm m)}{(2l + 1)(2l + 3)}},$$

$$\int Y^*_{l-1,m} \sin \theta e^{\pm i\varphi} Y_{l,m\mp 1} \, d\omega = \pm \sqrt{\frac{(l \mp m)(l + 1 \mp m)}{(2l + 1)(2l - 1)}}, \qquad (11\text{--}176)$$

$$\text{others} = 0.$$

These lead to the following selection rules:

$$\Delta l = \pm 1, \quad \Delta m = \pm 1 \text{ (for circularly polarized light).} \qquad (11\text{–}177)$$

The choice of the z-direction for the linearly polarized light and the xy-plane for the circularly polarized light is merely for convenience in making calculations. If the linearly polarized light is in the x-direction or the circularly polarized light is in the yz-plane, we may use the x-direction as the polar axis and the spherical harmonics may be transformed accordingly.

Both the linearly and circularly polarized light discussed above propagate in one fixed direction. We now consider a radiation field which is isotropic in space, such as the electromagnetic radiation in an enclosure which obeys the law of blackbody radiation. For an isotropic radiation the electric field vector may be oriented in any direction specified by direction cosines α, β, γ, which are random variables. Equation (11–149) is to be replaced by the following:

$$E_x = A\alpha \cos \omega t, \quad E_y = A\beta \cos \omega t, \quad E_z = A\gamma \cos \omega t, \qquad (11\text{–}178)$$

and all equations following Eq. (11–149) remain valid if $z_{nlm;NLM}$ is replaced by $(\alpha x + \beta y + \gamma z)_{nlm;NLM}$. In the expression for the transition probability, the matrix element appears in absolute value squared, which is

$$
\begin{aligned}
|(\alpha x &+ \beta y + \gamma z)_{nlm;NLM}|^2 \\
&= \alpha^2 |x_{nlm;NLM}|^2 + \beta^2 |y_{nlm;NLM}|^2 + \gamma^2 |z_{nlm;NLM}|^2 \\
&\quad + \alpha\beta[(x_{nlm;NLM})^*(y_{nlm;NLM}) + (y_{nlm;NLM})^*(x_{nlm;NLM})] \\
&\quad + \beta\gamma[(y_{nlm;NLM})^*(z_{nlm;NLM}) + (z_{nlm;NLM})^*(y_{nlm;NLM})] \\
&\quad + \gamma\alpha[(z_{nlm;NLM})^*(x_{nlm;NLM}) + (x_{nlm;NLM})^*(z_{nlm;NLM})]. \qquad (11\text{–}179)
\end{aligned}
$$

Since the direction cosines are random variables, the average values of $\alpha\beta$, $\beta\gamma$, and $\gamma\alpha$ are all zero while those of α^2, β^2, and γ^2 are all $\frac{1}{3}$. Thus

$$
\begin{aligned}
\overline{|(\alpha x + \beta y + \gamma z)_{nlm;NLM}|^2} &= \tfrac{1}{3}[|x_{nlm;NLM}|^2 + |y_{nlm;NLM}|^2 + |z_{nlm;NLM}|^2] \\
&\equiv \tfrac{1}{3}|r_{nlm;NLM}|^2. \qquad (11\text{–}180)
\end{aligned}
$$

The last step defines the quantity $|r_{nlm;NLM}|^2$ which will be used for convenience in writing. Equation (11–167) thus becomes

$$\frac{1}{\tau_{NLM \to nlm}} = \frac{4\pi^2 e^2}{3\hbar^2} U(\omega_{nN})|r_{nlm;NLM}|^2. \qquad (11\text{–}181)$$

The selection rules may be obtained from the matrix elements of x, y, and z. Since the matrix element of x (or y) may be expressed in terms

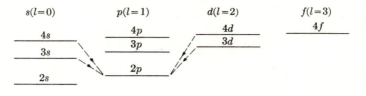

Allowed transitions to $2p$ state

FIGURE 11–19

of those of $(x + iy)$ and $(x - iy)$, the corresponding selection rules are the same as in Eq. (11–177). Therefore the selection rules for isotropic radiation are a combination of those for circularly polarized light in the xy-plane and those for linearly polarized light in the z-direction, i.e.,

$$\Delta l = \pm 1, \quad \Delta m = 0, \pm 1 \text{ (for isotropic radiation).} \qquad (11\text{–}182)$$

Referring to the energy level diagram reproduced in Fig. 11–19, the selection rule $\Delta l = \pm 1$ means that a transition can take place only between adjacent columns. A level in this diagram represents $(2l + 1)$ states with $m = -l, \ldots, l$. Between two levels a number of transitions may be possible satisfying the selection rule $\Delta m = 0, \pm 1$. For example, the line $3d \to 2p$ includes the following possible changes of m: $2 \to 1$; $1 \to 1, 0$; $0 \to 1, 0, -1$; $-1 \to 0, -1$; and $-2 \to -1$.

In spectral analysis, a number of intensity rules have been obtained empirically. Some of them may be derived with the aid of Eqs. (11–170) and (11–176). Since the states that differ in magnetic quantum numbers only have the same energy, we do not distinguish them in spectral analysis unless a magnetic field is applied. (The effect of the magnetic field will be discussed in Section 12–10.) Thus a single observed spectral line is actually the aggregation of a number of lines associated with different magnetic quantum numbers. Consider transitions between a given initial level (N, L) and levels of the same principal quantum number n. The selection rule requires the l values of the final level to be $L \pm 1$. Thus only two levels are available for allowed transition, giving rise to two spectral lines. The intensity of each of the two lines represents the total transition probability of all individual transitions between states of different magnetic quantum numbers. The matrix elements of all these transitions have nearly the same radial integrals. Thus the relative probabilities of these transitions may be expressed in terms of the quantities in Eqs. (11–170)

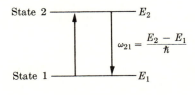

State 2 — E_2

$\omega_{21} = \dfrac{E_2 - E_1}{\hbar}$

State 1 — E_1

FIGURE 11–20

and (11–176). The total intensity of the two lines turns out to be proportional to $(2L + 1)$, which is the quantum weight of the initial level. This result agrees with the empirical Burger and Dorgelo rule.

(C) *The spontaneous emission.* The above derivation of the frequency, selection, polarization, and intensity rules are subject to the same criticisms we mentioned previously, i.e., they are valid for the absorption process; their validity for the induced emission depends on additional assumptions; and they have nothing to do with the spontaneous emission. Let us forget for a moment the incompleteness of the semiclassical theory in dealing with the emission processes and consider its success in dealing with the absorption process. If we have another piece of information, independent of the semiclassical theory, which gives us the relation between the emission and absorption processes, then we can combine this information with the correct result of absorption obtained in the semiclassical theory to derive the correct laws of emission. It so happens that the equilibrium condition governing an atom exposed to a blackbody radiation provides us with exactly such an information. In Fig. 11–20 we consider two of the quantum states of an atom designated by 1 and 2. An atom in state 1 may absorb radiation and be excited to state 2. An atom in state 2 may emit radiation and return to state 1. In order to establish equilibrium, the kind and amount of radiation emitted and absorbed by an ensemble of atoms must be the same. Set the emission rate equal to the absorption rate; we obtain

$$P_2 \Lambda_{2 \to 1} = P_1 \Lambda_{1 \to 2}, \tag{11–183}$$

where P_2 and P_1 are the equilibrium probabilities of finding the atomic system in states 2 and 1, and $\Lambda_{2 \to 1}$ and $\Lambda_{1 \to 2}$ are the corresponding transition probabilities per unit time. P_2 and P_1 are determined by the general laws of statistical mechanics:

$$P_2 \sim e^{-E_2/kT}, \qquad P_1 \sim e^{-E_1/kT}. \tag{11–184}$$

Therefore

$$\frac{P_1}{P_2} = e^{(E_2 - E_1)/kT} = e^{\hbar \omega_{21}/kT}. \tag{11–185}$$

According to the semiclassical theory of absorption, $\Lambda_{1\to2}$ is given by

$$\Lambda_{1\to2} = \frac{4\pi^2 e^2}{3\hbar^2} |r_{21}|^2 U(\omega_{21}), \qquad (11\text{--}186)$$

where $U(\omega)$ is now the spectral distribution function of the blackbody radiation given by Eq. (1–2) which is reproduced below:

$$U(\omega) = \frac{\omega^2}{\pi^2 c^3}\, \frac{\hbar\omega}{e^{\hbar\omega/kT} - 1}. \qquad (11\text{--}187)$$

From Eq. (11–187) we obtain

$$U(\omega) e^{\hbar\omega/kT} = \frac{\omega^2}{\pi^2 c^3}\, \hbar\omega + U(\omega).$$

Equation (11–183) thus gives the transition probability from the upper state 2 to the lower state 1,

$$\Lambda_{2\to1} = e^{\hbar\omega_{21}/kT}\, \frac{4\pi^2 e^2}{3\hbar^2} |r_{21}|^2 U(\omega_{21}) \qquad (11\text{--}188)$$

or

$$\Lambda_{2\to1} = \frac{4\pi^2 e^2}{3\hbar^2} |r_{21}|^2 U(\omega) + \frac{4e^2\omega^3}{3\hbar c^3} |r_{21}|^2. \qquad (11\text{--}189)$$

The first term of Eq. (11–189) is exactly the same as Eq. (11–186). This term represents a probability for the transition from state 2 to state 1 accompanied by emission, which is proportional to the intensity of the radiation field $U(\omega)$. Therefore it may be interpreted as the transition probability due to induced emission. The expression for induced emission thus obtained is the same as that of the semiclassical theory. The second term of Eq. (11–189) is independent of the radiation field and remains there when the field $U(\omega)$ is reduced to zero. This term represents a transition probability which is intrinsic to the atomic system and so may be interpreted as being due to the spontaneous emission. Thus we have derived the spontaneous emission* as well as the induced emission. The lifetime of spontaneous emission is given by

$$\frac{1}{\tau_{2\to1}} = \frac{4}{3}\, \frac{e^2\omega^3}{\hbar c^3} |r_{21}|^2, \qquad (11\text{--}190)$$

the order of magnitude of $\tau_{2\to1}$ for an atomic system being 10^{-8} sec. Because of the fact that the emitted radiation must be in equilibrium

* The expression for the spontaneous emission may also be obtained from a consideration of the radiation damping force. See Problem 11–8.

with the absorbed radiation, the frequency, selection, polarization, and intensity rules derived for absorption must apply equally well to the emission process. Thus by the help of the equilibrium condition for an atom in a blackbody radiation, the semiclassical theory succeeds in providing us with the essential information regarding the emission process as well as the absorption process, and thereby gives us a certain degree of understanding of the radiation process without having to go into the details of the quantum theory of radiation.

PROBLEMS

11-1. Before the time $t = 0$, the motion of a particle of mass M in a cubic box of finite volume a^3 is known to be described by an eigenfunction, the quantum numbers of which are $n_x = 1$, $n_y = 1$, $n_z = 1$. At the time $t = 0$, a weak perturbation potential is switched on. The perturbation potential is such that all matrix elements are equal, their values being denoted by λH. Determine the state of motion of the particle in the long run after the time $t = 0$.

11-2. In the previous problem the quantum numbers are changed to $n_x = 1,000,000$, $n_y = 0$, $n_z = 0$; other conditions remain the same. Determine the state of motion of the particle in the time period $0 < t < \infty$. What conclusion may one draw from this problem with respect to quantum statistical mechanics, which concerns the behavior of a system when $t \to \infty$?

11-3. The quadruply ionized boron atom $(Z = 5)$ B^{++++} is a hydrogenlike atom. Assume hypothetically that the nuclear charge, after the time $t = 0$, increases linearly in time so that after 10^{-20} sec the nucleus has changed to a carbon nucleus $(Z = 6)$. After the time $t = 10^{-20}$ sec, the nuclear charge is assumed to change no more. Before $t = 0$ the electron is known to be at the ground state. Give the quantum-mechanical prediction of the state of motion of the electron after $t = 10^{-20}$ sec. Repeat the same problem with the time interval 10^{-20} sec changed to 10^{-14} sec. What conclusion may one draw concerning the possibility of ionizing an atom by the β-decay of its nucleus?

11-4. The following problems are for those who are familiar with nuclear physics: (a) Derive the energy spectrum of the β-particles in a β-decay process. Assume the matrix element to be a constant. Note that the final state is specified by the momenta of three particles—the β-particle, the neutrino, and the nucleus, the sum of which is zero. (b) Discuss the energy dependence of the neutron cross sections in elastic scattering (n, n), inelastic scattering (n, n') and radiative capture (n, γ) processes. The nuclear physics part of the problem concerns the matrix element only, which is assumed to be a constant. (c) Discuss qualitatively the nuclear reaction through compound nucleus formation (resonance theory) from the point of view of the second-order perturbation theory.

11-5. Determine the differential cross section of scattering by a spherical potential well in the low energy limit.

11-6. Discuss the process of scattering by an electric dipole, the moment of which lies in the direction of the incident beam.

11-7. Show that the second term in the parentheses of Eq. (11–107) gives rise to a term which may be interpreted as a transient that fades away rapidly.

[*Hint:* $(e^{i\omega_{mn}t} - 1)$ is a complex number the representative point of which in the complex plane is on a circle centered at $x = -1$ and having a radius equal to unity. Note also that ω_{mn} in the denominator changes sign when n passes m.]

*11-8. Consider a charged linear harmonic oscillator in the high quantum number region. Derive the expression for the rate of spontaneous emission by considering the fact that the classical rate of energy loss by radiation is

$$\frac{dW}{dt} = \frac{2}{3}\frac{Q^2}{c^3}\ddot{x}^2, \tag{11-191}$$

where Q is the charge and x is the coordinate of the oscillator. [*Hint:* Express the matrix element $[x^2]_{nn}$ in terms of the matrix elements x_{mn} and remember that in the high quantum number region the matrix element x_{mn} changes little when the index changes a few units.

11-9. Equation (11-31) does not exclude $m = k$. From the result of Eq. (11-31), including $m = k$, calculate $\sum_{m=1}^{\infty} |a_m|^2$. Does the result agree with the fact that the total probability is always unity? *Note:* The first-order perturbation contributes only terms of the order of λ^2 to the total probability to which the second-order perturbation also contributes. When all contributions are considered, the total probability is always unity.

11-10. Discuss the case of a charged linear harmonic oscillator for which an electric field is switched on slowly but is turned off suddenly. Also the case in which the field is switched on suddenly but is turned off slowly.

11-11.# *Lorentz theory of electron-positron annihilation.* Consider a positronium reaching the Bohr radius. Assume the electron stops and plunges to the positron, radiating energy according to Eq. (11-191), the energy lost being supplied from the electrostatic field of the electron and thus from its mass. The Lorentz theory of electron thus forces us to give up Newton's dictum that particle is unbreakable and occasionally it can melt gradually. With the electron mass reduced, and thus its charge, the acceleration and radiation slow down, finally to zero and the masses of the electrons are entirely converted to radiation energy, completing the annihilation. Calculate the time required to reach this point, which is the classical theory of the lifetime of positronium. This is too short because the electron should go in circles. See the next problem.

*11-12.# The more realistic scenario of the above problem is a trajectory in many circles with the *Bohr velocity* as the initial condition. By accepting the veracity of the Lorentz theory, this seems to be the orthodox way to calculate the lifetime of positronium. On this attempt we lose the cherished concept of Newton's unbreakable particle, which is never proved; but in return we get rid of the notorious Coulomb singularity that bedevils high energy physics. A mathematical *point* is not a valid physical object; so is the infinity accompanying it. The concept of a melting electron is more acceptable. Compare the result with the experimental value of 10^{-9} sec and that of QED. (Since every particle is accompanied with an anti-particle, the total value of all singularities of the universe, positive and negative, added up to zero, which means singularity has no real existence.)

---

* Indicates more difficult problems.

\# Indicates new problems added to the expanded edition.

CHAPTER 12

GENERAL FORMULATION OF QUANTUM MECHANICS AND ITS APPLICATIONS

We have developed quantum mechanics in a way which brings out the physical meaning most clearly. It is limited to quantum-mechanical systems of one particle. For systems of more than one particle, a more general formulation is necessary. Because of its use of the noncommutative operators, which seems too abstract for beginners, the general formulation is postponed to this last chapter. We shall first restate the basic assumptions we have used so far and then derive a few theorems stated in terms of the operators. By a generalization of this logical system we obtain the general formulation of quantum mechanics. The general theory will then be applied to several physical systems to show the wide application of quantum mechanics. Many-particle systems, rigid bodies, radiation fields and matter waves will be discussed. The wave-particle duality will be further discussed from the viewpoint of quantization of wave fields to conclude this introductory volume on the quantum theory.

12–1 Dynamical quantities represented by operators. We have established quantum mechanics on the basis of three assumptions:

Assumption A. *Born's first assumption.* The quantum-mechanical state of a particle at a given time t_0 is described by a wave function $\Psi(x, y, z, t_0)$, the square of the absolute value of which, $|\Psi(x, y, z, t_0)|^2$, represents the probability density distribution of the position of the particle in coordinate space (x, y, z).

Assumption B. *Born's second assumption.* The square of the absolute value of the Fourier coefficients of $\Psi(x, y, z, t_0)$, $|a_{p_x p_y p_z}|^2$, represents the probabilities of finding the particle at time t_0 to have momentum values equal to (p_x, p_y, p_z), these values being related to the coefficients by

$$\Psi(x, y, z, t_0) = \sum_{p_x p_y p_z} a_{p_x p_y p_z} \frac{1}{\sqrt{\Omega}} e^{(i/\hbar)(p_x x + p_y y + p_z z)}. \qquad (12\text{–}1)$$

Assumption C. *Time-dependent Schrödinger equation.* The evolution of the wave function $\Psi(x, y, z, t)$ in time with a given initial condition $\Psi(x, y, z, t_0)$ is determined by the equation of motion

$$i\hbar \frac{\partial \Psi}{\partial t} = H\left(\frac{\hbar}{i} \frac{\partial}{\partial x}, \frac{\hbar}{i} \frac{\partial}{\partial y}, \frac{\hbar}{i} \frac{\partial}{\partial z}, x, y, z\right) \Psi(x, y, z, t), \qquad (12\text{–}2)$$

until a measurement is made at some later time t_1. The measurement causes a sudden and uncontrollable change of the wave function at t_1 and thus gives a new initial condition for the evolution of $\Psi(x, y, z, t)$ after t_1.

From these assumptions we now deduce a few theorems. The theorems will be expressed in terms of the operators, such as the operators in Eq. (12–2), the use of which was first made in Chapters 10 and 11. These theorems will bring out the physical significance of the operators. We first make a few remarks concerning operators in general. An *operator* **O** is a mathematical operation which may be applied to a function $f(x)$ and changes the function to another, $g(x)$. This relation may be represented *symbolically* by the equation

$$\boldsymbol{O}f(x) = g(x). \tag{12–3}$$

Throughout this chapter we shall use boldface italic letters to denote operators. A few simple examples of the operators will be mentioned here. "Multiplication by x" may be considered as an operator. Thus the equation

$$x(x^3 + 1) = x^4 + x \tag{12–4}$$

may be stated in the operator language as follows: the operator $\boldsymbol{O} = x$, operated on the function $f(x) = x^3 + 1$, changes $f(x)$ to $g(x) = x^4 + x$. The operation is multiplication. "Differentiation with respect to x" may also be considered as an operator. Thus the equation

$$\frac{d}{dx}(x^3 + 1) = 3x^2 \tag{12–5}$$

may be stated in the operator language as follows: the operator $\boldsymbol{O} = d/dx$, operated on $f(x) = x^3 + 1$, changes $f(x)$ to $g(x) = 3x^2$. The operation is differentiation. Operators may be *added* and *multiplied* to form new operators. Thus the operator $\boldsymbol{O} = 3(d/dx)^2 + 2x^2$ is defined by

$$\boldsymbol{O}f(x) = \left\{ 3\left(\frac{d}{dx}\right)^2 + 2x^2 \right\} f(x) = 3\frac{d^2 f(x)}{dx^2} + 2x^2 f(x). \tag{12–6}$$

There are many other kinds of operators.

One important class of operators is the *linear operator*, which is defined by the following equations:

$$\boldsymbol{O}[f_1(x) + f_2(x)] = \boldsymbol{O}f_1(x) + \boldsymbol{O}f_2(x),$$
$$\boldsymbol{O}[cf(x)] = c\boldsymbol{O}f(x), \tag{12–7}$$

where c is a constant. "Multiplication by x" and "differentiation with respect to x" are both linear operators, while "taking square" and "taking square root" are not, as may be verified readily. In quantum mechanics we consider mostly linear operators.

The most important class of operators used in quantum mechanics is the *Hermitian operator*, which is defined as follows: For any two well-behaved functions $f(x)$ and $g(x)$ vanishing at infinity, an operator O satisfying the equation

$$\int_{-\infty}^{\infty} g^*(Of)\, dx = \left[\int_{-\infty}^{\infty} f^*(Og)\, dx\right]^* = \int_{-\infty}^{\infty} (Og)^* f\, dx$$

is said to be Hermitian. The importance of the Hermitian operator lies in the fact that $\int_{-\infty}^{\infty} \psi^* O\psi\, dx$ is a real quantity as may be proved readily by the above equation.

THEOREM 1. The average values* of the coordinates x, y, z of a particle described by a normalized wave function $\Psi(x, y, z, t_0)$ are

$$\bar{x}(t_0) = \iiint \Psi^*(x, y, z, t_0)x\Psi(x, y, z, t_0)\, dx\, dy\, dz,$$

$$\bar{y}(t_0) = \iiint \Psi^*(x, y, z, t_0)y\Psi(x, y, z, t_0)\, dx\, dy\, dz, \qquad (12\text{–}8)$$

$$\bar{z}(t_0) = \iiint \Psi^*(x, y, z, t_0)z\Psi(x, y, z, t_0)\, dx\, dy\, dz.$$

Stated in the operator language, the average values of x, y, z are obtained by inserting the multiplication operators x, y, z between Ψ^* and Ψ and then performing integration.

This theorem follows immediately from assumption A. We here deliberately write the operator between Ψ^* and Ψ for a reason to be made clear later in Theorem 2. This order means that Ψ is to be multiplied by x (or y, z) first, then by Ψ^* before the final integration. By the definition it may be shown that the operators x, y, z are Hermitian. Therefore the average values are real, as they should be.

COROLLARY 1–1. The average value of any function $f(x, y, z)$ at t_0 in a quantum-mechanical state described by $\Psi(x, y, z, t_0)$ is

$$\overline{f_{t_0}(x, y, z)} = \iiint \Psi^*(x, y, z, t_0)f(x, y, z)\Psi(x, y, z, t_0)\, dx\, dy\, dz. \quad (12\text{–}9)$$

Here $f(x, y, z)$ may be regarded as a multiplication operator, i.e., multiplication by $f(x, y, z)$. This operator is also Hermitian.

---

* In some books the average value is called the expectation value.

THEOREM 2. The average values of momenta p_x, p_y, p_z of a particle described by $\Psi(x, y, z, t_0)$ are

$$\overline{p_x}(t_0) = \iiint \Psi^*(x, y, z, t_0) \frac{\hbar}{i} \frac{\partial}{\partial x} \Psi(x, y, z, t_0) \, dx \, dy \, dz,$$

$$\overline{p_y}(t_0) = \iiint \Psi^*(x, y, z, t_0) \frac{\hbar}{i} \frac{\partial}{\partial y} \Psi(x, y, z, t_0) \, dx \, dy \, dz, \qquad (12\text{–}10)$$

$$\overline{p_z}(t_0) = \iiint \Psi^*(x, y, z, t_0) \frac{\hbar}{i} \frac{\partial}{\partial z} \Psi(x, y, z, t_0) \, dx \, dy \, dz.$$

In other words the average values of p_x, p_y, p_z may be obtained by inserting the differentiation operators $(\hbar/i)(\partial/\partial x)$, $(\hbar/i)(\partial/\partial y)$, $(\hbar/i)(\partial/\partial z)$ between Ψ^* and Ψ and then performing integration.

Proof.

$$\iiint \Psi^*(x, y, z, t_0) \frac{\hbar}{i} \frac{\partial}{\partial x} \Psi(x, y, z, t_0) \, dx \, dy \, dz$$

$$= \iiint \left(\sum_{p_x' p_y' p_z'} a^*_{p_x' p_y' p_z'} \frac{1}{\sqrt{\Omega}} e^{-(i/\hbar)(p_x' x + p_y' y + p_z' z)} \right)$$

$$\times \frac{\hbar}{i} \frac{\partial}{\partial x} \left(\sum_{p_x p_y p_z} a_{p_x p_y p_z} \frac{1}{\sqrt{\Omega}} e^{(i/\hbar)(p_x x + p_y y + p_z z)} \right) dx \, dy \, dz$$

$$= \iiint \left(\sum_{p_x' p_y' p_z'} a^*_{p_x' p_y' p_z'} \frac{1}{\sqrt{\Omega}} e^{-(i/\hbar)(p_x' x + p_y' y + p_z' z)} \right)$$

$$\times \left(\sum_{p_x p_y p_z} p_x a_{p_x p_y p_z} \frac{1}{\sqrt{\Omega}} e^{(i/\hbar)(p_x x + p_y y + p_z z)} \right) dx \, dy \, dz. \qquad (12\text{–}11)$$

The dummy indices p_x', p_y', p_z' are used in the expression of Ψ^* to distinguish them from the other dummy indices p_x, p_y, p_z. Integrating with respect to x, y, z first and making use of the orthonormal property of the plane-wave functions, we reduce the double summation to a single summation as the cross terms vanish:

$$\iiint \sum_{p_x' p_y' p_z'} \sum_{p_x p_y p_z} \left(a^*_{p_x' p_y' p_z'} \frac{1}{\sqrt{\Omega}} e^{-(i/\hbar)(p_x' x + p_y' y + p_z' z)} \right) p_x$$

$$\times \left(a_{p_x p_y p_z} \frac{1}{\sqrt{\Omega}} e^{(i/\hbar)(p_x x + p_y y + p_z z)} \right) dx \, dy \, dz = \sum_{p_x p_y p_z} p_x a^*_{p_x p_y p_z} a_{p_x p_y p_z}.$$

$$(12\text{–}12)$$

The right-hand side, according to assumption B, is the average value of p_x over the momentum probability distribution. The theorem is thus proved.

Here, unlike in Eqs. (12–8), the order of writing Ψ, $(\hbar/i)(\partial/\partial x)$, and Ψ^* becomes important. It means that Ψ is to be differentiated first and then the result is to be multiplied by Ψ^* before integration. Equations (12–8) are so written that they have forms similar to those in Eqs. (12–10). By the definition it may be proved that the *momentum operators* $(\hbar/i)(\partial/\partial x)$, $(\hbar/i)(\partial/\partial y)$, $(\hbar/i)(\partial/\partial z)$ are Hermitian. Therefore the average values are all real, as they should be.

COROLLARY 2–1.   The average value of p_x^n may be obtained if in the right-hand side of Eq. (12–10) the operator $(\hbar/i)(\partial/\partial x)$ is replaced by the operator $[(\hbar/i)(\partial/\partial x)]^n$; i.e., successive operation of $(\hbar/i)(\partial/\partial x)$ for n times.

This may be proved by the fact that

$$\left(\frac{\hbar}{i}\frac{\partial}{\partial x}\right)^n e^{(i/\hbar)(p_x x + p_y y + p_z z)} = p_x^n e^{(i/\hbar)(p_x x + p_y y + p_z z)}. \tag{12–13}$$

Note that $[(\hbar/i)(\partial/\partial x)]^n$ is a Hermitian operator.

COROLLARY 2–2.   The average value of a function of p_x, $f(p_x)$, which is expressible in power series $\sum_{n=0}^{\infty} a_n p_x^n$, may be obtained by letting the operator $f((\hbar/i)(\partial/\partial x))$ take the place of $(\hbar/i)(\partial/\partial x)$ in the right-hand side of Eq. (12–10), $f((\hbar/i)(\partial/\partial x))$ being defined by

$$f\left(\frac{\hbar}{i}\frac{\partial}{\partial x}\right)\psi \equiv \sum_{n=0}^{\infty} a_n \left(\frac{\hbar}{i}\frac{\partial\psi}{\partial x}\right)^n.$$

This result follows directly from Corollary 2–1.

THEOREM 3.   The average value of any function $f(p_x, p_y, p_z, x, y, z)$ may be obtained as follows provided that the function may be expressed in a sum of two parts: one depending on x, y, z only, and the other depending on p_x, p_y, p_z, only the latter being expressible in power series of p_x, p_y, p_z:

$$\overline{f_{t_0}(p_x, p_y, p_z; x, y, z)} = \iiint \Psi^*(x, y, z, t_0) f\left(\frac{\hbar}{i}\frac{\partial}{\partial x}, \frac{\hbar}{i}\frac{\partial}{\partial y}, \frac{\hbar}{i}\frac{\partial}{\partial z}; x, y, z\right)$$

$$\times \Psi(x, y, z, t_0)\, dx\, dy\, dz. \tag{12–14}$$

This follows from Corollary 1–1 and Corollary 2–2. Note that the operator is Hermitian.

If $f(p_x, p_y, p_z; x, y, z)$ cannot be separated into two parts, assumptions A and B give no information regarding its average value. We are thus free to define an average value of this function by Eq. (12–14). This definition is consistent with the previous results and thus is a natural generalization of assumptions A and B. Equation (12–14) is thus assumed for any function $f(p_x, p_y, p_z; x, y, z)$. Furthermore, the operator is required to be Hermitian as the average value is always required to be real.

As an example we consider the energy as a function of momenta and coordinates:

$$H(p_x, p_y, p_z; x, y, z) = \frac{1}{2M} (p_x^2 + p_y^2 + p_z^2) + U(x, y, z). \quad (12\text{–}15)$$

This function consists of two parts and the momentum part is in the form of a power series. According to Theorem 3, the average value of energy is

$$\overline{H(p_x, p_y, p_z; x, y, z)} = \iiint \Psi^*(x, y, z, t_0) H \left(\frac{\hbar}{i} \frac{\partial}{\partial x}, \frac{\hbar}{i} \frac{\partial}{\partial y}, \frac{\hbar}{i} \frac{\partial}{\partial z}; x, y, z \right)$$

$$\times \Psi(x, y, z, t_0) \, dx \, dy \, dz. \quad (12\text{–}16)$$

Let the wave function be expressed as a superposition of energy eigenfunctions:

$$\Psi(x, y, z, t_0) = \sum_n b_n \psi_n(x, y, z) e^{-(i/\hbar) E_n t_0}. \quad (12\text{–}17)$$

Making use of the time-independent Schrödinger equation

$$H \left(\frac{\hbar}{i} \frac{\partial}{\partial x}, \frac{\hbar}{i} \frac{\partial}{\partial y}, \frac{\hbar}{i} \frac{\partial}{\partial z}; x, y, z \right) \psi_n = E_n \psi_n \quad (12\text{–}18)$$

and the orthonormal property of its solutions, we reduce Eq. (12–16) to the following:

$$\overline{H(p_x, p_y, p_z; x, y, z)} = \sum_n E_n b_n^* b_n. \quad (12\text{–}19)$$

In Section 4–3 we made a generalization of Born's second assumption: $b_n^* b_n$ represents the probability of finding the system to have energy value E_n. As a result, the average value is to be $\sum_n E_n b_n^* b_n$. Equation (12–19) shows that this generalization is consistent with Theorem 3. Later we shall show that this generalization may be derived from Theorem 3.

It may be shown that the energy operator is Hermitian. Therefore its average value is real as it should be. In quantum mechanics dynamical quantities are associated with Hermitian operators and their average values are all real as they should be.

We now consider another aspect of the time-independent Schrödinger equation which is the equation determining the Fourier coefficients of $\Psi(x, y, z, t)$ with respect to time. Equation (12–18) may be interpreted as follows: the energy eigenvalue E_n and the energy eigenfunction ψ_n satisfy an operator equation

$$H\psi_n = E_n\psi_n, \tag{12-20}$$

where H is the operator corresponding to energy as specified in Theorem 3. Let us also consider the fact that the plane-wave function satisfies

$$\left(\frac{\hbar}{i}\frac{\partial}{\partial x}\right)e^{(i/\hbar)(p_x x + p_y y + p_z z)} = p_x e^{(i/\hbar)(p_x x + p_y y + p_z z)}, \tag{12-21}$$

which may be interpreted to mean that the momentum eigenvalue p_x and eigenfunction $e^{(i/\hbar)(p_x x + p_y y + p_z z)}$ satisfy an operator equation similar to Eq. (12–20). These facts strongly suggest that the eigenvalues and eigenfunctions of any other physical quantities satisfy an operator equation similar to Eq. (12–20). This generalization will be discussed in the following theorems.

THEOREM 4.   The wave function $\psi_n(x, y, z)$, satisfying the operator equation

$$f\left(\frac{\hbar}{i}\frac{\partial}{\partial x}, \frac{\hbar}{i}\frac{\partial}{\partial y}, \frac{\hbar}{i}\frac{\partial}{\partial z}; x, y, z\right)\psi_n(x, y, z) = f_n\psi_n(x, y, z) \tag{12-22}$$

in which f_n is a constant, represents a quantum-mechanical state for which measurements of the quantity $f(p_x, p_y, p_z; x, y, z)$ (a function of momenta and coordinates) yield invariably a definite value f_n (instead of many different values).

Proof.   By Theorem 3 we may show that the average value of $f(p_x, p_y, p_z; x, y, z)$ in the state ψ_n is f_n. Now consider the square of $f(p_x, p_y, p_z; x, y, z)$, the average value of which is, according to Theorem 3,

$$\overline{f^2} = \iiint \psi_n^*(x, y, z)f\left(\frac{\hbar}{i}\frac{\partial}{\partial x}, \frac{\hbar}{i}\frac{\partial}{\partial y}, \frac{\hbar}{i}\frac{\partial}{\partial z}; x, y, z\right)$$

$$\times f\left(\frac{\hbar}{i}\frac{\partial}{\partial x}, \frac{\hbar}{i}\frac{\partial}{\partial y}, \frac{\hbar}{i}\frac{\partial}{\partial z}; x, y, z\right)\psi_n(x, y, z)\,dx\,dy\,dz$$

$$= \iiint \psi_n(x, y, z)f\left(\frac{\hbar}{i}\frac{\partial}{\partial x}, \frac{\hbar}{i}\frac{\partial}{\partial y}, \frac{\hbar}{i}\frac{\partial}{\partial z}; x, y, z\right)f_n\psi_n(x, y, z)\,dx\,dy\,dz$$

$$= \iiint \psi_n(x, y, z)f_n^2\psi_n(x, y, z)\,dx\,dy\,dz$$

$$= f_n^2. \tag{12-23}$$

Similarly the average values of any powers of f, i.e., f^m, may be shown to be f_n^m. In statistics the average value of the square (or any other power) of a quantity f is usually not the same as the square of the average of f. Here we have a situation where the average of any power of f is equal to the same power of the average value f_n. This can be the case only when the statistical distribution of the possible values of f is sharply peaked at a single value f_n; in other words, the probability of finding the value of f to be f_n is unity and the probabilities for any other values are zero. The theorem is thus proved.

The wave functions ψ_n and the constants f_n satisfying Eq. (12–22) are called eigenfunctions and eigenvalues of $f(p_x, p_y, p_z; x, y, z)$.

THEOREM 5. All normalized eigenfunctions of an operator f satisfying the boundary condition at infinity form a *complete orthonormal set* by which a well-behaved function vanishing at infinity may be expanded.

A special case of this theorem concerning energy eigenfunctions has been mentioned before in Sections 4–2 and 9–4. We omit the general proof.

THEOREM 6. The eigenvalues of $f(p_x, p_y, p_z; x, y, z)$ of a system are the only possible values of f that may be obtained upon an observation of f in any quantum-mechanical state of the system.

The proof will be given in conjunction with the following theorem.

THEOREM 7. The square of the absolute value of a coefficient of expansion, $C_n^* C_n$, of a wave function ψ in terms of an orthonormal set of eigenfunctions of $f(p_x, p_y, p_z; x, y, z)$, represents the probability of finding the value of f in the state ψ, upon an observation, to be f_n.

Proof.

$$\psi(x, y, z) = \sum_n C_n \psi_n(x, y, z). \qquad (12\text{–}24)$$

From Theorem 3 the average value of $f(p_x, p_y, p_z; x, y, z)$ in the state $\psi(x, y, z)$ is

$$\bar{f} = \iiint \psi^*(x, y, z) f\left(\frac{\hbar}{i}\frac{\partial}{\partial x}, \frac{\hbar}{i}\frac{\partial}{\partial y}, \frac{\hbar}{i}\frac{\partial}{\partial z}; x, y, z\right) \psi(x, y, z)\, dx\, dy\, dz$$

$$= \iiint \left[\sum_n C_n^* \psi_n^*(x, y, z)\right] f\left(\frac{\hbar}{i}\frac{\partial}{\partial x}, \frac{\hbar}{i}\frac{\partial}{\partial y}, \frac{\hbar}{i}\frac{\partial}{\partial z}; x, y, z\right)$$

$$\times \left[\sum_n C_n \psi_n(x, y, z)\right] dx\, dy\, dz$$

$$= \sum_n f_n C_n^* C_n. \qquad (12\text{–}25)$$

Similarly, the average value of the square of f is

$$\overline{f^2} = \sum_n f_n^2 C_n^* C_n. \tag{12-26}$$

Furthermore,

$$\overline{f^m} = \sum_n f_n^m C_n^* C_n. \tag{12-27}$$

Equations (12–25) through (12–27) may be satisfied only when the quantities $C_n^* C_n$ represent the probabilities of finding the value of f to be f_n. Theorem 7 is thus proved. Since the probability distribution is limited to the eigenvalues f_n only, these values represent the only possible values that may be obtained upon observation. Theorem 6 is thus proved. It may be remarked that the generalization of Born's second assumption in Section 4–3 is a special case of Theorem 7. Therefore it is not an independent assumption.

The above theorems show that the kinematics of quantum mechanics may be very conveniently formulated in the operator language. With each dynamical quantity expressed as a function of momenta and coordinates $f(p_x, p_y, p_z; x, y, z)$ we may associate an operator

$$f\left(\frac{\hbar}{i}\frac{\partial}{\partial x}, \frac{\hbar}{i}\frac{\partial}{\partial y}, \frac{\hbar}{i}\frac{\partial}{\partial z}; x, y, z\right),$$

which may be abbreviated by the boldface letter f, by means of which the average value of f in a state ψ may be calculated according to Theorem 3. The eigenvalues of f may be obtained by solving Eq. (12–22). As the eigenvalues usually form a discrete set, Eq. (12–22) leads to quantization of the dynamical quantity $f(p_x, p_y, p_z; x, y, z)$. Thus we have a *general method of quantization* applicable to any dynamical quantity f. (This method is thus the quantum-mechanical answer to the first problem mentioned at the end of Section 1–3. The quantum-mechanical answer to the second problem there is given by the time-dependent perturbation theory of Chapter 11.)

It is commonly stated in textbooks that a dynamical quantity may be represented by an operator. The physical meaning will be clearer if we state that the average value of a dynamical quantity may be calculated with the aid of an operator. The significant physical quantity is the average value (or its equivalent, i.e., the probability distribution over the eigenvalues). The operator is but a mathematical tool for obtaining this physical quantity. The situation is similar to the use of the *imaginary* numbers in electrical network theory with an ultimate aim of obtaining *real* physical results. If the student will keep in mind the average values behind the operators, the use of this abstract mathematical tool will no longer appear mysterious.

12–2 **The algebra of operators.** We have seen that operators may be added and multiplied to form new operators. Also it can be shown that operators obey the distributive law and the associative law. Thus the algebraic relations among the operators are very much the same as those among the numbers (integral, rational, irrational, or imaginary). However, there is one major difference, i.e., the commutative law of numbers is not obeyed by operators. Consider the following example. The operator formed by the product of two operators x and d/dx in the order $x \cdot (d/dx)$ is defined by

$$\left(x \cdot \frac{d}{dx} \right) f(x) = x \, \frac{df}{dx}. \tag{12–28}$$

If the order of the product is $(d/dx) \cdot x$, the operator is defined by

$$\left(\frac{d}{dx} \cdot x \right) f(x) = \frac{d}{dx} \, [xf(x)] = f(x) + x \, \frac{df}{dx}. \tag{12–29}$$

Evidently the two operators $x \cdot (d/dx)$ and $(d/dx) \cdot x$ are not the same. Thus the operators x and d/dx are not commutative. In special cases, operators may be commutative. For example, "multiplication by x" and "multiplication by y" are commutative. So are "differentiation with respect to x" and "differentiation with respect to y". When noncommuting operators are involved, the order of writing the product becomes important. In all previous examples, e.g., in the energy operator, the products do not involve noncommuting operators. Thus we need not pay, and actually have not paid, any attention to the order of the operators. Later we shall see more of the noncommuting operators. When two operators O_1 and O_2 are not commutative to each other, we may define a new operator $(O_1 O_2 - O_2 O_1)$, which is not vanishing, called the *commutator* of O_1 and O_2. For example, the commutator of d/dx and x may be found by taking the difference of Eqs. (12–29) and (12–28):

$$\left(\frac{d}{dx} \cdot x - x \cdot \frac{d}{dx} \right) f(x) = f(x).$$

Since $f(x)$ is arbitrary, the above equation implies that the operator

$$\left(\frac{d}{dx} \cdot x - x \cdot \frac{d}{dx} \right)$$

is equivalent to "multiplication by unity." Therefore the commutator is equal to unity:

$$\left(\frac{d}{dx} \cdot x - x \cdot \frac{d}{dx} \right) = 1. \tag{12–30}$$

Similarly, the commutator of the momentum operator $(\hbar/i)(\partial/\partial x)$ and the position operator x is

$$\left(\frac{\hbar}{i}\frac{\partial}{\partial x}\cdot x - x\cdot\frac{\hbar}{i}\frac{\partial}{\partial x}\right) = \frac{\hbar}{i},$$

or

$$p_x x - x p_x = \frac{\hbar}{i},$$

$$p_y y - y p_y = \frac{\hbar}{i},\tag{12–31}$$

$$p_z z - z p_z = \frac{\hbar}{i}.$$

On the other hand p_x, p_y, p_z commute among themselves. The same is true for x, y, z. Thus,

$$xy - yx = 0, \qquad p_x p_y - p_y p_x = 0,$$

$$yz - zy = 0, \qquad p_y p_z - p_z p_y = 0,\tag{12–32}$$

$$zx - xz = 0, \qquad p_z p_x - p_x p_z = 0.$$

One complication thus arises when we write down the operator f from the classical expression $f(p_x, p_y, p_z; x, y, z)$ if some products in the expression of f involve noncommuting operators. In the classical expression, the order of a product is immaterial. Here new assumptions are needed for determining the order of the quantum-mechanical operators.* One important consideration is that the operator so written must be Hermitian, as the average value is always required to be real. Usually, but not always, a symmetric average of the operators, e.g., $\frac{1}{2}(p_x x + x p_x)$, may be used as the quantum-mechanical counterpart of the classical product $x p_x$ (which is the same as $p_x x$). The justification for this assumption lies not only in the fact that this symmetric average reduces to the classical value in the classical limit but also that its deductions agree with experimental results.

The significance of the noncommuting property of the operators x and p_x may be brought out by the following theorem which relates it to the Heisenberg uncertainty relation.

THEOREM 8.   The commutation relation between p_x and x, i.e., $p_x x - x p_x = (\hbar/i)$, leads to the result that the root-mean-square devi-

---

* These are assumptions pertaining to specific quantum-mechanical systems, not to the general theory of quantum mechanics itself.

ations of position and momentum satisfy the following relation:

$$\Delta p_x \, \Delta x \geqq \frac{\hbar}{2}. \tag{12-33}$$

The same holds when x is replaced by y or z.

Proof. Consider the one-dimensional case for simplicity. By using the operators $\boldsymbol{p_x}$ and $\boldsymbol{x}$ we may calculate the average values $\overline{p}_x$ and $\overline{x}$. The mean-square deviation of position from the average $\overline{x}$ may be obtained by

$$(\Delta x)^2 = \int_{-\infty}^{\infty} \psi^*(\boldsymbol{x} - \overline{x})^2 \psi \, dx. \tag{12-34}$$

Similarly, the mean-square deviation of momentum from the average $\overline{p}_x$ is

$$(\Delta p_x)^2 = \int_{-\infty}^{\infty} \psi^*(\boldsymbol{p_x} - \overline{p}_x)^2 \psi \, dx. \tag{12-35}$$

Define two new operators:

$$\boldsymbol{\alpha} \equiv \boldsymbol{x} - \overline{x}, \qquad \boldsymbol{\beta} \equiv \boldsymbol{p_x} - \overline{p}_x. \tag{12-36}$$

Thus

$$\begin{aligned}
(\Delta x)^2 \, (\Delta p)^2 &= \int_{-\infty}^{\infty} \psi^* \boldsymbol{\alpha}^2 \psi \, dx \int_{-\infty}^{\infty} \psi^* \boldsymbol{\beta}^2 \psi \, dx \\
&= \int_{-\infty}^{\infty} (\boldsymbol{\alpha}\psi)^*(\boldsymbol{\alpha}\psi) \, dx \int_{-\infty}^{\infty} \psi^* \boldsymbol{\beta}^2 \psi \, dx.
\end{aligned} \tag{12-37}$$

Integrating by parts and remembering that ψ vanishes at infinity, we transform the second integral and obtain the following result:

$$(\Delta x)^2 \, (\Delta p_x)^2 = \int_{-\infty}^{\infty} (\boldsymbol{\alpha}\psi)^*(\boldsymbol{\alpha}\psi) \, dx \int_{-\infty}^{\infty} (\boldsymbol{\beta}\psi)^*(\boldsymbol{\beta}\psi) \, dx. \tag{12-38}$$

The above equation may be obtained directly from the first equation in Eq. (12-37) by making use of the Hermitian properties of $\boldsymbol{\alpha}$ and $\boldsymbol{\beta}$. We now make use of a mathematical relation known as the Schwarz inequality, which states that for any two complex functions f and g,

$$\int_{-\infty}^{\infty} |f|^2 \, dx \int_{-\infty}^{\infty} |g|^2 \, dx \geqq \left| \int_{-\infty}^{\infty} f^*g \, dx \right|^2. \tag{12-39}$$

This relation may be derived from the following self-evident inequality

$$\int_{-\infty}^{\infty} \left| f - g \frac{\int_{-\infty}^{\infty} fg^* \, dx}{\int_{-\infty}^{\infty} |g|^2 \, dx} \right|^2 dx \geqq 0. \tag{12-40}$$

By Eq. (12–39) we change Eq. (12–38):

$$(\Delta x)^2 (\Delta p_x)^2 \geqq \left| \int_{-\infty}^{\infty} (\boldsymbol{\alpha}\psi)^*(\boldsymbol{\beta}\psi) \, dx \right|^2$$

$$= \left| \int_{-\infty}^{\infty} \psi^* \boldsymbol{\alpha}\boldsymbol{\beta}\psi \, dx \right|^2$$

$$= \left| \int_{-\infty}^{\infty} \psi^* \tfrac{1}{2}(\boldsymbol{\alpha}\boldsymbol{\beta} - \boldsymbol{\beta}\boldsymbol{\alpha})\psi \, dx + \int_{-\infty}^{\infty} \psi^* \tfrac{1}{2}(\boldsymbol{\alpha}\boldsymbol{\beta} + \boldsymbol{\beta}\boldsymbol{\alpha})\psi \, dx \right|^2$$

$$= \tfrac{1}{4} \left| \int_{-\infty}^{\infty} \psi^* (\boldsymbol{\alpha}\boldsymbol{\beta} - \boldsymbol{\beta}\boldsymbol{\alpha})\psi \, dx \right|^2 + \tfrac{1}{4} \left| \int_{-\infty}^{\infty} \psi^* (\boldsymbol{\alpha}\boldsymbol{\beta} + \boldsymbol{\beta}\boldsymbol{\alpha})\psi \, dx \right|^2.$$

$$(12\text{–}41)$$

In the last expression the two cross terms cancel because of the following relation:

$$\left[\int_{-\infty}^{\infty} \psi^* (\boldsymbol{\alpha}\boldsymbol{\beta} \pm \boldsymbol{\beta}\boldsymbol{\alpha})\psi \, dx \right]^* = \pm \left[\int_{-\infty}^{\infty} \psi^* (\boldsymbol{\alpha}\boldsymbol{\beta} \pm \boldsymbol{\beta}\boldsymbol{\alpha})\psi \, dx \right], \quad (12\text{–}42)$$

which may be verified by such operations as are shown in Eqs. (12–37) and (12–38) or by the Hermitian properties of the operators $(\boldsymbol{\alpha}\boldsymbol{\beta} + \boldsymbol{\beta}\boldsymbol{\alpha})$ and $i(\boldsymbol{\alpha}\boldsymbol{\beta} - \boldsymbol{\beta}\boldsymbol{\alpha})$. Equation (12–41) means that

$$(\Delta x)^2 (\Delta p_x)^2 \geqq \tfrac{1}{4} \left| \int_{-\infty}^{\infty} \psi^* (\boldsymbol{\alpha}\boldsymbol{\beta} - \boldsymbol{\beta}\boldsymbol{\alpha})\psi \, dx \right|^2$$

$$= \tfrac{1}{4} \left| \int_{-\infty}^{\infty} \psi^* (x\boldsymbol{p}_x - \boldsymbol{p}_x x)\psi \, dx \right|^2$$

$$= \tfrac{1}{4}\hbar^2.$$

Therefore

$$\Delta x \, \Delta p_x \geqq \frac{\hbar}{2}. \qquad (12\text{–}43)$$

[Cf. Eq. (3–25).]

The above proof may be reformulated in a more general way in which no explicit use of the differential operator $\boldsymbol{p}_x = (\hbar/i)(\partial/\partial x)$ is made. From the Hermitian property of x and $\boldsymbol{p}_x$, and the commutation relation $\boldsymbol{p}_x x - x\boldsymbol{p}_x = \hbar/i$, the same result may be derived. Equation (12–43) represents the uncertainty relation. The above theorem shows that the uncertainty relation is borne out by any wave function ψ and therefore is satisfied in any quantum-mechanical state. Previously we have derived it only in special quantum-mechanical states.

This theorem shows that the mathematical origin of the uncertainty relation lies in the noncommutativity of the operators $\boldsymbol{p}_x$ and x. From the mathematical point of view, the extension from classical mechanics to quantum mechanics corresponds to an extension from a commutative algebra to a noncommutative algebra. And the requirement that the uncertainty relation be incorporated in a mechanical theory is fulfilled by the noncommutativity of the operators. This consideration also enables us to gain some insight into the meaning of the noncommutativity of the operators. It also guides us, as we shall see later, in formulating quantum-mechanical theories for physical systems not yet discussed.

The above theorem implies that momentum eigenfunctions ($\Delta p = 0$) cannot be position eigenfunctions ($\Delta x = 0$) at the same time and vice versa. This leads to the following theorem concerning simultaneous eigenfunctions.

THEOREM 9. The necessary and sufficient condition that simultaneous eigenfunctions exist for two operators O_1 and O_2 is that they commute,

$$O_1 O_2 - O_2 O_1 = 0. \tag{12-44}$$

Proof. If simultaneous eigenfunctions ψ_n exist for O_1 and O_2, i.e.,

$$O_1 \psi_n = O_{1n} \psi_n, \qquad O_2 \psi_n = O_{2n} \psi_n,$$

then

$$O_1 O_2 \psi_n = O_1(O_{2n} \psi_n) = O_{2n} O_1 \psi_n = O_{2n} O_{1n} \psi_n,$$

$$O_2 O_1 \psi_n = O_2(O_{1n} \psi_n) = O_{1n} O_2 \psi_n = O_{1n} O_{2n} \psi_n.$$

Therefore

$$(O_1 O_2 - O_2 O_1)\psi_n = 0.$$

As any wave function may be expanded in a series of ψ_n, the above equation holds also for any arbitrary wave function ψ. Therefore

$$O_1 O_2 - O_2 O_1 = 0.$$

Thus the condition is proved to be necessary. To prove the condition to be sufficient we consider a set of eigenfunctions of O_1, satisfying

$$O_1 \psi_{1n} = O_{1n} \psi_{1n}.$$

From the condition $O_1 O_2 = O_2 O_1$, we have

$$O_1(O_2 \psi_{1n}) = O_2 O_1 \psi_{1n} = O_2 O_{1n} \psi_{1n} = O_{1n}(O_2 \psi_{1n}).$$

Therefore $(O_2 \psi_{1n})$ is a wave function which is an eigenfunction of O_1 with eigenvalue O_{1n}. In the nondegenerate case there is only one eigenfunction for eigenvalue O_{1n} which is ψ_{1n}. Therefore $(O_2 \psi_{1n})$ and ψ_{1n} must be equivalent and they can differ only by a multiplication constant O_{2n}:

$$O_2 \psi_{1n} = O_{2n} \psi_{1n}.$$

The last equation states that ψ_{1n} is an eigenfunction of O_2 with eigenvalue O_{2n}. The theorem is thus proved. We omit the proof for the degenerate case which may be carried out similarly with slight modification.

12–3 Angular momentum in quantum mechanics. The theorems of the last two sections will now be applied to a special case for illustration. We consider the dynamical quantity angular momentum. In classical mechanics the angular momentum $\vec{M}$ is a vector quantity defined as a function of momenta and coordinates:

$$\vec{M} = \vec{r} \times \vec{p}. \tag{12–45}$$

The three components are

$$M_x = yp_z - zp_y,$$
$$M_y = zp_x - xp_z, \tag{12–46}$$
$$M_z = xp_y - yp_x.$$

Actually there are three functions, M_x, M_y, M_z, of momenta and coordinates. As these functions are not the kind that can be separated in two parts, one containing momenta and the other containing coordinates, we cannot rely on Born's two assumptions to determine their average values (or their equivalent). However, they may be treated by the generalized point of view of Theorem 3. Accordingly, the following operators are associated with the angular momentum components:

$$M_x \rightarrow \frac{\hbar}{i} \left(y \frac{\partial}{\partial z} - z \frac{\partial}{\partial y} \right),$$

$$M_y \rightarrow \frac{\hbar}{i} \left(z \frac{\partial}{\partial x} - x \frac{\partial}{\partial z} \right), \tag{12–47}$$

$$M_z \rightarrow \frac{\hbar}{i} \left(x \frac{\partial}{\partial y} - y \frac{\partial}{\partial x} \right).$$

The order of the two operators in the six products is immaterial, since the two always commute. By means of these operators their average values in a state ψ may be obtained by Eq. (12–14). Also the quantization of angular momentum may be worked out by solving Eq. (12–22), the eigenvalues of which are the allowed angular momentum values. Incidentally, it may be shown that these operators are Hermitian.

Let us first consider $\boldsymbol{M}_z$. Making a transformation from rectangular to spherical coordinates we may express $\boldsymbol{M}_z$ as follows:

$$\boldsymbol{M}_z = \frac{\hbar}{i} \frac{\partial}{\partial \varphi}. \tag{12–48}$$

The eigenvalue equation is thus

$$\frac{\hbar}{i} \frac{\partial \psi}{\partial \varphi} = M_z \psi, \tag{12-49}$$

where M_z is a constant representing the eigenvalue of $\boldsymbol{M_z}$. This equation may be solved readily. The eigenfunction takes the form

$$\psi = C(r, \theta)e^{(i/\hbar)M_z\varphi}, \tag{12-50}$$

where $C(r, \theta)$ is an arbitrary function of the other two spherical coordinates. As φ is a variable with a period of 2π, ψ will not be a single-valued function unless

$$\frac{i}{\hbar} M_z 2\pi = 2\pi i m, \qquad m = 0, \pm 1, \pm 2, \ldots,$$

or

$$M_z = m\hbar, \qquad m = 0, \pm 1, \pm 2, \ldots. \tag{12-51}$$

Thus the z-component of the angular momentum is quantized to integral units of $\hbar$. The corresponding eigenfunctions are

$$\psi_m = C(r, \theta)e^{im\varphi}, \qquad m = 0, \pm 1, \pm 2, \ldots. \tag{12-52}$$

As M_x and M_y are on the same footing with M_z, the rule of quantization for M_x and M_y is the same as M_z. However, we may not conclude that they are all quantized at the same time. First, we show that $\boldsymbol{M_x}$, $\boldsymbol{M_y}$, and $\boldsymbol{M_z}$ do not commute to one another. To do this we calculate $\boldsymbol{M_x M_y}$ and $\boldsymbol{M_y M_x}$:

$$\boldsymbol{M_x M_y} = (yp_z - zp_y)(zp_x - xp_z) = yp_z zp_x - yp_z xp_z - zp_y zp_x + zp_y xp_z, \tag{12-53}$$

$$\boldsymbol{M_y M_x} = (zp_x - xp_z)(yp_z - zp_y) = zp_x yp_z - zp_x zp_y - xp_z yp_z + xp_z zp_y. \tag{12-54}$$

The order of multiplication is preserved in all products. The four negative terms in Eqs. (12–53, 54) are all products of commuting operators while the four positive terms contain noncommuting operators. Taking the difference of the two equations, we have

$$\boldsymbol{M_x M_y} - \boldsymbol{M_y M_x} = yp_x(p_z z - zp_z) + xp_y(zp_z - p_z z)$$

$$= \frac{\hbar}{i} yp_x - \frac{\hbar}{i} xp_y$$

$$= -\frac{\hbar}{i} \boldsymbol{M_z}. \tag{12-55}$$

Two similar equations may be obtained by permutation of x, y, z. The three together specify the commutation rules:

$$M_x M_y - M_y M_x = -\frac{\hbar}{i} M_z,$$

$$M_y M_z - M_z M_y = -\frac{\hbar}{i} M_x, \qquad (12\text{–}56)$$

$$M_z M_x - M_x M_z = -\frac{\hbar}{i} M_y.$$

These equations may be summarized in vector notation,

$$\vec{M} \times \vec{M} = -\frac{\hbar}{i} \vec{M}. \qquad (12\text{–}57)$$

Since the three operators do not commute, we conclude from Theorem 9 that any eigenfunction of M_z cannot be at the same time an eigenfunction of M_x or M_y. For a quantum state described by a wave function of the form of Eq. (12–52), M_z has a definite value but M_x (or M_y) has not. Upon observation of M_x (or M_y), the values obtained are limited to the eigenvalues of M_x, which are also given by Eq. (12–51). Each eigenvalue has a certain probability of being observed. Similarly, wave functions may be constructed such that M_x has a definite value. For these wave functions, M_z (or M_y) does not have a definite value but has a distribution of probable values limited to its eigenvalues.

We now consider a new physical quantity, the square of the angular momentum vector M^2:

$$M^2 = M_x^2 + M_y^2 + M_z^2. \qquad (12\text{–}58)$$

The operator associated with this quantity is

$$M^2 = -\hbar^2 \left(y \frac{\partial}{\partial z} - z \frac{\partial}{\partial y} \right)^2 - \hbar^2 \left(z \frac{\partial}{\partial x} - x \frac{\partial}{\partial z} \right)^2 - \hbar^2 \left(x \frac{\partial}{\partial y} - y \frac{\partial}{\partial x} \right)^2.$$

$$(12\text{–}59)$$

By straightforward transformation to the spherical coordinates, we obtain

$$M^2 = -\hbar^2 \left[\frac{1}{\sin \theta} \frac{\partial}{\partial \theta} \left(\sin \theta \frac{\partial}{\partial \theta} \right) + \frac{1}{\sin^2 \theta} \frac{\partial^2}{\partial \varphi^2} \right]. \qquad (12\text{–}60)$$

The eigenvalue equation is thus

$$-\hbar^2 \left[\frac{1}{\sin \theta} \frac{\partial}{\partial \theta} \left(\sin \theta \frac{\partial}{\partial \theta} \right) + \frac{1}{\sin^2 \theta} \frac{\partial^2}{\partial \varphi^2} \right] \psi = M^2 \psi, \qquad (12\text{–}61)$$

where M^2 is a constant representing the eigenvalue of $\boldsymbol{M}^2$. This equation is identical with the Schrödinger equation for a space rotator, Eq. (8–4), if in the latter $2Ma^2E$ is replaced by M^2. The eigenvalues and eigenfunctions of Eq. (8–4) are already determined in Section 8–1 and may be taken over for the equation (12–61) if the substitution $2Ma^2E \rightarrow M^2$ is made. The eigenvalues of M^2 are thus, according to Eq. (8–36),

$$M^2 = l(l+1)\hbar^2, \qquad l = 0, 1, 2, \ldots . \qquad (12\text{–}62)$$

The eigenfunctions are, according to Eq. (8–35),

$$\psi_{lm} = C(r)\,Y_{lm} = C(r)e^{im\varphi}P_l^m\,(\cos\theta), \qquad \begin{aligned} &l = 0, 1, 2, \ldots, \\ &m = -l, -l+1, \ldots, 0, \ldots l, \end{aligned}$$

$$(12\text{–}63)$$

where $C(r)$ is an arbitrary function of the radius vector r. There are $(2l+1)$ independent eigenfunctions, with different m values, corresponding to one eigenvalue of M^2 designated by a given value of l. Comparison of Eqs. (12–52) and (12–63) leads immediately to the conclusion that the eigenfunctions of M^2 are also eigenfunctions of M_z, but the $2l+1$ eigenfunctions for a given l correspond to different eigenvalues of M_z from $-l\hbar$ to $+l\hbar$. Thus Eq. (12–63) actually gives the simultaneous eigenfunctions of both operators M^2 and M_z. According to Theorem 9, the two operators M^2 and M_z must commute. This is verified immediately by using Eqs. (12–48) and (12–60). Since M_x, M_y, M_z are on an equal footing, M^2 must commute with M_x and M_y. The reader is urged to verify these relations by using Eqs. (12–56) [not by Eqs. (12–48) and (12–60)].

In central force problems, the energy or Hamiltonian operator may be written as follows:

$$H = -\frac{\hbar^2}{2M}\left\{\frac{1}{r^2}\frac{\partial}{\partial r}\left(r^2\frac{\partial}{\partial r}\right) + \frac{1}{r^2}\left[\frac{1}{\sin\theta}\frac{\partial}{\partial\theta}\left(\sin\theta\frac{\partial}{\partial\theta}\right) + \frac{1}{\sin^2\theta}\frac{\partial^2}{\partial\varphi^2}\right]\right\} + U(r).$$

$$(12\text{–}64)$$

From Eqs. (12–48), (12–60), and (12–64) it may be verified readily that the three operators $\boldsymbol{M}_z$, $\boldsymbol{M}^2$, and $\boldsymbol{H}$ commute with one another. From Theorem 9 it follows that there exist simultaneous eigenfunctions for all three operators. In fact, the reader may verify that the wave functions of the hydrogen atom ψ_{nlm} are actually simultaneous eigenfunctions of $\boldsymbol{M}_z$, $\boldsymbol{M}^2$, and $\boldsymbol{H}$, the energy eigenvalues being $-(MZ^2e^4/2\hbar^2n^2)$, M^2 eigenvalues being $l(l+1)\hbar^2$, and M_z eigenvalues being $m\hbar$. The three indices n, l, m, thus designate the three eigenvalues of $\boldsymbol{H}, \boldsymbol{M}^2, \boldsymbol{M}_z$. As the electron in the hydrogen atom has three degrees of freedom, its motion is specified by three constants of motion. In the old quantum theory three quantum conditions are imposed and thus the three constants of motion are quantized. In quantum mechanics the three commutation relations, Eqs. (12–31), serving the same purpose as the three quantum conditions (we shall not discuss in detail the relation between commutation relation and quantum condition here), lead to the simultaneous quantization of the three constants of motion. Other dynamical quantities like M_x and M_y are also quantized but not simultaneously with M_z. Their eigenfunctions may be expressed as linear superpositions of the eigenfunctions of M_z. It may also be concluded that in a system of three degrees of freedom it requires three observations to determine the quantum-mechanical state completely. The observation of energy determines the quantum number n, but there are n^2 degenerate states having the same energy value. A second observation of M^2 determines the quantum number l, but there still remain $(2l + 1)$ degenerate states for the same n and l. A third observation of M_z determines the quantum number m. The quantum-mechanical state ψ_{nlm} is thus completely determined without degeneracy. (It may be remarked that in some special cases one measurement may serve the purpose of determining two quantum numbers, e.g., the energy measurement in the alkali atoms determines both n and l. Also all s-states are completely determined, once n and l are known, M_z being zero automatically.)

In classical mechanics the magnitude of the angular momentum is defined as the length of the angular momentum vector; it equals the maximum value of a component of it. In quantum mechanics the length of the angular momentum vector is $\sqrt{l(l + 1)}\,\hbar$ while the maximum value of a component is $l\hbar$. The two are not the same. The reason is that the three components are represented by noncommuting operators so that there exists no simultaneous eigenfunctions of all three components.

12–4 Position eigenfunctions and the principle of superposition.

We have considered the eigenvalue problems of momentum, energy, and angular momentum. It is natural to extend this discussion to the simplest kind of all operators—the position operator. The coordinate x is associated with a multiplication operator $\boldsymbol{x}$. The eigenvalues and eigenfunctions of $\boldsymbol{x}$ are determined by the following equation:

$$\boldsymbol{x}\psi_n = x_n\psi_n, \qquad (12\text{–}65)$$

where x_n is a constant. The eigenfunction ψ_n is a function of x such that, after multiplying it by the variable x, it changes to itself multiplied by a constant x_n. A function $\Delta_n(x)$, narrowly peaked at $x = x_n$, defined by

$$\Delta_n(x) = \begin{cases} 0, & \text{for} \quad x_n > x > x_n + \epsilon, \\ \dfrac{1}{\epsilon}, & \text{for} \quad x_n \le x \le x_n + \epsilon, \end{cases} \qquad (12\text{–}66)$$

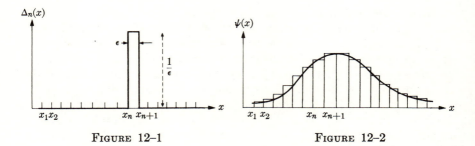

FIGURE 12–1   FIGURE 12–2

and shown graphically in Fig. 12–1, approximately satisfies Eq. (12–65), as may be verified readily by substitution. Therefore it may be considered as an approximation for the position eigenfunction. The approximation becomes better when the peak width ϵ becomes narrower. Before we let ϵ approach zero, we first define the so-called *Dirac δ-function*, $\delta(x)$, by the following equations:

$$\delta(x) = 0, \quad \text{for} \quad x \neq 0,$$
$$\int_{-\infty}^{\infty} \delta(x) \, dx = 1. \tag{12–67}$$

When $\epsilon \to 0$ the function $\Delta_n(x)$ approaches $\delta(x - x_n)$ as its limit. Thus $\delta(x - x_n)$ may be considered as the position eigenfunction with position eigenvalue equal to x_n. Since a solution $\delta(x - x_n)$ may be found for any value of x_n, the coordinate x is not quantized, its eigenvalues forming a continuous spectrum from $-\infty$ to ∞.

Since any wave function may be expanded in a series of eigenfunctions of any operator, it should be possible to do so with the position eigenfunctions. We shall carry out this expansion by using the approximate form $\Delta_n(x)$. We first note that $\Delta_n(x)$ and $\Delta_m(x)$ are orthogonal to each other if $x_n \neq x_m$. This result may be verified readily. Next we find the normalization constant of $\Delta_n(x)$, which turns out to be $\sqrt{\epsilon}$. Thus $\sqrt{\epsilon}\Delta_n(x)$'s form an orthonormal set. An arbitrary wave function $\psi(x)$ may now be expanded in terms of this set:

$$\psi(x) = \sum_n a_n \sqrt{\epsilon} \, \Delta_n(x); \tag{12–68}$$

the coefficients of expansion may be calculated as follows:

$$a_n = \int_{-\infty}^{\infty} \psi(x) \sqrt{\epsilon} \, \Delta_n(x) \, dx$$
$$= \sqrt{\epsilon} \, \overline{\psi(x_n)}, \tag{12–69}$$

where $\overline{\psi(x_n)}$ is a constant representing the average of $\psi(x)$ over the range $x_n \to x_{n+1}$. From Eqs. (12–68) and (12–69) we have

$$\psi(x) = \sum_n \overline{\psi(x_n)}\, \epsilon\, \Delta_n(x). \tag{12–70}$$

The right-hand side of Eq. (12–70) is plotted in Fig. 12–2. It is a zigzag line. When ϵ is made smaller, the line approaches the smooth curve $\psi(x)$ as its limit. The expansion is thus verified. Furthermore, the condition that $\sum_n a_n^* a_n = 1$ leads to

$$\sum_n \epsilon |\overline{\psi(x_n)}|^2 = 1. \tag{12–71}$$

When $\epsilon \to 0$, Eq. (12–71) becomes

$$\int_{-\infty}^{\infty} |\psi(x)|^2\, dx = 1, \tag{12–72}$$

which is actually satisfied because $\psi(x)$ is normalized. The above expansion may be easily written in terms of the δ-function; this is left to the reader.

The meaning of the expansion coefficient specified by Theorem 7 leads to the conclusion that $\epsilon |\overline{\psi(x_n)}|^2$ represents the probability of finding the particle in the region $x_n \le x \le x_n + \epsilon$. When ϵ is small, this reduces to $|\psi(x_n)|^2\, dx$, which is actually the probability of finding the particle at x_n in dx according to Born's first assumption. Therefore Born's first assumption may be stated in a similar fashion as his second assumption; the square of the absolute value of the coefficient of expansion in terms of position eigenfunctions represents the relative probability of finding the position equal to the corresponding position eigenvalue. In Chapter 2 the mathematical expressions of the two assumptions are not similar in form, leaving the impression that they are different in nature. We now see that they are of the same kind and may be stated as two special cases of a general statement.

The statement of Born's first assumption as given in assumption A is based on a consideration of the *geometric* property of the wave function (the amplitude of a wave propagating in space-time). The introduction of the position eigenfunctions makes it possible to consider the wave function from its *algebraic* property. Thus the wave function ψ, or more correctly the ensemble of numbers $\overline{\psi(x_1)}, \overline{\psi(x_2)}, \ldots, \overline{\psi(x_n)}, \ldots$, in the limit $\epsilon \to 0$, may be regarded, according to Eq. (12–69), as a set of coefficients of expansion. While the geometric property has strong appeal to intuition, it is useful only for systems of three degrees of freedom. The algebraic approach is more useful for more complicated systems.

When wave functions are expanded in series of orthonormal functions, e.g., the position eigenfunctions, they may be regarded as vectors in a space of infinitely many, continuous dimensions (Hilbert space). The orthonormal set plays the role of a set of orthogonal unit vectors and the expansion coefficients play the role of the components of the vectors. The addition of two wave functions may be regarded as the addition of two vectors, which is achieved by adding the corresponding components. Thus the wave functions may be studied by the algebraic properties of vectors in Hilbert space. A change of the wave function, such as its evolution from time t_1 to t_2, may be considered as a transformation of a vector in Hilbert space. Expanding a wave function $\psi(x)$ in terms of another orthonormal set, e.g., the momentum eigenfunctions, may be considered as expressing a vector in another set of unit vectors. The momentum eigenfunctions

$$\psi_{p_x} = \frac{1}{\sqrt{\Omega}}\, e^{(i/\hbar)p_x x}, \qquad p_x = -\infty, \ldots, \infty, \qquad (12\text{–}73)$$

may thus be regarded as the equations of transformation connecting the new unit vectors (momentum eigenfunctions) with the old unit vectors (position eigenfunctions). In fact, the most general formulation of quantum mechanics is based on the concepts of vectors and transformations in Hilbert space which we do not discuss here.

When a wave function ψ is expressed in a series of energy eigenfunctions,

$$\psi = \sum_n a_n \psi_n, \qquad (12\text{–}74)$$

we interpret $a_n^* a_n$ as the probability of finding the energy value to be E_n. This is also the probability of finding $f(E)$ to be $f(E_n)$ and finding other characteristics to be those of ψ_n. These may be summarized by saying that the state ψ has a probability $a_n^* a_n$ of being in the state ψ_n. In other words, the state ψ may be regarded as being partly in the state ψ_1, partly in ψ_2, etc., with relative probabilities $a_1^* a_1$, $a_2^* a_2$, etc. This situation may be referred to by saying that the state ψ is a *superposition of states ψ_1, ψ_2,* etc. (this is the definition of superposition). The two assumptions of Born and their generalization (Theorem 7) may be restated thus: *any quantum state ψ may be considered as a superposition of eigenstates of any operator f, the relative probabilities of the eigenstates being given by the squares of the absolute values of the coefficients of expansion of ψ in terms of the eigenfunctions of f.* This statement is a derived theorem in our presentation of quantum mechanics. However, because of its fundamental importance, it may be taken as a starting point in developing a general formulation of quantum mechanics (see Dirac, *The Principles of Quantum Mechanics*).

When used in this manner, the statement is referred to as the *principle of superposition*.*

The physical importance of the principle of superposition lies in the fact that it provides a mathematical scheme capable of describing the interference phenomena. This is due to the fact that the probability is always related to a square. The diffraction pattern of an electron may be considered as resulting from the interference of wave functions with regard to probability distribution of position. Nevertheless, the principle implies that wave functions may also interfere when we consider probability distribution of other dynamical quantities, thus giving rise to a variety of interference phenomena.

12–5 Equations of motion in the operator form. The previous formulation of quantum mechanics based on assumptions A, B, C has its equation of motion in the form of the time-dependent Schrödinger equation. We shall now deduce its equivalent in the operator formalism.

Consider any function of momentum and position $F(p_x, p_y, p_z; x, y, z)$, the corresponding operator

$$F\left(\frac{\hbar}{i}\frac{\partial}{\partial x}, \frac{\hbar}{i}\frac{\partial}{\partial y}, \frac{\hbar}{i}\frac{\partial}{\partial z}; x, y, z\right)$$

being Hermitian and denoted by $\boldsymbol{F}$. Its average value in a state $\Psi(x, y, z, t)$ is

$$\overline{F}(t) = \iiint\limits_{-\infty}^{\infty} \Psi^*(x, y, z, t)\boldsymbol{F}\Psi(x, y, z, t)\,dx\,dy\,dz. \qquad (12\text{--}75)$$

The average value is a function of time. This function may be differentiated with respect to time:

$$\frac{d\overline{F}(t)}{dt} = \iiint\limits_{-\infty}^{\infty} \left(\frac{\partial\Psi^*}{\partial t}\boldsymbol{F}\Psi + \Psi^*\boldsymbol{F}\frac{\partial\Psi}{\partial t}\right)dx\,dy\,dz. \qquad (12\text{--}76)$$

Making use of the assumption C, we have

$$\frac{d\overline{F}(t)}{dt} = \frac{1}{i\hbar}\iiint\limits_{-\infty}^{\infty}[-(\boldsymbol{H}\Psi)^*\boldsymbol{F}\Psi + \Psi^*\boldsymbol{F}\boldsymbol{H}\Psi]\,dx\,dy\,dz. \qquad (12\text{--}77)$$

---

* From the pedagogical point of view it seems desirable to introduce the mathematics and physics of the principle of superposition through the two assumptions of Born rather than to propose the principle axiomatically as a basic assumption.

By the Hermitian property of H we may change the first term as follows:

$$\frac{d\bar{F}(t)}{dt} = \frac{1}{i\hbar} \iiint\limits_{-\infty}^{\infty} (-\Psi^* HF\Psi + \Psi^* FH\Psi)\, dx\, dy\, dz$$

$$= \iiint\limits_{-\infty}^{\infty} \Psi^* \frac{i}{\hbar} (HF - FH)\Psi\, dx\, dy\, dz. \tag{12-78}$$

Let us define an operator $\dot{F}$ by

$$\iiint\limits_{-\infty}^{\infty} \Psi^* \dot{F}\Psi\, dx\, dy\, dz \equiv \frac{d\bar{F}(t)}{dt}. \tag{12-79}$$

$\dot{F}$ is thus an operator related to the time rate of change of the dynamical quantity $F(p_x, p_y, p_z; x, y, z)$. Equations (12–78) and (12–79) lead immediately to

$$\dot{F} = \frac{i}{\hbar} (HF - FH). \tag{12-80}$$

The right-hand side gives the explicit form of the operator $\dot{F}$. Once the time rate of change of F is known, the initial condition determines the future of F completely. Therefore Eq. (12–80) may be regarded as the equation of motion of the dynamical quantity F in the operator form. It may be noted that the equations of motion of all dynamical quantities have the same form of Eq. (12–80). We shall consider this result as a theorem derived from assumptions A, B, C. This theorem is stated below.

THEOREM 10. The equation of motion for a dynamical quantity F in the operator form is

$$\dot{F} = \frac{i}{\hbar} (HF - FH) \tag{12-81}$$

where $\dot{F}$ is defined by

$$\iiint\limits_{-\infty}^{\infty} \Psi^* \dot{F}\Psi\, dx\, dy\, dz = \frac{d\bar{F}(t)}{dt}. \tag{12-82}$$

The resemblance of quantum mechanics (in the operator form) to classical mechanics will now be demonstrated by considering two special cases: $F = x$ and $F = p_x$ (or the y, z counterparts). For $F = x$ we have

$$\dot{x} = \frac{i}{\hbar} (Hx - xH)$$

$$= \frac{i}{\hbar} \left\{ \left[-\frac{\hbar^2}{2M} \nabla^2 + U(x, y, z) \right] \cdot x - x \cdot \left[-\frac{\hbar^2}{2M} \nabla^2 + U(x, y, z) \right] \right\}$$

$$= \frac{i}{\hbar} \left\{ -\frac{\hbar^2}{2M} \left[\left(\frac{\partial^2}{\partial x^2} \right) \cdot x - x \left(\frac{\partial^2}{\partial x^2} \right) \right] \right\}, \tag{12–83}$$

since $U(x, y, z)$, as well as $\partial^2/\partial y^2$, $\partial^2/\partial z^2$, commutes with x. The last expression may be evaluated as follows: For an arbitrary function $f(x)$,

$$\left[\left(\frac{\partial^2}{\partial x^2} \right) \cdot x - x \left(\frac{\partial^2}{\partial x^2} \right) \right] f(x) = \frac{\partial^2}{\partial x^2} [x f(x)] - x \frac{\partial^2 f(x)}{\partial x^2}$$

$$= 2 \frac{\partial}{\partial x} f(x).$$

Therefore

$$\left(\frac{\partial^2}{\partial x^2} \right) \cdot x - x \cdot \left(\frac{\partial^2}{\partial x^2} \right) = 2 \frac{\partial}{\partial x} . \tag{12–84}$$

Equation (12–83) thus becomes

$$\dot{x} = \frac{\hbar}{iM} \frac{\partial}{\partial x} . \tag{12–85}$$

This may be expressed by

$$\dot{x} = \frac{1}{M} p_x, \tag{12–86}$$

or

$$\dot{x} = \frac{\partial H}{\partial p_x} , \tag{12–87}$$

where the differentiation of an operator function with respect to an operator may be defined in a similar manner as the differentiation of ordinary functions. Consider next $F = p_x$:

$$\dot{p}_x = \frac{i}{\hbar} (H p_x - p_x H)$$

$$= \frac{i}{\hbar} \left\{ \left[-\frac{\hbar^2}{2M} \nabla^2 + U(x, y, z) \right] \frac{\hbar}{i} \frac{\partial}{\partial x} - \frac{\hbar}{i} \frac{\partial}{\partial x} \left[-\frac{\hbar^2}{2M} \nabla^2 + U(x, y, z) \right] \right\}$$

$$= -\frac{\partial U(x, y, z)}{\partial x} , \tag{12–88}$$

or

$$\dot{p}_x = -\frac{\partial H}{\partial x} . \tag{12–89}$$

Equations (12–86, 87, 88, 89) are well known in classical mechanics if the boldface quantities are replaced by ordinary numbers. Actually, the classical *canonical equations of motion*,

$$\dot{x} = \frac{\partial H}{\partial p_x}, \qquad \dot{p}_x = -\frac{\partial H}{\partial x}, \qquad (12\text{–}90)$$

are identical with the quantum-mechanical equation in operator form, i.e., Eqs. (12–87) and (12–89). Equation (12–86) corresponds to the classical relation between velocity and momentum; Eq. (12–88) corresponds to the Newton second law. This resemblance strongly suggests that quantum mechanics is very much closer to classical mechanics when expressed in the operator form. Because of this similarity in the equations of motion, a classical theory may easily be translated into a quantum-mechanical theory. This is the basis of a general formulation of quantum mechanics which will be discussed in the next section.

12–6 General formulation of quantum mechanics. In previous chapters we developed quantum mechanics for the one-particle system from assumptions A, B, C. In this chapter we have derived Theorems 1–10 from assumptions A, B, C. In order to apply quantum mechanics to other systems, such as a system of many particles, it has to be generalized. Before doing so we first reformulate the previous theory in terms of the operators. This operator formalism should include the previous theory as a special case and at the same time should be easily generalizable to apply to other systems. For logical consistency, the reader is urged to forget temporarily the logical system built upon assumptions A, B, C, and to start the whole theory again from the very beginning. In the general formulation the assumptions are to be labeled by Greek letters, and the theorems that follow are to be labeled by Roman numerals, so that they may be distinguished from the previous assumptions which are labeled by Latin letters and the previous theorems which are labeled by Arabic numerals.

We first make the observation that quantum mechanics reduces to classical mechanics in a special case. Therefore every classical quantity, such as position, momentum, energy, etc., is to have a quantum-mechanical counterpart. Furthermore, the major difference between classical and quantum mechanics lies in the kinematics. Every classical quantity has a definite value at a given time while its quantum-mechanical counterpart has not, and instead, has a distribution of possible values. In particular, the position information of a particle at any time is to be specified by a function $\Psi(x, y, z, t)$ such that $\Psi^*\Psi$ represents the probability distribution of position. Other quantities may also be specified by probability distributions among the possible values. From the previous discussions it seems plausible that dynamical quantities may be represented by operators acting on $\Psi(x, y, z, t)$. We thus make the following assumption.

Assumption α. A particle at time t is to be described by a function of its coordinates $\Psi(x, y, z, t)$, the square of the absolute value of which represents the probability distribution of the coordinates of the particle, and by means of which the average values of any other dynamical quantities F may be calculated according to the following equation:

$$\bar{F} = \int \Psi^* F \Psi \, d\tau, \qquad (12\text{–}91)$$

where F is a Hermitian operator* associated with the quantity F. (The explicit form of F is to be specified later.)

This assumption implies that for any classical quantity F there is a quantum-mechanical quantity which may be related to an operator. Theorems 3, 4, 6, 7 suggest that the quantity F may have a set of discrete eigenvalues and thus may be quantized. They further suggest that the function $\Psi(x, y, z, t)$, representing a state, may provide information on the probability distribution of possible values of F (which are the eigenvalues) obtained in observation. Besides the dynamical quantities of the type $F(p_x, p_y, p_z; x, y, z)$, we may also associate operators with dynamical quantities of the type $\dot{F}(p_x, p_y, p_z; x, y, z)$. The meaning of the operators $\dot{F}$ is somewhat different from that of F because the measurement of a time rate of change requires two observations in which the first disturbs the quantum state. The eigenvalues of $\dot{F}$ and their probability distribution thus do not have physical significance and, in fact, do not appear in the mathematical theory at all. Therefore we are free to define the physical meaning of $\dot{F}$ subject only to the condition that it reduces to the classical value in the limit. It is thus defined that the average value of $\dot{F}$, which is the only significant quantity that enters the mathematical theory, equals the time rate of change of the average value of F;† this definition agrees with Eq. (12–79).

The plausibility of assumption α is demonstrated by the one-particle quantum mechanics developed in the previous sections. It may be mentioned that assumption α actually specifies a system of kinematics that allows the quantization of dynamical quantities and the existence of a probability distribution over a set of probable values. No dynamical law is involved here. Still, the kinematics specified by assumption α is too broad. The kinematics we need is more specific as the quantization is specified by the Planck constant. Therefore an additional assumption is required to introduce the Planck constant. The equation for this pur-

---

* Assumed to be Hermitian so that its eigenvalues are all real.

† This is the only quantity we need to derive the Schrödinger equation later in Theorem X.

pose, i.e., the quantum condition, is introduced by the following assumption.

Assumption β. The operators for coordinates and momenta obey the following commutation relations:

$$xy - yx = 0, \qquad yz - zy = 0, \qquad zx - xz = 0; \qquad (12\text{–}92)$$

$$p_x p_y - p_y p_x = 0, \qquad p_y p_z - p_z p_y = 0, \qquad p_z p_x - p_x p_z = 0; \qquad (12\text{–}93)$$

$$p_x x - x p_x = \frac{\hbar}{i}, \qquad p_y y - y p_y = \frac{\hbar}{i}, \qquad p_z z - z p_z = \frac{\hbar}{i}. \qquad (12\text{–}94)$$

To gain some physical insight into these commutation relations we turn to the relation between the commutation relation and the uncertainty relation stated in Theorem 8 (also Theorem 9). Although we cannot derive Eqs. (11–94) from the uncertainty relation (this is not surprising, because the uncertainty relation itself is not an exact law), we may satisfy ourselves by the fact that the kinematics based on assumptions α and β obeys the uncertainty relation. If we take the viewpoint that the uncertainty principle is the manifestation of a basic law in kinematics, we may say that this law is embodied in assumption β. Although we have not discussed in Chapter 6 the possibility of simultaneous measurement of two coordinates, or of two momenta, it can be shown that such measurements are possible, so that the assumptions of Eqs. (11–92) and (11–93) are justified. It may also be mentioned again that Theorem 8 may be proved directly by making use of the commutation relation without the explicit use of the differential operator for the momentum operator.

The quantum kinematics is now complete and the quantum dynamics will be introduced by the following assumption.

Assumption γ. The algebraic relations among the operators representing various dynamical quantities are exactly the same as those among the corresponding classical quantities. (In case the order of operators in a product may not be reversed because of the noncommutative nature of the operators, an additional assumption for the particular system involved will be required to determine the order of the operators.)

According to this assumption, the classical relation $E = (1/2M) \times (p_x^2 + p_y^2 + p_z^2) + U(x, y, z)$ demands that the operator for energy be related to those of momenta and coordinates by $H = (1/2M) \times (p_x^2 + p_y^2 + p_z^2) + U(x, y, z)$. Similarly the classical canonical equations of motion, Eqs. (12–90), demand that the quantum-mechanical equations of motion be Eqs. (12–87) and Eq. (12–89). Thus this assumption not only specifies the functional relations among all operators but also pronounces the equations of motion. At first look, to make such a broad claim, consisting of many apparently independent and irrelevant parts, seems rather greedy. The assumption for the expression of the energy operator and the assumption for the equation of motion seem to be two independent assumptions which should be introduced separately. On the other hand, if we remember that quantum mechanics reduces to classical mechanics as a special case, assumption γ appears to be the most reasonable assumption one may possibly make. Still, the operators and classical quantities are mathematically different, and the relations between classical quantities are no justification for their being valid for the operators. The justification lies solely in the fact that the system of quantum mechanics so formulated is able to reproduce classical mechanics as a special case, as will be shown in the next section.

We may remark that the revolution of quantum mechanics is actually a revolution in kinematics. No new dynamical laws are introduced to *supplant* the classical laws. The quantum dynamical laws are a natural extension of the classical laws in order to make them applicable in the new kinematics.

From assumptions α, β, γ, we shall now derive a number of theorems.

THEOREM I.   The operators corresponding to the coordinates x, y, z are simply multiplication operators $\boldsymbol{x}$, $\boldsymbol{y}$, $\boldsymbol{z}$.

This theorem follows from assumption α immediately.

THEOREM II.   The operators for the momentum components p_x, p_y, p_z are differential operators

$$\frac{\hbar}{i}\frac{\partial}{\partial x}, \quad \frac{\hbar}{i}\frac{\partial}{\partial y}, \quad \frac{\hbar}{i}\frac{\partial}{\partial z}.$$

*Proof.** From Eqs. (12–94), we have

$$\boldsymbol{p}_x\boldsymbol{x} - \boldsymbol{x}\boldsymbol{p}_x = \frac{\hbar}{i}. \tag{12–95}$$

Thus for any function $f(x)$, we have

$$\boldsymbol{p}_x\boldsymbol{x}f(x) = \boldsymbol{x}\boldsymbol{p}_xf(x) + \frac{\hbar}{i}f(x). \tag{12–96}$$

Consider a special case $f(x) = 0$ for $-\infty < x < \infty$. This function is represented by a line coinciding with the x-axis. For this $f(x)$, Eq. (12–96) reduces to

$$\boldsymbol{p}_x0 = \boldsymbol{x}\boldsymbol{p}_x0, \tag{12–97}$$

---

* This proof [Peter Fong, *Am. J. Phys.* **29**, 852 (1961)] is designed to show how the commutation relation leads to a differential operator; it is thus longer than that of Dirac as given in his book *The Principles of Quantum Mechanics*.

where $\boldsymbol{p}_x 0$ is a function of x obtained by the operation of $\boldsymbol{p}_x$ on $f(x) = 0$ for $-\infty < x < \infty$. For Eq. (12–97) to be true for any value of x it must be that

$$\boldsymbol{p}_x 0 = 0. \tag{12–98}$$

This means that the operator $\boldsymbol{p}_x$, when operated on the function represented by a line coinciding with the x-axis, turns this function into another, which is again a line coinciding with the x-axis. Next consider another special case, $f(x) = 1$ for $-\infty < x < \infty$. This function is represented by a horizontal line at a distance of unity above the x-axis. Equation (12–96) leads to

$$\boldsymbol{p}_x x = x\boldsymbol{p}_x 1 + \frac{\hbar}{i}. \tag{12–99}$$

$\boldsymbol{p}_x x$ and $\boldsymbol{p}_x 1$ represent two functions of x resulting respectively from the operation of $\boldsymbol{p}_x$ on $f(x) = x$ and $f(x) = 1$. Equation (12–99) tells us how they are related but not the explicit form of either. We then consider $f(x) = x$ in Eq. (12–96):

$$\boldsymbol{p}_x x^2 = x\boldsymbol{p}_x x + \frac{\hbar}{i} x. \tag{12–100}$$

By means of Eq. (12–99) we have

$$\boldsymbol{p}_x x^2 = x^2 \boldsymbol{p}_x 1 + \frac{\hbar}{i} 2x. \tag{12–101}$$

Thus the function $\boldsymbol{p}_x x^2$ is related to the function $\boldsymbol{p}_x 1$, which remains undetermined. By repeated applications of the above procedure, we have

$$\boldsymbol{p}_x x^n = x^n \boldsymbol{p}_x 1 + \frac{\hbar}{i} n x^{n-1}. \tag{12–102}$$

Since $\boldsymbol{p}_x$ is assumed to be linear, we can easily generalize Eq. (12–102) to the following where a polynomial of x denoted by $F(x)$ replaces x^n in Eq. (12–102):

$$\boldsymbol{p}_x F(x) = F(x)\boldsymbol{p}_x 1 + \frac{\hbar}{i} \frac{d}{dx} F(x). \tag{12–103}$$

Similarly, for any $f(x)$ expressible in a power series of x, we have

$$\boldsymbol{p}_x f(x) = f(x)\boldsymbol{p}_x 1 + \frac{\hbar}{i} \frac{d}{dx} f(x). \tag{12–104}$$

Throughout the above derivation, the function $\boldsymbol{p}_x 1$ remains undetermined. This represents a function of x resulting from the operation of $\boldsymbol{p}_x$ on the function $f(x)$ represented by a horizontal line at a distance of unity above the x-axis. Let us denote this function by $X(x)$. Equation (12–104) thus becomes

$$p_x f(x) = \left[\frac{\hbar}{i} \frac{d}{dx} + X(x) \right] f(x), \qquad (12\text{–}105)$$

which means that the operator p_x is a differential operator plus a multiplication operator:

$$p_x = \frac{\hbar}{i} \frac{d}{dx} + X(x). \qquad (12\text{–}106)$$

The form of $X(x)$ is still undetermined. It cannot be determined by the commutation relation. In fact, Eq. (12–106) is the general solution of an operator p_x satisfying Eq. (11–95). This may be verified readily.

$$\left[\frac{\hbar}{i} \frac{d}{dx} + X(x) \right] \cdot x - x \cdot \left[\frac{\hbar}{i} \frac{d}{dx} + X(x) \right] = \left[\frac{\hbar}{i} \frac{d}{dx} \right] \cdot x - x \cdot \left[\frac{\hbar}{i} \frac{d}{dx} \right] = \frac{\hbar}{i} \cdot$$

$$(12\text{–}107)$$

Equation (12–106) may be expressed in a different form:

$$p_x = e^{-(i/\hbar)\int^x X(x)\,dx} \cdot \frac{\hbar}{i} \frac{d}{dx} \cdot e^{+(i/\hbar)\int^x X(x)\,dx}. \qquad (12\text{–}108)$$

This may be verified easily upon operation on any arbitrary function $f(x)$, the exponentials being all multiplication operators. By choosing different forms of $X(x)$, we have different expressions of the operator p_x. However, it can be shown that these expressions are physically equivalent. We shall demonstrate this by the following example. Consider the eigenvalues of momentum. The eigenvalue equation is

$$p_x \psi(x) = p_x \psi(x).$$

By Eq. (12–108), we write

$$e^{-(i/\hbar)\int^x X(x)\,dx} \cdot \frac{\hbar}{i} \frac{d}{dx} \cdot e^{+(i/\hbar)\int^x X(x)\,dx} \psi(x) = p_x \psi(x),$$

or

$$\frac{\hbar}{i} \frac{d}{dx} \left[\psi(x) e^{(i/\hbar)\int^x X(x)\,dx} \right] = p_x \left[\psi(x) e^{(i/\hbar)\int^x X(x)\,dx} \right]. \qquad (12\text{–}109)$$

If we set $X(x) = 0$, the operator of p_x is

$$p_x = \frac{\hbar}{i} \frac{d}{dx}, \qquad (12\text{–}110)$$

and the eigenvalue equation is

$$\frac{\hbar}{i} \frac{d}{dx} \psi(x) = p_x \psi(x).$$ (12–111)

Equation (12–111) is the same as Eq. (12–109) except for a change in the phase of the wave function. Thus the inclusion of $X(x)$ in Eq. (12–108) merely changes the phase of the eigenfunctions. The eigenvalues remain the same. The reader is urged to verify the same conclusion in the energy eigenvalue problem. It may also be shown that in dynamical problems the use of Eqs. (12–108) and (12–110) are completely equivalent except for a trivial change of phase of the eigenfunctions.* As the Schrödinger equation leaves the phase factor of an eigenfunction undetermined, the inclusion of $X(x)$ has no physical significance, and $X(x)$ may thus be dropped. We therefore conclude that the operator for momentum p_x is $(\hbar/i)(d/dx)$. Although this proof is worked out for the one-dimensional case, it may be generalized easily for the three-dimensional case.

THEOREM III. The operator associated with an arbitrary function of momenta and position $f(p_x, p_y, p_z; x, y, z)$ is

$$f\left(\frac{\hbar}{i} \frac{\partial}{\partial x}, \frac{\hbar}{i} \frac{\partial}{\partial y}, \frac{\hbar}{i} \frac{\partial}{\partial z}; x, y, z\right)$$

if all factors in any product appearing in $f(p_x, p_y, p_z; x, y, z)$ are commutative to one another.

This follows immediately from Theorems I and II and assumption γ.

THEOREM IV. (Identical with Theorem 3.)

This follows from assumption α and Theorem III.

THEOREMS V, VI, VII, VIII. (Identical with Theorems 4, 5, 6, 7.)

They follow from Theorem IV.

Thus the present formulation has included the kinematics of the previous formulation, i.e., assumptions A, B, and their derivatives. On the other hand, the present formulation is broader, since previously in Theorem 3 we have to make a generalization of assumptions A and B.

We shall now show that the present formulation also includes the dynamics of the previous formulation. In other words, we can derive the time-

---

* The exact proof of the equivalence of Eq. (12–108) and Eq. (12–110) is to be found in the transformation theory which we do not discuss here. In this theory Eq. (12–108) is merely a transformation of Eq. (12–110) as a result of a change of the phases of the position eigenfunctions.

dependent Schrödinger equation, which is assumption C of the previous formulation. First we establish Theorem 10 in the general formulation.

THEOREM IX. (Identical with Theorem 10.)

Proof. The quantum-mechanical equation of motion for a general dynamical quantity $F(p_x, p_y, p_z; x, y, z)$ may be derived from the equation

$$\dot{F} = \frac{\partial F}{\partial p_x} \dot{p}_x + \frac{\partial F}{\partial p_y} \dot{p}_y + \frac{\partial F}{\partial p_z} \dot{p}_z + \frac{\partial F}{\partial x} \dot{x} + \frac{\partial F}{\partial y} \dot{y} + \frac{\partial F}{\partial z} \dot{z}, \quad (12\text{--}112)$$

which may be obtained directly from the classical equation according to assumption γ. Using the canonical equations of motion, we may write

$$\dot{F} = \begin{vmatrix} \dfrac{\partial F}{\partial x} & \dfrac{\partial H}{\partial x} \\[2mm] \dfrac{\partial F}{\partial p_x} & \dfrac{\partial H}{\partial p_x} \end{vmatrix} + \begin{vmatrix} \dfrac{\partial F}{\partial y} & \dfrac{\partial H}{\partial y} \\[2mm] \dfrac{\partial F}{\partial p_y} & \dfrac{\partial H}{\partial p_y} \end{vmatrix} + \begin{vmatrix} \dfrac{\partial F}{\partial z} & \dfrac{\partial H}{\partial z} \\[2mm] \dfrac{\partial F}{\partial p_z} & \dfrac{\partial H}{\partial p_z} \end{vmatrix}. \quad (12\text{--}113)$$

The expression in the right-hand side is defined, in classical mechanics, as the *Poisson bracket* of F and H, and is denoted by $[F, H]$. It may be verified (by the rules of differentiation) that Poisson brackets satisfy the following algebraic relations:

$$[A, B] = -[B, A],$$

$$[A, c] = 0, \quad c \text{ being a constant,}$$

$$[(A + B), C] = [A, C] + [B, C], \quad (12\text{--}114)$$

$$[AB, C] = [A, C]B + A[B, C].$$

Therefore, for any two polynomial functions of position and momentum, $f_1(p_x, p_y, p_z; x, y, z)$ and $f_2(p_x, p_y, p_z; x, y, z)$, their Poisson bracket may be reduced to a sum of elementary Poisson brackets of coordinates and momenta, each multiplied by a polynomial function of $p_x, p_y, p_z; x, y, z$. The elementary Poisson brackets may be evaluated readily according to their definitions. The results are

$$[x, p_x] = 1, \quad [y, p_y] = 1, \quad [z, p_z] = 1;$$

$$[x, y] = 0, \quad [y, z] = 0, \quad [z, x] = 0; \quad (12\text{--}115)$$

$$[p_x, p_y] = 0, \quad [p_y, p_z] = 0, \quad [p_z, p_x] = 0.$$

Thus Eqs. (12–114) and (12–115) together furnish a method of evaluating the Poisson bracket of f_1 and f_2, and we do not have to calculate it directly according to its definition. Now we make the important observation that the commutator of two operators A and B obeys the same algebraic rules of Eqs. (12–114). (The proof is left to the reader.) Therefore the process of reducing the commutator $(AB - BA)$, where A and B are functions of operators x, p_x, etc., to elementary commutators $(p_x x - x p_x)$ and similar ones, is exactly the same as the reduction of Poisson bracket to elementary Poisson brackets. Furthermore, a comparison of Eqs. (12–92, 93, 94) with Eqs. (12–115) shows that the values of the elementary commutators equal those of the elementary Poisson bracket multiplied by $i\hbar$. Therefore the expression of the commutator $(AB - BA)$ in terms of p_x, p_y, p_z, x, y, z is exactly the same as that of the Poisson bracket multiplied by $i\hbar$:

$$\frac{1}{i\hbar}(AB - BA) = [A, B]. \tag{12–116}$$

Equation (12–113) thus becomes

$$\dot{F} = \frac{i}{\hbar}(HF - FH). \tag{12–117}$$

This form of the equation of motion is identical with Eq. (12–81). Thus Theorem 10 is now proved in the general formulation.

THEOREM X. The $\Psi(x, y, z, t)$ function in assumption α satisfies the following equation of motion:

$$i\hbar \frac{\partial \Psi}{\partial t} = H\Psi, \tag{12–118}$$

where H is the Hamiltonian or the energy operator.

Proof. First we consider the quantum-mechanical equation of motion for momentum. By the definition of $\dot{p}_x$ and setting F equal to p_x in Eq. (12–117) we derive Eq. (12–79):

$$\frac{d}{dt}\overline{p_x}(t) = \frac{i}{\hbar}\iiint \Psi^*(Hp_x - p_x H)\Psi\, dx\, dy\, dz. \tag{12–119}$$

Again, working backward from Eq. (12–78), we obtain Eq. (12–77). On the other hand, the left-hand side of Eq. (12–119) is by definition given by Eq. (12–76). Comparing Eqs. (12–76) and (12–77) we arrive at

$$i\hbar \frac{\partial \Psi}{\partial t} = H\Psi + J\Psi, \tag{12–120}$$

where $J\Psi$ is a quantity which must satisfy

$$\iiint (-J^*\Psi^* p_x \Psi + \Psi^* p_x J\Psi)\, dx\, dy\, dz = 0. \tag{12–121}$$

Now repeat the same procedure for the equation of motion of x. We derive again Eq. (12–120), but $J\Psi$ must satisfy

$$\iiint (-J^*\Psi^* x\Psi + \Psi^* x J\Psi)\, dx\, dy\, dz = 0. \qquad (12\text{–}122)$$

The same may be repeated for any other quantities and Eq. (12–120) will be reproduced, but $J\Psi$ has to satisfy a new equation every time a new dynamical quantity is considered. The only way that $J\Psi$ may satisfy all these equations is

$$J\Psi = 0. \qquad (12\text{–}123)$$

Therefore

$$i\hbar \frac{\partial \Psi}{\partial t} = H\Psi. \qquad (12\text{–}124)$$

The theorem is thus proved.

The quantum-mechanical equations of motion of *all* dynamical quantities, expressed by Eq. (12–117), lead to the *same* equation of motion of the wave function Ψ, Eq. (12–124). This gives us the advantage of not having to solve each of the equations of motion of the dynamical quantities individually. The solution of Eq. (12–124) alone will give us information on all dynamical quantities. In fact, the practical applications of quantum mechanics are usually carried out by solving the Schrödinger equation.

We have now shown that the general formulation includes assumptions A, B, C. Therefore it includes all deductions of the previous formulation, i.e., all theorems labeled with Arabic numerals.

It may be remarked that the general formulation here described is not the most general form. In the most general formulation, which we do not discuss in this volume, the quantum-mechanical state is described by a vector in a Hilbert space and dynamical variables are represented by operators operating on the state vector; in other words, by transformations in the Hilbert space. What we have developed in this chapter is a special representation known as the *position representation* or the *Schrödinger representation*, in which the Hilbert space is represented by a special coordinate system, the unit vectors of which are the position eigenfunctions. In the most general formulation other eigenfunctions may also be used as unit vectors leading to formulations in other representations. For example, in the *momentum representation* the momentum eigenfunctions are used as unit vectors (see Problems 12–8 and 12–9). Another point may also be mentioned here. Our formulation is based on the so-called *Schrödinger picture*, in which the state vector depends on time explicitly while the operators are independent of time. An alternate is the so-called *Heisenberg picture*, in which the state vector is independent of time and the evolution of the system in time is represented by the explicit time-dependence of the operators. We shall not discuss these generalizations.

The general formulation of quantum mechanics in this section, though developed for a one-particle system, may very easily be generalized to apply to other mechanical systems. The generalization is based on the substitution for assumption β of the following assumption which is a natural extension of assumption β and includes it as a special case.

Assumption β'. In a general mechanical system describable by a Hamiltonian, the operators for the canonical coordinates and their conjugate momenta obey the following commutation rules:

$$p_i q_i - q_i p_i = \frac{\hbar}{i},$$
$$p_i p_k - p_k p_i = 0, \qquad \begin{cases} i = 1, 2, \ldots n, \\ k = 1, 2, \ldots n; \end{cases} \qquad (12\text{–}125)$$
$$q_i q_k - q_k q_i = 0,$$

in deriving the Schrödinger equation the substitution of momenta by differential operators must be carried out in Cartesian coordinates [this point will be illustrated in an example later in Section 12–9(c)].

In some cases this generalization may be justified by an extension of the reasoning in the one-particle case. In others this is taken as a basic assumption. It follows from assumptions β' and γ that the quantum-mechanical equation of motion is again Eq. (12–117) or, equivalently, the Schrödinger equation.

The assumptions α, β', γ form the basis for establishing a general theory of quantum mechanics which is applicable to any mechanical system describable by a Hamiltonian. In the remainder of this chapter we shall apply this general theory to many-particle systems, rigid bodies, charged particles in a magnetic field, radiation fields, and matter waves. Before we make these diversified applications, we first prove in the next section that the general theory based on assumptions α, β', γ reduces to classical mechanics as a limiting case, and thereby establish the fact that quantum mechanics in the general theory is a natural extension of classical mechanics.

12–7 Classical mechanics as a limiting case of quantum mechanics. If one has any feeling of mystery regarding the general theory of quantum mechanics due to the abstractness of its mathematical apparatus, the mystery disappears when we find that classical mechanics may be formulated by using the same mathematical apparatus and, in fact, is a special case of quantum mechanics.

That quantum mechanics reduces to classical mechanics in the special case of a one-particle system has been discussed before on many occasions. We now consider the general case of any system describable by a Hamiltonian. In Section 12–6 the constant h is identified with the Planck constant. Let us assign to h a new value. A new system of mechanics is then obtained which is of a similar structure. We now show that this system of mechanics approaches classical mechanics as h approaches zero.

When h approaches zero, x and p_x become commutative. As a result, all functions of x and p_x commute with one another. According to Theorem 9, it becomes possible to find simultaneous eigenfunctions for all dynamical quantities. For such an eigenfunction $\Phi(x, y, z, t)$ at a given time t, all coordinates x, y, z, and momenta p_x, p_y, p_z, have definite values. Therefore any function $f(p_x, p_y, p_z; x, y, z)$ has a definite value. Such a kinematics is just what the classical kinematics claims to be.

Now consider the dynamics. In the equation of motion

$$\dot{F} = \frac{i}{\hbar}\,(HF - FH), \qquad (12\text{--}126)$$

the commutator is zero but $\hbar$ is also zero. The right-hand side of Eq. (12–126) is thus an indeterminate form which may be evaluated by the limiting process of letting $\hbar$ approach zero gradually. We already know that the right-hand side of Eq. (12–126) is independent of $\hbar$, being equal to the Poisson bracket of F and H. Therefore even when $\hbar \rightarrow 0$, we still have

$$\dot{F} = [F, H]. \qquad (12\text{--}127)$$

The right-hand side is a function of x, p_x, etc., and, when it operates on Φ, it yields a definite value equal to that obtained by replacing x, p_x, etc., in the Poisson bracket operator by the eigenvalues of x, p_x, etc., for the state Φ. As a result, Φ is an eigenfunction of $\dot{F}$ with an eigenvalue $\dot{F}$ as just described, i.e.,

$$\dot{F} = [F, H]. \qquad (12\text{--}128)$$

The above equation means that the eigenvalues of $\dot{F}$, x, and p_x for the state Φ satisfy an equation exactly the same as the classical equation of motion. Therefore the change in time of the dynamical quantities of the system described by Φ follows exactly the laws of classical mechanics. Thus the limiting form of quantum mechanics, when $\hbar \rightarrow 0$, is classical mechanics. In applications to cases where $\hbar$ may be regarded as small, classical mechanics is valid.

The above argument makes no use of any explicit form of the operators and is thus quite general. Using the Schrödinger representation, i.e., $x = x$ and $p_x = (\hbar/i)(\partial/\partial x)$, we see that the wave function Φ becomes more and more localized when $\hbar \rightarrow 0$. When $\hbar$ is infinitesimally small, it becomes possible to construct a wave function Φ localized in a small region Δx within which there still exists an infinitely large number of wave crests, so that it may be considered as a plane wave within Δx. The momentum value is thus almost definite (like a plane wave). Such a wave packet has almost definite position and momentum values, and therefore has almost definite values of any other dynamical quantities. As $\hbar \rightarrow 0$, the exactness may be indefinitely improved. The classical picture is thus reproduced.

From the above discussion we realize that the operator formulation is not particularly abstruse, for classical mechanics may also be reformulated in its terms. The extension from classical mechanics to quantum mechanics is thus quite natural, this being achieved by introducing a quantum condition which changes the value of h from infinitesimal to 6.625×10^{-27} erg·sec. Classical and quantum mechanics may thus be regarded as theories of a similar structure, differing only in the magnitude of a parameter h.

The fact that classical mechanics may be derived without any use of the explicit forms of the operators is a good illustration of the point that the relations among the operators, including the commutation relations, completely determine the behavior of the system. The use of an explicit form, such as $p_x = (\hbar/i)(\partial/\partial x)$ in the Schrödinger representation, is but *one way* of working out the results already contained in the algebra of the operators. This is also illustrated by the fact that the angular momentum eigenvalues, Eqs. (12–51, 62), may be obtained from the commutation relation, Eq. (12–57), without the use of the explicit form of the operators in the Schrödinger representation, Eqs. (12–47) (this will not be discussed in the present volume). In the general formulation, the algebra of the operators plays the dominant role.

We noted that the classical equation of motion, Eq. (12–128), is derived as a result of the assumption that the quantum-mechanical equation of motion, Eq. (12–127), has the same form as the classical equation (assumption γ). This is the only justification for assuming that quantum-mechanical quantities satisfy the same relations as the classical quantities.

In the following sections we shall apply the general method of Section 12–6, i.e., applying assumptions α, β', γ, to a number of mechanical systems to which the previous method of Chapter 2 is not applicable.

12–8 Quantization of many-particle systems. Quantum mechanics as developed in Chapter 2 is a theory for just one particle moving in a specified potential. It tells us nothing about quantum mechanics of two particles. In order to generalize the theory along the line of Chapter 2, we have to conceive of a wave in a six-dimensional space. The intuitive advantage of the wave picture is thus lost. On the other hand, we may easily treat the two-particle system by the general formulation of Section 12–6. To find the quantum-mechanical laws of a system by inference from the classical theory of the system is a process called *the quantization of a classical system*. The technique of quantization according to Section 12–6 consists in introducing assumptions α, β', γ. We shall now consider the quantization of a two-particle system.

A two-body system has six coordinates, x_1, y_1, z_1; x_2, y_2, z_2, and six momenta, p_{x1}, p_{y1}, p_{z1}; p_{x2}, p_{y2}, p_{z2}, for the two particles. There are many other dynamical quantities which are functions of the coordinates and momenta. According to assumption α, the system is to be described by a wave function $\Psi(x_1, y_1, z_1; x_2, y_2, z_2; t)$ and each dynamical quantity is to be associated with an operator. It seems reasonable and natural to assume that each particle taken individually must obey the commutation relations of the one-particle system. This is exactly what assumption β' asserts. Thus we obtain two sets of commutation relations, each like Eqs. (12–92, 93, 94), for the two particles. From Theorems I and II we conclude that the operators for the coordinates are multiplication operators x_1, y_1, z_1; x_2, y_2, z_2 and those for the momenta are differentiation operators

$$\frac{\hbar}{i}\frac{\partial}{\partial x_1}, \quad \frac{\hbar}{i}\frac{\partial}{\partial y_1}, \quad \frac{\hbar}{i}\frac{\partial}{\partial z_1}; \quad \frac{\hbar}{i}\frac{\partial}{\partial x_2}, \quad \frac{\hbar}{i}\frac{\partial}{\partial y_2}, \quad \frac{\hbar}{i}\frac{\partial}{\partial z_2}.$$

The operator for a general dynamical quantity $F(p_{x1}, p_{y1}, p_{z1}, p_{x2}, p_{y2}, p_{z2}; x_1, y_1, z_1, x_2, y_2, z_2)$ is thus

$$F\left(\frac{\hbar}{i}\frac{\partial}{\partial x_1}, \frac{\hbar}{i}\frac{\partial}{\partial y_1}, \frac{\hbar}{i}\frac{\partial}{\partial z_1}, \frac{\hbar}{i}\frac{\partial}{\partial x_2}, \frac{\hbar}{i}\frac{\partial}{\partial y_2}, \frac{\hbar}{i}\frac{\partial}{\partial z_2}; x_1, y_1, z_1, x_2, y_2, z_2\right).$$

The eigenvalue equations may thus be written which lead to quantization of dynamical quantities. Assumption γ, by virtue of Theorem X, leads to the time-dependent Schrödinger equation which determines the evolution of the $\Psi(x_1, y_1, z_1, x_2, y_2, z_2, t)$ function. The Hamiltonian of the two-particle system is

$$H = \frac{1}{2M_1}(p_{x_1}^2 + p_{y_1}^2 + p_{z_1}^2) + \frac{1}{2M_2}(p_{x_2}^2 + p_{y_2}^2 + p_{z_2}^2)$$
$$+ U(x_1, y_1, z_1, x_2, y_2, z_2). \quad (12\text{–}129)$$

Therefore the Schrödinger equation is   .

$$i\hbar\,\frac{\partial\Psi}{\partial t}$$

$$= H\left(\frac{\hbar}{i}\frac{\partial}{\partial x_1}, \frac{\hbar}{i}\frac{\partial}{\partial y_1}, \frac{\hbar}{i}\frac{\partial}{\partial z_1}, \frac{\hbar}{i}\frac{\partial}{\partial x_2}, \frac{\hbar}{i}\frac{\partial}{\partial y_2}, \frac{\hbar}{i}\frac{\partial}{\partial z_2}; x_1, y_1, z_1, x_2, y_2, z_2\right)\Psi$$

$$(12\text{–}130)$$

or

$$i\hbar\,\frac{\partial\Psi}{\partial t} = \left[-\frac{\hbar^2}{2M_1}\nabla_1^2 - \frac{\hbar^2}{2M_2}\nabla_2^2 + U(x_1, y_1, z_1, x_2, y_2, z_2)\right]\Psi. \quad (12\text{–}131)$$

Equation (12–131) may be solved by Fourier expansion of Ψ in the time variable, the expansion coefficients ψ_n being determined by the time-independent Schrödinger equation

$$H\psi_n = E_n\psi_n. \tag{12–132}$$

Since Eq. (12–132) is also the eigenvalue equation of energy, the constant E_n is identified with the energy eigenvalue. Once Eq. (12–131) is solved and Ψ is determined by its initial value at t_0, the average value or the probability distribution of the eigenvalues of any dynamical quantity F at any time t may be obtained from the operator F by Theorems IV through VIII. The quantum-mechanical problem of the two-body system is thus completely solved.

We shall illustrate the above general procedure by two special cases. First we consider a system of two distinct and independent particles, the Hamiltonian of which is

$$H = \frac{1}{2M_1}\,(p_{x_1}^2 + p_{y_1}^2 + p_{z_1}^2) + \frac{1}{2M_2}\,(p_{x_2}^2 + p_{y_2}^2 + p_{z_2}^2)$$
$$+ U_1(x_1, y_1, z_1) + U_2(x_2, y_2, z_2). \tag{12–133}$$

The time-independent Schrödinger equation is thus

$$\frac{1}{2M_1}\,\nabla_1^2\psi(x_1, y_1, z_1, x_2, y_2, z_2) + \frac{1}{2M_2}\,\nabla_2^2\psi(x_1, y_1, z_1, x_2, y_2, z_2)$$
$$+ \frac{1}{\hbar^2}\,[E_n - U_1(x_1, y_1, z_1) - U_2(x_2, y_2, z_2)]\psi(x_1, y_1, z_1, x_2, y_2, z_2) = 0. \tag{12–134}$$

Particular solutions of this equation may be found by the method of separation of variables. The solutions are of the following form:

$$\psi_n(x_1, y_1, z_1, x_2, y_2, z_2) = \psi_1(x_1, y_1, z_1)\psi_2(x_2, y_2, z_2), \tag{12–135}$$

$$E_n = E_1 + E_2, \tag{12–136}$$

where ψ_1, ψ_2 and E_1, E_2 satisfy

$$\nabla_1^2\psi_1(x_1, y_1, z_1) + \frac{2M_1}{\hbar^2}\,[E_1 - U_1(x_1, y_1, z_1)]\psi_1(x_1, y_1, z_1) = 0, \tag{12–137}$$

$$\nabla_2^2\psi_2(x_2, y_2, z_2) + \frac{2M_2}{\hbar^2}\,[E_2 - U_2(x_2, y_2, z_2)]\psi_2(x_2, y_2, z_2) = 0. \tag{12–138}$$

Actually ψ_1, ψ_2 and E_1, E_2 are the eigenfunctions and eigenvalues of the

two particles taken individually. Therefore we conclude that the eigenfunctions of the two-independent-particle system are the products of the
eigenfunctions of the individual particles, while the energy eigenvalues
are the sum of the individual energy eigenvalues.

Next we consider a system of two particles with central-force interaction.
It includes as special cases the hydrogen atom, which consists of an
electron and a proton, and the positronium, which consists of an electron and a positron. In each case the force between the two bodies is
central (in fact, Coulomb force). We first write the time-dependent
Schrödinger equation:

$$\frac{1}{2M_1}\,\nabla_1^2\psi_n + \frac{1}{2M_2}\,\nabla_2^2\psi_n + \frac{1}{\hbar^2}\,[E_n - U(x_1, y_1, z_1, x_2, y_2, z_2)]\psi_n = 0.$$

$$(12\text{–}139)$$

The potential is no longer separable into two parts and the method of
separation of variables used in the first example is no longer applicable.
However, it is a well-known method in solving the central force problems
to introduce the relative coordinates (x, y, z) and the coordinates of the
center of mass (X, Y, Z), these being defined by

$$x = x_2 - x_1, \qquad y = y_2 - y_1, \qquad z = z_2 - z_1;$$

$$(12\text{–}140)$$

$$X = \frac{M_1x_1 + M_2x_2}{M_1 + M_2}, \qquad Y = \frac{M_1y_1 + M_2y_2}{M_1 + M_2}, \qquad Z = \frac{M_1z_1 + M_2z_2}{M_1 + M_2}.$$

$$(12\text{–}141)$$

Equations (12–140, 141) represent a transformation from the six original
coordinates $x_1, y_1, z_1, x_2, y_2, z_2$ to the six new coordinates x, y, z, X, Y, Z.
By a straightforward though tedious calculation, we obtain

$$\frac{1}{2M_1}\left(\frac{\partial^2}{\partial x_1^2} + \frac{\partial^2}{\partial y_1^2} + \frac{\partial^2}{\partial z_1^2}\right) + \frac{1}{2M_2}\left(\frac{\partial^2}{\partial x_2^2} + \frac{\partial^2}{\partial y_2^2} + \frac{\partial^2}{\partial z_2^2}\right) = \frac{1}{2(M_1 + M_2)}$$

$$\times \left(\frac{\partial^2}{\partial X^2} + \frac{\partial^2}{\partial Y^2} + \frac{\partial^2}{\partial Z^2}\right) + \frac{1}{2[M_1M_2/(M_1 + M_2)]}\left(\frac{\partial^2}{\partial x^2} + \frac{\partial^2}{\partial y^2} + \frac{\partial^2}{\partial z^2}\right).$$

$$(12\text{–}142)$$

Since the central force is the only force acting in the system, the potential
depends on the relative coordinates only:

$$U(x_1, y_1, z_1, x_2, y_2, z_2) = V(x, y, z). \qquad (12\text{–}143)$$

Substituting Eqs. (12–142, 143) in Eq. (12–139), we find that the latter may be solved by separation of variables. The solutions are found to be

$$\psi_n(x_1, y_1, z_1, x_2, y_2, z_2) = \psi_1(X, Y, Z)\psi_2(x, y, z), \qquad (12\text{–}144)$$

$$E_n = E_1 + E_2, \qquad (12\text{–}145)$$

where ψ_1, ψ_2 and E_1, E_2 satisfy

$$\frac{\partial^2 \psi_1}{\partial X^2} + \frac{\partial^2 \psi_1}{\partial Y^2} + \frac{\partial^2 \psi_1}{\partial Z^2} + \frac{2(M_1 + M_2)}{\hbar^2} E_1\psi_1 = 0, \qquad (12\text{–}146)$$

$$\frac{\partial^2 \psi_2}{\partial x^2} + \frac{\partial^2 \psi_2}{\partial y^2} + \frac{\partial^2 \psi_2}{\partial z^2} + \frac{2[M_1 M_2/(M_1 + M_2)]}{\hbar^2} [E_2 - V(x, y, z)]\psi_2 = 0. \qquad (12\text{–}147)$$

Equation (12–146) is the Schrödinger equation of a free particle of mass $(M_1 + M_2)$, the coordinates of which are those of the center of mass (X, Y, Z). Equation (12–147) is that of a particle of mass $M_1M_2/(M_1 + M_2)$, the coordinates of which are the relative coordinates (x, y, z). In classical mechanics the solution of a system of two particles with central force is such that the coordinates of the center of mass (X, Y, Z) change like those of a particle of mass $(M_1 + M_2)$, while the relative coordinates (x, y, z) change like those of a particle of mass $M_1M_2/(M_1 + M_2)$, which is the *reduced mass*. Thus the quantum-mechanical results are very similar to the classical results. The energy of the quantum-mechanical system accordingly may be regarded as consisting of two parts: the energy of the center of mass E_1 and that of the relative motion E_2. In the previous discussion of the hydrogen atom we considered only the relative motion. This may be justified because of the fact that the motion of the center of mass is an independent and constant part which does not need to be considered. Nevertheless, the discussion here shows that when considering the relative motion, we have to use the reduced mass for M instead of the mass of the electron. Since the electron is 1836 times lighter than the proton, the reduced mass of the electron in any atom is nearly the same as the electron mass itself. Yet spectroscopic data are accurate enough to detect and verify this small reduced-mass effect. It may be mentioned that the reduced-mass effect was considered in Bohr's theory and the deductions were verified by experiments. Although this effect is small for electrons in atoms, it becomes appreciable for the electron in positronium. Here the reduced mass is one-half of the electron mass. Thus the energy levels and other quantities for positronium change accordingly when compared with those of the hydrogen atom.

So far we have considered two distinct particles. The case of identical particles is complicated by the so-called *exchange degeneracy* and the *statistics* (Bose-Einstein or Fermi-Dirac), which will not be discussed in this volume.

The above theory of the two-particle system may be generalized to the N-particle system, the time-dependent Schrödinger equation of which is easily found to be

$$i\hbar \frac{\partial}{\partial t} \Psi(x_1, y_1, z_1, \ldots, x_N, y_N, z_N, t)$$

$$= \left[-\frac{\hbar^2}{2M_1} \nabla_1^2 - \frac{\hbar^2}{2M_2} \nabla_2^2 \cdots - \frac{\hbar^2}{2M_N} \nabla_N^2 + U(x_1, y_1, z_1, \ldots, x_N, y_N, z_N) \right]$$

$$\times \Psi(x_1, y_1, z_1, \ldots, x_N, y_N, z_N, t), \qquad (12\text{–}148)$$

where $\nabla_1^2 = (\partial^2/\partial x_1^2) + (\partial^2/\partial y_1^2) + (\partial^2/\partial z_1^2)$, etc. This is the equation of motion of the N-particle system.

12–9 Quantization of the motion of a rigid body. We have not yet introduced the concept of a rigid body in quantum mechanics. From the discussion in Chapter 2 we have no idea of how a rigid body behaves in quantum mechanics. Its quantization will have to be considered from the general theory of Section 12–6. The following three special cases will be discussed.

(A) *The plane rotator.* In classical mechanics a plane rotator, or a rigid rotator in a plane, is a rigid body rotating about a fixed axis. The Hamiltonian of this system is

$$H = \frac{1}{2I} M_z^2, \qquad (12\text{–}149)$$

where I is the moment of inertia of the rigid body about the fixed axis, assumed to be the z-axis, and M_z is the angular momentum about the z-axis. To quantize this system we make use of assumptions α, β', γ. In the classical theory the canonical variables are M_z and φ. As this system has one degree of freedom, it is completely specified by one coordinate, φ. In quantum mechanics the wave function is thus a function of φ and assumption β' leads to the replacing of M_z by an operator as follows:

$$M_z \rightarrow \frac{\hbar}{i} \frac{\partial}{\partial \varphi}. \qquad (12\text{–}150)$$

The plausibility of this assumption may be seen in its similarity to Eq. (12–48), which gives the operator for the z-component of the angular momentum of a particle. The quantization of energy of the plane rotator may be carried out by solving the following equation obtained from Eqs. (12–149) and (12–150):

$$-\frac{\hbar^2}{2I} \frac{\partial^2}{\partial \varphi^2} \psi_m(\varphi) = E_m \psi_m(\varphi).$$

Such an equation, except for the constants, has been solved in Chapter 8. The eigenfunctions and eigenvalues are

$$\psi_m(\varphi) = \frac{1}{\sqrt{2\pi}} \, e^{im\varphi}, \qquad m = 0, \pm 1, \pm 2, \dots ; \qquad (12\text{–}151)$$

$$E_m = \frac{\hbar^2}{2I} \, m^2, \qquad m = 0, \pm 1, \pm 2, \dots . \qquad (12\text{–}152)$$

Thus the energy is quantized to the square of an integer.

It may be interesting to note that the same result of quantization was obtained by using the Wilson-Sommerfeld quantum condition in the old quantum theory:

$$\oint M_z \, d\varphi = mh, \qquad (12\text{–}153)$$

$$M_z = m\hbar, \qquad (12\text{–}154)$$

$$E = \frac{M_z^2}{2I} = \frac{\hbar^2}{2I} \, m^2. \qquad (12\text{–}155)$$

The justification of the assumption of Eq. (12–150) lies solely in the experimental verification of its deductions, such as Eq. (12–152). Whether such a quantized system actually existed in nature is a different problem. However, we know at least one system which may be considered as a special case of the plane rotation and thus may test the validity of the theory. This is a particle of mass M constrained to move along a circle of radius a without any other force acting on it. The reader is urged to solve the quantum-mechanical problem of this system. The result of the energy eigenvalues is as follows:

$$E_m = \frac{\hbar^2}{2Ma^2} \, m^2, \qquad m = 0, \pm 1, \pm 2, \dots . \qquad (12\text{–}156)$$

This system may also be considered as a special case of the plane rotator with a moment of inertia Ma^2. The quantized plane rotator leads to the energy eigenvalues Eq. (12–152) which are the same as Eq. (12–156). The procedure of quantizing the plane rotator is thus consistent with the mechanics of a particle. That Eq. (12–152) agrees with Eq. (15–156) is due to the fact that the assumption in Eq. (12–150) is so made that it reduces to Eq. (12–48) as a special case when the plane rotator consists of a single particle.

The physical meaning of the wave function is that $|\psi(\varphi)|^2$ represents the probability (when ψ is normalized) of finding the rotator at angular position φ. As the rotator is specified by one coordinate only, the foregoing is the complete kinematical description of the plane rotator in quantum mechanics.

(B) *The rigid rotator.* In classical mechanics the rigid rotator is a rotating rigid body satisfying two conditions: (1), two of its principal moments of inertia are equal; and (2), it does not rotate about its third principal axis. Referring to the principal axes, these conditions may be written as follows:

$$I_1 = I_2, \tag{12–157}$$

$$\omega_3 = 0. \tag{12–158}$$

In other words, it is a symmetric top which does not spin. The energy is

$$\begin{aligned} E &= \tfrac{1}{2}(I_1\omega_1^2 + I_2\omega_2^2 + I_3\omega_3^2) \\ &= \tfrac{1}{2}I_1\omega^2 \\ &= \frac{1}{2I_1}\,M^2, \end{aligned} \tag{12–159}$$

where M^2 is the square of the total angular momentum. The Hamiltonian may thus be written, dropping the index of I, which is now trivial,

$$H = \frac{1}{2I}\,M^2. \tag{12–160}$$

The standard procedure of quantizing the rigid rotator proceeds by making use of the two canonical momenta. This may be worked out as a special case of a more general problem—the symmetric top to be described in the next section. Here we shall take a short cut by assuming that M^2 is replaced by the same operator for the square of the angular momentum of a single particle as given by Eq. (12–47). This procedure is a natural extension of the quantization of the plane rotator. The quantization of energy may be carried out by solving the following equation obtained from Eqs. (12–47) and (12–60):

$$-\frac{\hbar^2}{2I}\left[\frac{1}{\sin\theta}\frac{\partial}{\partial\theta}\left(\sin\theta\frac{\partial}{\partial\theta}\right) + \frac{1}{\sin^2\theta}\frac{\partial^2}{\partial\varphi^2}\right]\psi_n(\theta,\varphi) = E_n\psi_n(\theta,\varphi). \tag{12–161}$$

This eigenvalue problem is already solved in Chapter 8. The eigenfunctions and eigenvalues are

$$\psi_{lm} = Y_{lm}(\theta,\varphi), \tag{12–162}$$

$$E_{lm} = \frac{\hbar^2}{2I}\,l(l+1). \tag{12–163}$$

The rule of energy quantization is thus different from that of a plane rotator. The justification of this procedure of quantization lies in the experimental verification of Eq. (12–163). In molecular spectroscopy we have known that a diatomic molecule has a part of its energy quantized to the rule of Eq. (12–163). This part of energy may thus be regarded as due to the quantized rotational motion of the diatomic molecule (without spin about the axis of the molecule). These molecules may be considered in classical mechanics as either rigid rotators or plane rotators. The experimental results show that they are not plane rotators. Actually the diatomic molecule problem may be solved without regard to quantization of a rigid body, since it is a two-particle problem. By the method of Section 12–8 it may be separated in two parts. The part for the relative motion may be solved by the method of Section 8–1 which leads to the same result of energy quantization. Thus the method of quantization here employed for the rigid rotator is consistent with quantum mechanics of two-particle systems. This conclusion is also expected in view of the fact that the operator for M^2 in Eq. (12–160) is assumed to be the same as in Eq. (12–47).

The physical meaning of the wave function $\psi(\theta, \varphi)$ is that the normalized $|\psi(\theta, \varphi)|^2$ represents the probability of finding the third axis of the rigid rotator in the angular position (θ, φ). As the rigid rotator is a system of two degrees of freedom, two coordinates θ and φ, determining the position of the axis, completely determine the position of the rigid rotator. The wave function thus provides a complete kinematical description of the rigid rotator in quantum mechanics.

(C) *The symmetric top.* A rigid body with two of its principal moments of inertia equal to each other is a *symmetric top.* Referring to the principal axes,

$$I_1 = I_2. \tag{12–164}$$

It differs from the rigid rotator in that it is allowed to spin about the third of the principal axes ξ, η, ζ. A symmetric top has three degrees of freedom, its position being completely determined by the three Euler angles θ, φ, ψ, where θ is the angle between the moving ζ-axis and the fixed z-axis (co-latitude), φ is the angle between the moving ξ-axis and the line of intersection of the xy-plane and $\xi\eta$-plane (angle of spin), and ψ is the angle between the fixed x-axis and the same line of intersection (longitude). (See Fig. 12–3.) [In a sense, it is a combination of a rigid rotator (θ, ψ) and a plane rotator (φ).] In classical mechanics the kinetic energy of the symmetric top is

$$T = \tfrac{1}{2}I_1(\omega_1^2 + \omega_2^2) + \tfrac{1}{2}I_3\omega_3^2, \tag{12–165}$$

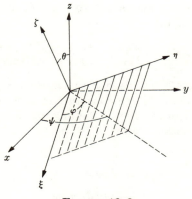

FIGURE 12–3

where I_1, I_2, I_3 are the principal moments of inertia and ω_1, ω_2, ω_3 are the components of the angular velocity with respect to the principal axes ξ, η, ζ. We shall use the three Euler angles as the generalized coordinates. The angular velocity ω_1, ω_2, ω_3 may be expressed in terms of θ, ψ, φ as follows:

$$\omega_1 = \dot{\psi} \sin \theta \sin \varphi + \dot{\theta} \cos \varphi,$$
$$\omega_2 = \dot{\psi} \sin \theta \cos \varphi - \dot{\theta} \sin \varphi, \qquad (12\text{–}166)$$
$$\omega_3 = \dot{\psi} \cos \theta + \dot{\varphi}.$$

The kinetic energy is thus

$$T = \tfrac{1}{2}I_1(\dot{\theta}^2 + \dot{\psi}^2 \sin^2 \theta) + \tfrac{1}{2}I_3(\dot{\psi} \cos \theta + \dot{\varphi})^2. \qquad (12\text{–}167)$$

The corresponding canonical momenta are

$$p_\theta = \frac{\partial T}{\partial \dot{\theta}} = I_1 \dot{\theta},$$

$$p_\varphi = \frac{\partial T}{\partial \dot{\varphi}} = I_3(\dot{\psi} \cos \theta + \dot{\varphi}), \qquad (12\text{–}168)$$

$$p_\psi = \frac{\partial T}{\partial \dot{\psi}} = I_1 \sin^2 \theta \dot{\psi} + I_3 \cos \theta(\dot{\psi} \cos \theta + \dot{\varphi}).$$

The kinetic energy may be expressed in terms of the generalized momenta,

$$T = \frac{p_\theta^2}{2I_1} + \frac{p_\varphi^2}{2I_3} + \frac{(p_\psi - p_\varphi \cos \theta)^2}{2I_1 \sin^2 \theta}. \qquad (12\text{–}169)$$

For the free rotation of a symmetric top the Hamiltonian consists of the kinetic energy only. Thus

$$H(p_\theta, p_\varphi, p_\psi, \theta, \varphi, \psi) = \frac{p_\theta^2}{2I_1} + \frac{p_\varphi^2}{2I_3} + \frac{(p_\psi - p_\varphi \cos \theta)^2}{2I_1 \sin^2 \theta}. \qquad (12\text{-}170)$$

The Hamiltonian does not contain φ and ψ; thus φ and ψ are cyclic variables. Therefore p_φ and p_ψ are constants of motion. If we choose the z-axis lying in the direction of the total angular momentum, the solutions of the equations of motion are found to be

$$\theta = \theta_0,$$
$$\varphi = 2\pi\nu_K t + \varphi_0, \qquad (12\text{-}171)$$
$$\psi = 2\pi\nu_J t + \psi_0.$$

It means that the top is spinning about its axis of symmetry (ζ-axis) with a frequency ν_K and also has a precessional motion about the axis of the total angular momentum (z-axis) with a frequency ν_J.

We now proceed to quantize this classical system, i.e., to find the quantum-mechanical equations of motion and their solutions. Assumptions α, β', γ are introduced. The Hamiltonian in canonical variables is already known. However, it is not in the Cartesian form:

$$H = \frac{1}{2M} (p_\lambda^2 + p_\mu^2 + p_\nu^2). \qquad (12\text{-}172)$$

According to assumption β', we have to transform Eq. (12-170) to a form of Eq. (12-172) first, then we may replace p_λ, p_μ, p_ν respectively by

$$\frac{\hbar}{i} \frac{\partial}{\partial \lambda}, \quad \frac{\hbar}{i} \frac{\partial}{\partial \mu}, \quad \frac{\hbar}{i} \frac{\partial}{\partial \nu}$$

to obtain the Hamiltonian operator which operates on wave function $\psi(\lambda, \mu, \nu)$. The quantized equations of motion may thus be formulated in the coordinate system (λ, μ, ν). If the original coordinates are preferred, the quantum-mechanical equation may be transformed back to the (θ, φ, ψ) system. However, the result of this lengthy procedure of obtaining an equation of motion in the (θ, φ, ψ) system may be shown to be obtainable directly from $H(p_\theta, p_\varphi, p_\psi, \theta, \varphi, \psi)$ by replacing p_θ, p_φ, p_ψ respectively with

$$\frac{\hbar}{i} \frac{\partial}{\partial \theta}, \quad \frac{\hbar}{i} \frac{\partial}{\partial \varphi}, \quad \frac{\hbar}{i} \frac{\partial}{\partial \psi}$$

provided the Hamiltonian is written in a particular manner.* We shall

---

* B. Podolsky, *Phys. Rev.* **32**, 812 (1928).

not prove this general theorem nor give the specifications of the particular form of the Hamiltonian; we shall merely write down the appropriate Hamiltonian for the symmetric top in the (θ, φ, ψ) coordinate system:

$$H(p_\theta, p_\varphi, p_\psi; \theta, \varphi, \psi) = \frac{1}{2I_1} \frac{1}{\sin\theta} p_\theta \sin\theta\, p_\theta$$

$$+ \frac{1}{2I_3} p_\varphi^2 + \frac{(p_\psi - p_\varphi \cos\theta)^2}{2I_1 \sin^2\theta}. \qquad (12\text{–}173)$$

The Hamiltonian operator is thus

$$H\left(\frac{\hbar}{i}\frac{\partial}{\partial\theta}, \frac{\hbar}{i}\frac{\partial}{\partial\varphi}, \frac{\hbar}{i}\frac{\partial}{\partial\psi}; \theta, \varphi, \psi\right)$$

$$= -\frac{\hbar^2}{2I_1}\frac{1}{\sin\theta}\frac{\partial}{\partial\theta}\left(\sin\theta\frac{\partial}{\partial\theta}\right) - \frac{\hbar^2}{2I_1}\left(\frac{I_1}{I_3} + \frac{\cos^2\theta}{\sin^2\theta}\right)\frac{\partial^2}{\partial\varphi^2}$$

$$- \frac{\hbar^2}{2I_1}\frac{1}{\sin^2\theta}\frac{\partial^2}{\partial\psi^2} + \frac{\hbar^2}{2I_1}\frac{2\cos\theta}{\sin^2\theta}\frac{\partial^2}{\partial\psi\,\partial\varphi}. \qquad (12\text{–}174)$$

From this Hamiltonian operator the time-dependent Schrödinger equation may be written down, the wave function being a function of θ, φ, ψ. The solution of this equation determines the quantum-mechanical behavior of this system, which we shall not describe in detail. It may be remarked that the Hamiltonian of Eq. (12–173) is identical with Eq. (12–170) in classical mechanics. They become different in quantum mechanics because $\sin\theta$ and p_θ no longer commute. Because of this noncommutative nature of the quantum-mechanical quantities, one classical expression of the Hamiltonian corresponds to infinitely many different forms of the Hamiltonian in quantum mechanics. The question thus arises: Which one of the forms leads to the correct equation in quantum mechanics? The answer to this question, according to assumption β', is that in Cartesian coordinates the Hamiltonian should be just like Eq. (12–172) instead of something like

$$H = \frac{1}{2M}\left(\frac{1}{f(\lambda)} p_\lambda f(x) p_\lambda + \frac{1}{g(\mu)} p_\mu g(\mu) p_\mu + \frac{1}{h(\nu)} p_\nu h(\nu) p_\nu\right). \qquad (12\text{–}175)$$

This actually is an assumption justified by the fact that its application in the one-particle system leads to the correct result. Once the Hamiltonian operator is specified in one coordinate system, its forms in other systems are uniquely determined by straightforward transformation. The Podolsky theorem is just a short cut for writing down the result of the transformation without going through the intermediate steps. The reader may notice a similar situation when he writes the Schrödinger equation of a particle in spherical coordinates. The same equation may be obtained from the classical Hamiltonian in (r, θ, φ) system only when the Hamiltonian is arranged in a way similar to that of Eq. (12–173). The above discussion explains why in assumption β' the Cartesian coordinate system is insisted upon.

We shall consider one important application of the quantized equation of motion, i.e., the quantization of energy. The eigenvalue equation is

$$H\psi_E = E\psi_E. \tag{12–176}$$

From Eq. (12–174) we obtain

$$\frac{1}{\sin\theta}\frac{\partial}{\partial\theta}\left(\sin\theta\frac{\partial\psi_E}{\partial\theta}\right) + \left(\frac{I_1}{I_3} + \frac{\cos^2\theta}{\sin^2\theta}\right)\frac{\partial^2\psi_E}{\partial\varphi^2} + \frac{1}{\sin^2\theta}\frac{\partial^2\psi_E}{\partial\psi^2}$$

$$- \frac{2\cos\theta}{\sin^2\theta}\frac{\partial^2\psi_E}{\partial\psi\,\partial\varphi} + \frac{2I_1}{\hbar^2}E\psi_E = 0. \tag{12–177}$$

This equation may be solved by separation of variables. Assume that

$$\psi_E = \Theta(\theta)e^{iM\psi}e^{iK\varphi}. \tag{12–178}$$

As both ψ and φ are coordinates with a period 2π, the constants M and K must take integral value in order that ψ_E be a single-valued function:

$$M = 0, \pm1, \pm2, \ldots,$$
$$K = 0, \pm1, \pm2, \ldots. \tag{12–179}$$

M and K thus become two quantum numbers. The equation of $\Theta(\theta)$ may be obtained by substitution. Its acceptable solutions may be expressed in terms of the hypergeometric functions, and the eigenvalues of energy are given by the following expression:

$$E_{JK} = \frac{\hbar^2}{2}\left[\frac{J(J+1)}{I_1} + K^2\left(\frac{1}{I_3} - \frac{1}{I_1}\right)\right],$$
$$J = 0, 1, 2, \ldots, \tag{12–180}$$
$$K = 0, \pm1, \pm2, \ldots, \pm J,$$
$$M = 0, \pm1, \pm2, \ldots, \pm J.$$

It may be remarked that the rigid rotator is a special case of the symmetric top for which the wave function is independent of φ. Equation (12–177) reduces exactly to Eq. (12–161) if φ in the latter is changed to ψ. The eigenvalues are related accordingly. As mentioned previously, the quantization of the rigid rotator may be carried out as a special case of the symmetric top.

In molecular spectroscopy the rotational motion of a molecule as a whole may be treated approximately as the rotation of a rigid body. (Actually the molecule is not rigid, for it executes vibrational motion in addition to rotation.) Some molecules possess geometric symmetry such that two (or even three) of their principal moments of inertia are equal. Examples are CO_2, NH_3, CH_3Cl. They may be considered as symmetric tops. According to quantum mechanics, their rotational energy is quantized according to Eq. (12–180). The analysis of the rotational bands of the spectra of these molecules agrees with Eq. (12–180) and thus provides experimental verification of the theory.

Many other molecules have no geometrical symmetry and thus $I_1 \neq I_2 \neq I_3$. The classical theory for such a rigid body is more complicated than that for the symmetric top and the quantum theory is even more complicated. They will not be discussed here.

12–10 Quantization of the motion of a charged particle in a magnetic field.

When a light source is placed in a magnetic field, each spectral line is split up in a number of components. This phenomenon is called the Zeeman effect. The quantum-mechanical theory of the Zeeman effect is based on the quantization of the motion of an electron in a magnetic field.

We again start from the classical theory, express the equation of motion in the Hamiltonian form, and then quantize it according to assumptions α, β', γ. (We simplify the problem by neglecting the effect of the *electron spin*.) The motion of a charged particle in a magnetic field cannot be described by a scalar potential; thus the method of Chapter 2 is not applicable.

A magnetic field $\vec{H}$ may be related to a vector potential $\vec{A}$ by

$$\vec{H} = \nabla \times \vec{A}. \tag{12–181}$$

$\vec{A}$ may always be chosen to satisfy the following condition:

$$\frac{\partial A_x}{\partial x} + \frac{\partial A_y}{\partial y} + \frac{\partial A_z}{\partial z} = 0.$$

This condition is conventionally imposed on $\vec{A}$ and thus is assumed here. In particular, a uniform magnetic field H along the z-direction may be represented by

$$A_x = -\frac{H}{2}\, y, \qquad A_y = \frac{H}{2}\, x, \qquad A_z = 0, \tag{12–182}$$

as may be verified readily by substitution in Eq. (12–181). Without the magnetic field, the Hamiltonian of a charged particle is

$$H = \frac{1}{2M}\left(p_x^2 + p_y^2 + p_z^2\right) + q\varphi(x, y, z), \tag{12–183}$$

where the potential energy $U(x, y, z)$ is replaced by the product of the charge q and the scalar potential φ. The effect of the magnetic field, as will be proved presently, may be accounted for if in the Hamiltonian we make the substitution

$$p_x \rightarrow p_x - \frac{q}{c}A_x,$$

$$p_y \rightarrow p_y - \frac{q}{c}A_y, \tag{12–184}$$

$$p_z \rightarrow p_z - \frac{q}{c}A_z,$$

where c is the velocity of light. For an electron,

$$q = -e = -4.80 \times 10^{-10} \text{ esu}. \tag{12–185}$$

The new Hamiltonian is thus

$$H = \frac{1}{2M}\left[\left(p_x - \frac{q}{c}A_x\right)^2 + \left(p_y - \frac{q}{c}A_y\right)^2 + \left(p_z - \frac{q}{c}A_z\right)^2\right] + q\varphi(x, y, z). \tag{12–186}$$

We shall show that the equations of motion derived from this Hamiltonian are exactly those of a charged particle under the influence of a magnetic field described by $\vec{A}$, as well as an electric field described by φ. It may be mentioned that in Eq. (12–186), p_x, p_y, p_z are generalized momenta conjugate to x, y, z, but no longer equal to mass times the velocity components.

The canonical equations of motion are

$$\frac{dx}{dt} = \frac{\partial H}{\partial p_x} = \frac{1}{M}\left(p_x - \frac{q}{c}A_x\right), \tag{12–187}$$

$$\frac{dp_x}{dt} = -\frac{\partial H}{\partial x} = -\frac{\partial U}{\partial x}$$
$$+ \frac{q}{Mc}\left[\left(p_x - \frac{q}{c}A_x\right)\frac{\partial A_x}{\partial x} + \left(p_y - \frac{q}{c}A_y\right)\frac{\partial A_y}{\partial x} + \left(p_z - \frac{q}{c}A_z\right)\frac{\partial A_z}{\partial x}\right]. \tag{12–188}$$

Equation (12–188) may be rewritten with the help of Eq. (12–187):

$$\frac{dp_x}{dt} = -\frac{\partial U}{\partial x} + \frac{q}{c}\left(v_x\frac{\partial A_x}{\partial x} + v_y\frac{\partial A_y}{\partial x} + v_z\frac{\partial A_z}{\partial x}\right). \tag{12–189}$$

Since

$$\frac{dA_x}{dt} = \frac{\partial A_x}{\partial t} + \frac{\partial A_x}{\partial x} v_x + \frac{\partial A_x}{\partial y} v_y + \frac{\partial A_x}{\partial z} v_z, \qquad (12\text{--}190)$$

we may calculate $M(d^2x/dt^2)$ from Eqs. (12–187, 189, 190):

$$M \frac{d^2x}{dt^2} = -q \frac{\partial \varphi}{\partial x} + \frac{q}{c} \left(v_x \frac{\partial A_x}{\partial x} + v_y \frac{\partial A_y}{\partial x} + v_z \frac{\partial A_z}{\partial x} \right)$$

$$- \frac{q}{c} \left(\frac{\partial A_x}{\partial t} + v_x \frac{\partial A_x}{\partial x} + v_y \frac{\partial A_x}{\partial y} + v_z \frac{\partial A_x}{\partial z} \right)$$

$$= -q \frac{\partial \varphi}{\partial x} - \frac{q}{c} \frac{\partial A_x}{\partial t} + \frac{q}{c} v_y \left(\frac{\partial A_y}{\partial x} - \frac{\partial A_x}{\partial y} \right) - \frac{q}{c} v_z \left(\frac{\partial A_x}{\partial z} - \frac{\partial A_z}{\partial x} \right)$$

$$= q E_x + \frac{q}{c} (\vec{v} \times \vec{H})_x. \qquad (12\text{--}191)$$

This is the equation of motion of a particle of mass M and charge q under the influence of the Lorentz force of an electric field E and a magnetic field H. Therefore Eq. (12–186) is the correct Hamiltonian.

We now quantize this mechanical system according to assumptions α, β', γ. The wave function is to be a function of x, y, z, and the operators for p_x, p_y, p_z are

$$\frac{\hbar}{i} \frac{\partial}{\partial x}, \qquad \frac{\hbar}{i} \frac{\partial}{\partial y}, \qquad \frac{\hbar}{i} \frac{\partial}{\partial z}. \overset{*}{} \qquad (12\text{--}192)$$

The Hamiltonian operator is thus

$$H = \frac{1}{2M} \left[\left(\frac{\hbar}{i} \frac{\partial}{\partial x} - \frac{q}{c} A_x \right)^2 + \left(\frac{\hbar}{i} \frac{\partial}{\partial y} - \frac{q}{c} A_y \right)^2 + \left(\frac{\hbar}{i} \frac{\partial}{\partial z} - \frac{q}{c} A_z \right)^2 \right]$$

$$+ U(x, y, z). \qquad (12\text{--}193)$$

It is noted that

$$\vec{p} \cdot \vec{A} = \vec{A} \cdot \vec{p}, \qquad (12\text{--}194)$$

since

$$\vec{p} \cdot \vec{A} - \vec{A} \cdot \vec{p} = \frac{\hbar}{i} \left(\frac{\partial A_x}{\partial x} + \frac{\partial A_y}{\partial y} + \frac{\partial A_z}{\partial z} \right) = 0. \qquad (12\text{--}195)$$

---

* As p_x is not equal to Mv_x it is difficult from the theory of Chapter 2 to understand why Mv_x is not replaced by $(\hbar/i)(\partial/\partial x)$ but p_x is. The present substitution follows necessarily from the general theory of Section 12–6.

We may thus rewrite Eq. (12–193):

$$H = \frac{1}{2M}\, p^2 + U(x, y, z) - \frac{q}{Mc}\, \vec{A} \cdot \vec{p} + \frac{q^2}{2Mc^2}\, A^2. \qquad (12\text{–}196)$$

The first two terms consist of the Hamiltonian of the field-free system. The last two terms, representing the effect of the magnetic field, may be considered as a perturbation and the problem may be solved by the perturbation theory.

The above general method will be applied to a special case, i.e., the Zeeman effect. The field is assumed to be weak, so that we may neglect $(q^2/2Mc^2)A^2$ (of the second order of H) in comparison with $(q/Mc)\vec{A} \cdot \vec{p}$ (of the first order of H). The uniform external magnetic field H being along the z-axis, we use Eq. (12–182) and obtain

$$-\frac{q}{Mc}\, \vec{A} \cdot \vec{p} = \frac{qH}{2Mc}\, (yp_x - xp_y). \qquad (12\text{–}197)$$

By Eq. (12–47) we have

$$-\frac{q}{Mc}\, \vec{A} \cdot \vec{p} = -\frac{qH}{2Mc}\, M_z. \qquad (12\text{–}198)$$

Equation (12–196) may thus be written as follows:

$$H = H^{(0)} + \lambda H^{(1)},$$

where

$$H^{(0)} = \frac{1}{2M}\, p^2 + U(x, y, z),$$

$$H^{(1)} = M_z,$$

$$\lambda = -\frac{qH}{2Mc}. \qquad (12\text{–}199)$$

The problem is thus expressed in a form suitable for solution by the perturbation method. There is only one difference. The perturbation Hamiltonian now contains an operator instead of a function of x, y, z. However, the perturbation theory of Chapter 10 is so formulated that it applies equally well for perturbation Hamiltonians containing operators, provided that in the matrix element

$$H^{(1)}_{mn} = \iiint \psi_m^* H^{(1)} \psi_n \, d\tau \qquad (12\text{–}200)$$

$H^{(1)}\psi_n$ is interpreted as the result of the operation of $H^{(1)}$ on ψ_n, instead of simple multiplication. The reader is urged to go back to Chapter 10 and verify this generalization for himself.

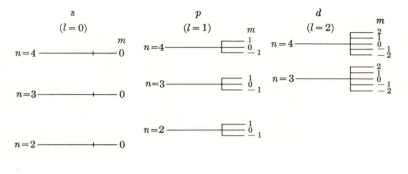

FIGURE 12–4

$n=1$ ——————|———— 0

In the Zeeman effect, the valence electron (which causes the emission of light) of an atom moves under the central-force field of the nucleus shielded by the inner electrons, and the external magnetic field. The unperturbed wave functions of the electron are determined by the Hamiltonian $H^{(0)}$. Because of the central force the wave functions are of the following form:

$$\psi_{nlm}(r, \theta, \varphi) = R_{nl}(r) Y_{lm}(\theta, \varphi). \tag{12–201}$$

The perturbed energy eigenvalues, to the first order, are thus

$$E_{nlm} = E_{nlm}^{(0)} - \frac{qH}{2Mc} \iiint \psi_{nlm}^* M_z \psi_{nlm} \, d\tau. \tag{12–202}$$

Since $Y_{lm}(\theta, \varphi)$ is an eigenfunction of M_z [which is represented by the operator $(\hbar/i)(\partial/\partial\varphi)$] with eigenvalue $m\hbar$, we have

$$E_{nlm} = E_{nlm}^{(0)} - m \frac{q\hbar H}{2Mc}. \tag{12–203}$$

Since ψ_{nlm} is an eigenfunction of $H^{(0)}$ and also $H^{(1)}$, it is an eigenfunction of $H^{(0)} + \lambda H^{(1)}$, i.e., of H. Thus ψ_{nlm}'s are themselves the first-order perturbed eigenfunctions. The first-order perturbation is thus solved.

The energy value now depends on the magnetic quantum number m as well as the other two quantum numbers n, l. The $(2l + 1)$-fold degeneracy of the energy level designated by (n, l) is thus removed by the perturbation of the magnetic field. The energy level diagram is represented schematically in Fig. 12–4. There is no split for the s terms ($l = 0$). The p terms split into 3 levels $(2l + 1 = 3)$. The d terms split into 5 levels, etc. The spacings of the multiplet levels for any term (n, l) are independent of n, l, m, being a constant equal to $(e\hbar/2Mc)H$. Equation (12–203) may also be interpreted to mean that the state ψ_{nlm} has a magnetic moment $m(e\hbar/2Mc)$ along the z-direction. This value suggests that the z-component of the magnetic moment is quantized to integral multiples of the Bohr magneton. It may be remembered that in Bohr's theory each orbit has a magnetic moment equal to an integral multiple of the Bohr magneton. When the orientation of the orbit is quantized the z-component becomes quantized also. Although the picture of a circulating electron is no longer valid in quantum mechanics, there is a corresponding quantity, the probability current density, expressed by $(e\hbar/2Mi)(\psi^*\nabla\psi - \psi\nabla\psi^*)$, which may be interpreted as currents contributing to the magnetic moment of the atom. Actually, the magnetic moment of a state ψ_{nlm} may be calculated from the probability current density, and the result turns out to be exactly the same as we obtained here from an energy consideration.

If the external field is placed along the x-axis we may use the x-axis as the polar axis of the spherical coordinates for writing the spherical harmonics. The results will be the same. The choice of the polar axis is completely arbitrary and we usually choose the most convenient one. It is also possible, when the field is along the x-axis, to solve the problem by using spherical harmonics referred to the z-axis. Then the eigenfunctions will no longer be Y_{lm}, but will be linear combinations of them. These combinations represent a transformation from spherical harmonics referring to the z-axis to those referring to the x-axis. The physical meaning of the solutions and the eigenvalues are not changed.

12–11 Quantization of the radiation field. We can only mention a few general ideas of the elaborate theory of quantization of radiation field. The radiation field in vacuum is specified by two vector field quantities $\vec{E}(x, y, z)$ and $\vec{H}(x, y, z)$, which are governed by the Maxwell equations,

$$\nabla \cdot \vec{E} = 0,$$

$$\nabla \cdot \vec{H} = 0,$$

$$\nabla \times \vec{E} = -\frac{1}{c} \frac{\partial \vec{H}}{\partial t}, \qquad (12\text{–}204)$$

$$\nabla \times \vec{H} = \frac{1}{c} \frac{\partial \vec{E}}{\partial t}.$$

The last two equations, specifying the time rate of change, may be regarded as the equations of motion. Solutions of these equations, in the form of three-dimensional waves, describes the propagation of the electromagnetic waves. A wave exhibits the properties of interference, diffraction, and polarization. The electromagnetic theory of light asserts that light is a form of electromagnetic wave and explains successfully the propagation of light and the phenomena of interference, diffraction, and polarization.

Before we proceed to quantize this physical system we first note a difference between this system and those we discussed before. In a system of N particles the number of independent variables, being equal to $3N$, is always finite. The number of variables of an electromagnetic field, however, is infinite. The value of the electric (or magnetic) field, say E_x, at a point (x_1, y_1, z_1) is a dependent variable. For the infinitely many points in space there are infinitely many variables of E_x. The same is true for E_y, E_z, H_x, H_y, H_z. Actually $E_x(x, y, z)$ (or E_y, etc.) represents a collection of variables each labeled by three indices, (x, y, z). To complicate the situation further, these variables are continuous, instead of discrete as in the case of a many-particle problem.

In spite of the fact that the classical theory of the electromagnetic wave is theoretically complete and experimentally verified, we have to re-examine the theory in view of both the development of the quantum theory and the appearance of new experimental facts. Consider the field values of E and H at (x_1, y_1, z_1). In the spirit of the operational philosophy, which quantum theory helped to develop, they are meaningful only when they can be ascertained by some experimental operations. To measure the electric field we have to use a test charge. Since the test charge is a particle the mechanical description of which is subject to the uncertainty relation, the measured results of E and H are subject to some kind of uncertainty. In fact Heisenberg has shown that the uncertainties of E_x and H_y are subject to a relation similar to, but not exactly the same as, the uncertainty relation of a particle. This seems to suggest that we may not be justified in assuming definite values for E_x, H_y, etc. at a given point (x_1, y_1, z_1) and given time t_1 as it is tacitly assumed in the classical wave theory. We may have to assume that as in the quantum theory of a particle, E_x at (x_1, y_1, z_1) at time t_1 is to be specified by a probability distribution of values which may be obtained through an operator. Also, these operators may satisfy some kind of commutation relations as do the operators p_x and x of a particle. We have seen that a classical system may be quantized by assigning operators to dynamical quantities. The quantized system reduces to the classical system as a limiting case, but has the additional feature that many physical quantities are quantized. Similarly, by assigning operators to the field quantities E_x at (x_1, y_1, z_1), etc., assuming certain commutation relations between them, and considering the Maxwell equations (or their equivalent) as the operator equations of motion, we may establish a quantum theory of the radiation field which includes the classical wave theory as a limiting case and has the additional feature that physical quantities, such as energy and momentum, are quantized.

The need for quantization may also be made evident by the following argument. We know that the radiation field interacts with mechanical systems of charged particles. Since the latter are quantized, their energy and momentum can change only discontinuously. The energy and momentum of the field thus can change by discontinuous steps only. It may well be that the radiation field itself is quantized.

On the experimental side, many discoveries made early in the century show that the energy of a radiation field is quantized (photoelectric effect) and so is its momentum (Compton effect). Thus the quantization of radiation fields is not only suggested by theoretical reasoning but also demanded by experimental evidence.

The so-called corpuscular theory of light is a phenomenological theory describing the photoelectric effect and Compton effect. These effects may actually be explained by the quantization of energy and momentum. A quantized theory is thus able to explain the particle-like properties of light. On the other hand, in the classical limit, the quantized theory reduces to the classical wave theory, and therefore may explain the wave-like properties of light (which appear in the region of high quantum numbers). Thus the quantized theory holds the promise of explaining both the particle-like and wavelike properties of radiation and thereby resolves the paradox of wave-particle duality of light.

An elaborate quantum theory of radiation has been developed which belongs to an advanced course in quantum mechanics. In the following we shall consider only a very simple case. This discussion also demonstrates the wide applicability of the general method of quantum mechanics to physical systems which are quite different from those previously discussed.

We concern ourselves with the problem of quantizing a plane electromagnetic wave in the z-direction which may be represented by the following vector potential:

$$A_x = c \sqrt{\frac{4\pi}{\Omega\omega}} [q(t) \cos kz - p(t) \sin kz],$$

$$A_y = 0, \qquad\qquad\qquad (12\text{–}205)$$

$$A_z = 0,$$

where c is the velocity of light, Ω is the normalization volume, ω is the angular frequency of the wave, k is $2\pi/\lambda$, and $q(t)$, $p(t)$ are two functions of time.* The electric field E and magnetic field H are in the x- and y-direc-

---

* We are free to write in Eq. (12–205) a constant factor $c\sqrt{4\pi/\Omega\omega}$ which may be considered as taken out of $q(t)$ and $p(t)$. The reason for writing this factor here will be made clear in Eq. (12–221) later.

tions respectively, their magnitudes being given by

$$\vec{E} = -\frac{1}{c}\frac{\partial \vec{A}}{\partial t},$$

$$\vec{H} = \nabla \times \vec{A}. \tag{12–206}$$

For this case Maxwell's equations are equivalent to the following equation for the vector potential:

$$\frac{1}{c^2}\frac{\partial^2 A_x}{\partial t^2} - \frac{\partial^2 A_x}{\partial x^2} = 0, \tag{12–207}$$

or

$$\left[\cos kz\left(\frac{1}{c^2}\ddot{q} + k^2 q\right) - \sin kz\left(\frac{1}{c^2}\ddot{p} + k^2 p\right)\right] = 0, \tag{12–208}$$

or

$$\ddot{q} + \omega^2 q = 0,$$

$$\ddot{p} + \omega^2 p = 0. \tag{12–209}$$

Equations (12–209) may be regarded as the equations of motion for the two variables $q(t)$ and $p(t)$ which specify the plane wave completely. The solutions are simply

$$q = q_0 \cos(\omega t - \alpha),$$

$$p = p_0 \cos(\omega t - \beta). \tag{12–210}$$

It may be verified easily that the electric and magnetic fields as given by Eq. (12–206) are sinusoidal waves. Let us restrict ourselves to a wave in the $+z$-direction. This requires the solution A_x to take the form of $f(ct - z)$ instead of $f(ct + z)$. Thus,

$$\frac{1}{c}\frac{\partial A_x}{\partial t} + \frac{\partial A_x}{\partial z} = 0, \tag{12–211}$$

or

$$\left[\cos kz\left(\frac{1}{c}\dot{q} - kp\right) + \sin kz\left(-\frac{1}{c}\dot{p} - kq\right)\right] = 0, \tag{12–212}$$

or

$$\dot{q} = \omega p,$$

$$\dot{p} = -\omega q. \tag{12–213}$$

The equations of motion (12–213) imply the previous equations (12–209).

This set of equations of motion may be considered as the canonical equations of motion of the Hamiltonian

$$H = \frac{\omega}{2}(p^2 + q^2), \tag{12-214}$$

since

$$\dot{q} = \frac{\partial H}{\partial p} = \omega p, \qquad \dot{p} = -\frac{\partial H}{\partial q} = -\omega q. \tag{12-215}$$

Once our problem is formulated in the Hamiltonian form, we may proceed to quantize it by replacing the canonical variables p, q with operators $\boldsymbol{p}$, $\boldsymbol{q}$ which satisfy the following commutation relation according to assumption β':

$$\boldsymbol{p}\boldsymbol{q} - \boldsymbol{q}\boldsymbol{p} = \frac{\hbar}{i}. \tag{12-216}$$

The dynamical quantity $H(p, q)$ may thus be quantized. The eigenvalues of $H(p, q)$ may be obtained by a comparison with the linear harmonic oscillator, the Hamiltonian and eigenvalues of which are

$$H = \frac{1}{2M}(p^2 + M^2\omega^2 q^2), \tag{12-217}$$

$$E_n = \hbar\omega(n + \tfrac{1}{2}), \qquad n = 0, 1, 2, \ldots. \tag{12-218}$$

By letting $M = 1/\omega$ we obtain the Hamiltonian

$$H = \frac{\omega}{2}(p^2 + q^2), \tag{12-219}$$

which is identical with Eq. (12–214), and the eigenvalues

$$E_n = \hbar\omega(n + \tfrac{1}{2}), \qquad n = 0, 1, 2, \ldots, \tag{12-220}$$

which are thus the eigenvalues of Eq. (12–214). We now calculate the total energy of the plane wave:

$$\mathcal{E} = \iiint \frac{1}{8\pi}(E^2 + H^2)\, d\tau$$

$$= \frac{1}{4\pi}\overline{E^2}\Omega$$

$$= \frac{1}{4\pi}c^2\frac{4\pi}{\Omega\omega}\frac{\omega^2}{c^2}\frac{1}{2}(p^2 + q^2)\Omega \tag{12-221}$$

$$= \frac{\omega}{2}(p^2 + q^2)$$

$$= H.$$

Thus the eigenvalues of the total energy of the plane wave are those given by Eq. (12–220), which shows that the energy is quantized and can change only by an integral multiple of $h\nu$. Hence in the photoelectric effect, a plane wave can transfer energy to the electrons only in whole units of $h\nu$,* and therefore the wave appears as a group of particles each having energy $h\nu$. The first Einstein equation for the photon, $E = h\nu$, is thus *derived* from the quantum theory of radiation. However, it must be remembered that the concept of photon in the quantum theory of radiation is quite different from that of a light particle in the classical corpuscular theory of light.

The reader is urged to derive the eigenvalues of the momentum of a plane electromagnetic wave (Problem 12–7). The result is equivalent to the second Einstein equation, $p = h/\lambda$.

We have seen in Chapter 4 that for a linear harmonic oscillator in the high quantum number region, a superposition of energy eigenfunctions results in the formation of a wave packet for which the coordinate q and momentum p have *nearly* definite values and they change in time accord-ing to the solutions of the classical equation of motion. Similarly, for the quantized plane wave in the high quantum number region, a superposition of energy eigenstates results in a quantum state in which the variables p and q have *nearly* definite values and they change in time according to the classical solutions, Eqs. (12–210). For this quantum state, the energy value is not definite but nearly definite. Also E and H values at every point in space are not definite but specified with a small amount of un-certainty. These nearly definite values (or their average values) satisfy the classical equations of motion, i.e., the Maxwell equations. Therefore, the classical wave theory is *derived* as a special case of the quantum theory of radiation. The quantum theory thus includes both the corpuscular theory and the wave theory as its special cases.

12–12 Quantization of the matter wave. The wave-particle duality of the electromagnetic radiation may be accounted for by the quantization of the radiation field. In Chapter 3 we show that the wave-particle duality of matter may be accounted for by the quantization of the motion of a particle. Thus duality is very closely related to quantization. In this section we shall further demonstrate this relation by considering the quantization of the matter wave.

Suppose there existed an isolated civilization where physicists had performed experiments of the Davison and Germer type (electron diffraction) but not the Thomson and Millikan type (showing the existence of the electron). They

---

*This is justified from a quantum-mechanical consideration of the inter-action between the radiation field and the electron, which we do not discuss here.

would naturally conclude that the negative electricity was a wave instead of a stream of particles. Furthermore, they would try to find an equation of motion for this wave. The diffraction experiments had established the de Broglie relation. They naturally wanted the group velocity of the wave to equal the actual velocity of movement of a localized negative charge cloud. Their effort to find the equation of motion for the wave of negative electricity would be closely parallel to ours in Chapter 2 and the resulting equation for the wave amplitude χ would be

$$\nabla^2 \chi - \frac{2M}{\hbar^2} U\chi = -\frac{2Mi}{\hbar} \frac{\partial}{\partial t} \chi, \qquad (12\text{-}222)$$

in close analogy with our Schrödinger equation. Here M was a universal constant (not yet identified with the electron mass because they knew no electrons). From this equation of motion they developed a wave theory of negative electricity which was verified by experiments of the Davison and Germer type. For a time they seemed quite satisfied with the theory. Then some experimenter reduced the intensity of a beam of negative electricity and observed on a screen single scintillations, one at a time, instead of a steady diffraction pattern. When the beam intensity was increased gradually, they found that the scintillations became so numerous that their individuality was lost and they formed a continuous and steady pattern which showed dark and bright rings (or spots) typical of a diffraction phenomenon. They thus concluded that the negative electricity had a corpuscular structure which was unnoticeable at high intensity. Their theorists were baffled for a time, but soon found a theory which accounted for the corpuscular property by quantizing the equation of motion of χ. The procedure was similar to that of quantizing the Maxwell equations. The value of χ at a given point (x_1, y_1, z_1) was no longer considered to be definite; its average value was assumed to be obtainable from an operator. A set of commutation relations were introduced (between χ and χ^*). The total intensity of the wave was then found to be quantized to integral units which made the wave appear like a group of particles. That such a theory might be worked out successfully did not surprise us in view of our knowledge of the quantization of the electromagnetic wave. Since the equation of motion of χ was identical with our time-dependent Schrödinger equation, we might, for our own convenience, call this procedure the quantization of the Schrödinger equation. (The quantization actually has been worked out in our civilization, but we shall not discuss it further.)

When a delegate of physicists from this civilization visited us they would insist that the true law of the negative electricity was the quantized Schrödinger equation. The reader probably would be inclined to think that the true law of the electrons was the quantized equation of motion of an N-particle system [the N-particle Schrödinger equation (12-148)]. Both might claim that their theories were able to explain both the particle-like and the wavelike properties of the negative electricity (or electrons).

Jordan and Klein* have shown that the two approaches are actually identical. The quantized Schrödinger equation leads to results physically and mathe-

---

* P. Jordan and O. Klein, *Zeit. f. phys.* **45,** 751 (1927).

matically identical with those of the N-particle Schrödinger equation. (The *statistics* of the particles, Fermi-Dirac or Bose-Einstein, is accounted for by proper choice of the commutation relations.) With the benefit of hindsight, we may now conclude that there is only one theory, the quantum theory of matter, which has two continuous approximations, i.e., the classical corpuscular theory and the classical wave theory. Historically, it is usually the simpler special theories which come to be established first. In our civilization the corpuscular theory of matter (classical mechanics) was developed first; in the other the wave theory (the classical wave equation being the Schrödinger equation) was first. However, these special theories are merely approximate theories and they have to be generalized. The process of quantization enables us to establish the general (quantum) theory starting from a continuous approximation (classical theory). Both classical theories may be taken as the starting point; by quantization both lead to the same quantum theory of matter. It is therefore no surprise that the quantized particle theory and the quantized wave theory turn out to be identical.

And here lies the explanation of the wave-particle duality of matter.

PROBLEMS

12-1. Construct the operators for the total energy, the components of total linear momentum, and the components of total angular momentum of the N-particle system.

12-2. State and prove the quantum-mechanical theorems of conservation of energy, conservation of linear momentum, and conservation of angular momentum for the one-particle system. Give the physical interpretation and discuss their relations with the classical theorems.

12-3. Repeat the last problem for the N-particle system.

12-4. A linear harmonic oscillator is perturbed by $\lambda H^{(1)} = \lambda v$, where v is the velocity of the oscillator. What are the consequences? In particular derive the perturbed energy eigenvalues and eigenfunctions.

*12-5. A linear harmonic oscillator is exposed to a sound wave. Resonance may occur. Develop the quantum theory of resonance analogous to the semiclassical theory of radiation.

*12-6. Two linear harmonic oscillators of the same frequency are placed on the x-axis with a distance d apart. Both oscillate along the x-axis. A weak spring of a natural (unstretched) length d connects the two oscillators. In classical mechanics the energy of the system shifts back and forth from one oscillator to the other. Develop the quantum theory for this system. *Note:* The exchange of energy between a mechanical system and a radiation field is analogous to that of the present problem, and the quantum theory of radiation (including interaction with mechanical systems) is developed in a similar manner.

12-7. Determine the eigenvalues of momentum of a plane electromagnetic wave.

12–8. In the momentum representation, the momentum operator p_x is assumed to be a multiplication operator p_x. What is the form of the position operator x in order that the commutation relation may be satisfied?

*12–9. Find the momentum distribution of a linear harmonic oscillator in an energy eigenstate by the momentum representation (see Problem 12–8) and compare the results with Problem 4–5. [*Hint:* Use the result of the last problem to write the Schrödinger equation in momentum representation, the solution of which may be obtained immediately by using the mathematical results of Chapter 4.]

12–10. Let O_0 be the lowest eigenvalue of an operator O. Show that for two operators F and G,

$$(F + G)_0 \geq F_0 + G_0.$$

The equality sign holds when F and G commute.

*12–11. An α-particle of a definite momentum $\vec{p}$ enters a Wilson cloud chamber containing two water molecules assumed to be at fixed positions. Show that the probability of finding the two molecules ionized simultaneously is large only when the line joining the centers of the molecules is parallel to the momentum vector $\vec{p}$. *Note:* By a straightforward generalization it may be shown that the droplets forming the track lie on a straight line with a straggling consistent with the uncertainty relation governing a classical trajectory. Comparing this problem with Problem 6–6 we notice the following difference: In Problem 6–6 we consider the α-particle as an isolated system and the water molecule as an apparatus performing observation and thereby causing an uncontrollable amount of disturbance on the heretofore isolated system, the result of the observation being expressed by a reduction of the wave packet. In this problem we consider the water molecules and the α-particle as forming one physical system; this isolated system of three particles evolves causally according to the dictation of the 3-particle Schrödinger equation. Although we are able to predict the properties of the track formed, we have not reinstated determinism because we know only the probabilities of forming various tracks. It still requires an act of observation by an agent outside this system to find which of the possible tracks is actually formed. This process of including an observing apparatus as a part of the system may be extended *ad infinitum.* Yet the reduction of the wave packet cannot be eliminated. Indeterminism is thus an inherent part of quantum mechanics. The theory worked out in this problem, nevertheless, provides the mathematical formalism for the physical concepts employed in Problem 6–6; the reduction of a plane wave to a localized wave packet by the ionization of a molecule, heretofore assumed on physical ground, is now deduced explicitly.

*12–12.# Proof that Theorem II of Chapter 12 is correct for all normalizable wave functions. (This work has been done by a reader of this book and published in *Am. J. Physics.*)

---

* Indicates more difficult problems.

# Indicates new problem added to the expanded edition.

CHAPTER 13

MATRIX MECHANICS

Classical physics had great successes in the 19th century but ran into a grave crisis in the 20th. Bohr started the *old quantum theory* by restricting the classical theory by two *ad hoc* assumptions. The success led to the correspondence principle, which affirmed the classical theory in general but modified it in the low quantum number region.

Old quantum theory reached maturation with the Sommerfeld quantum condition. It was still a patched-up job. Then came the revolution in *quantum mechanics*. There were two versions: Heisenberg's matrix mechanics and Schrödinger's wave mechanics.

In the preceding chapters we follow the Schrödinger line, in which the truly quantum element, the Planck constant h enters through the de Broglie waves. The quantum number n enters through the boundary conditions. Thus the "nh" of the Sommerfeld quantum condition is no longer arbitrary but their genetic connection to classical mechanics is ambiguous. This is a general feature of the theory. Newton's law appears in the Ehrenfest theorem (Sections 2–6) as if a mere coincidence. The correspondence principle is not explicit. The wave feature dominates the particle aspect of mechanics.

In matrix mechanics the logic lineage to classical mechanics is clearer but the mathematical tool of matrix is more esoteric, which leads to difficulty for beginning students. In this new chapter of the expanded edition we follow the same approach used in Chapter 2 by reformulating classical mechanics in a new mathematical form, but now the matrix, not the wave, leaving an undetermined parameter H. By identifying the H with the Planck constant h, we get matrix mechanics. In this way the essential physics, the introduction of the indivisible quantum of action h into the otherwise continuous physical theory is separated from the purely mathematical complexities and can be understood clearly.

One important mathematical result is the derivation of the Sommerfeld quantum condition from the matrix commutation rule of momentum and position. The latter is an abstract, esoteric mathematical relation; the former the summation of a large variety of empirical information in atomic physics, which has now found its rightful position in the structure of the theory. It closes a missing link and wraps up a loose end of the fabric of the quantum theory. It offers satisfaction and delight to the pursuers of truth.

13–1 Introduction. Heisenberg's introduction of matrix mechanics[1] is a breakthrough in the development of the quantum theory. His original formulation, though imaginative and insightful, is difficult for students to grasp because it is not based on logical reasoning. This presents a conceptual difficulty which, together with others, makes quantum mechanics a difficult subject.

Actually many elements of quantum mechanics are already present in classical mechanics, and can be studied and understood within the logical framework of classical mechanics. The truly quantum-mechanical elements can then be isolated, which would be simpler and thus can be grasped more easily. For example, in the treatment of Chapter 2 the Schrödinger equation with the Planck constant h replaced by an undetermined constant H that is required to be small only, can be established in the framework of classical mechanics without any reference to the quantum phenomena. Quantum mechanics can then be introduced by the simple quantum condition that fixes the value of H at h.

$$H = h. \qquad (13\text{–}1)$$

Many of the consequences of the Schrödinger equation can then be understood in terms of classical mechanics and the conceptual difficulties will be reduced.

In this chapter we present a similar treatment for matrix mechanics. We first show that the mathematical framework of matrix mechanics with the constant h replaced by an undetermined constant H that is required to be small only, can be established as a mathematical substitute for classical mechanics without any reference to the quantum phenomena. Matrix mechanics can then be introduced by the simple quantum condition that identifies a constant matrix $\|\mathbf{H}\|$ related to H with the identity matrix multiplied by the Planck constant h.

$$\|\mathbf{H}\| = h\|\mathbf{I}\|. \qquad (13\text{–}2)$$

The analogy with the above mentioned treatment of the Schrödinger theory is obvious and the advantages are similar. The matrix $\|\mathbf{H}\|$ can then be shown to come from the evaluation of the commutator of momentum and coordinate matrices, i.e.,

$$\|\mathbf{H}\| \equiv 2\pi i(\|\mathbf{P}\| \cdot \|\mathbf{x}\| - \|\mathbf{x}\| \cdot \|\mathbf{p}\|). \qquad (13\text{–}3)$$

The quantum condition (13–2) is thus identical with the Heisenberg quantum condition,

$$\|\mathbf{P}\| \cdot \|\mathbf{x}\| - \|\mathbf{x}\| \cdot \|\mathbf{p}\| = \frac{h}{2\pi i}\|\mathbf{I}\|. \qquad (13\text{–}4)$$

This reformulation is not an empty exercise but leads to an important conclusion. We can now show that the quantum condition (13–2) or (13–4) is generally equivalent to the Sommerfeld quantum condition, thus establishing a logical connection of matrix mechanics with the old quantum theory and demonstrating explicitly that Eq. (13–2) or (13–4) performs the role of specifying the rule of quantization. Together with Eq. (13–3), this result gives us some physical insight into the mathematical quantity the commutator of two non-commuting matrices, which is usually thought of as not subject to direct physical interpretation. The elucidation of these relations should make matrix mechanics less mysterious, more transparent and easier to understand.

13–2 Matrix formulation as a substitute for classical mechanics. Consider classical mechanics with the linear harmonic oscillator as an example. The coordinate of the system is represented by a Fourier series that contains only two terms,

$$x = \frac{A}{2}e^{i\omega t} + \frac{A}{2}e^{-i\omega t} \tag{13–5}$$

where A is the classical amplitude. The square of x and any other functions of x can be evaluated by this Fourier series

$$x^2 = \frac{A^2}{4}e^{i2\omega t} + \frac{A^2}{2} + \frac{A^2}{4}e^{-i2\omega t}. \tag{13–6}$$

The momentum can be evaluated also from the Fourier series

$$p = im\omega\frac{A}{2}e^{i\omega t} - im\omega\frac{A}{2}e^{-i\omega t}. \tag{13–7}$$

Any functions of x and p can be evaluated and the results expressed as Fourier series. The canonical equations of motion may be regarded as equations that determine the Fourier series (13–5) and (13–7). The angular frequency ω is determined by the constants of the system and the amplitude A is determined by the initial condition x_0 ($x_0 = A$). For a variety of initial conditions x_0 we have a variety of series of the form (13–5), the totality of which covering all possible initial conditions (classically a continuum) represents the complete solution of the problem.

We now propose a mathematical substitute for the Fourier series in this problem by using matrices. For one solution with a fixed initial condition x_0 we write the Fourier coefficients as a row of a matrix,

$$\cdots \quad 0 \quad 0 \quad 0 \quad \frac{A_n}{2} \quad 0 \quad \frac{A_n}{2} \quad 0 \quad 0 \quad 0 \quad \cdots.$$

For all possible x_0 from zero to infinity we write similar rows and put them together to form a matrix with the rule that the middle zero be placed at the diagonal element of the matrix

$$\frac{1}{2} \left\| \begin{array}{ccccccccc} 0 & 0 & 0 & A_{n-1} & 0 & A_{n-1} & 0 & 0 & 0 \\ 0 & 0 & 0 & 0 & A_n & 0 & A_n & 0 & 0 \\ 0 & 0 & 0 & 0 & 0 & A_{n+1} & 0 & A_{n+1} & 0 \end{array} \right\| . \qquad (13\text{--}8)$$

The matrix then contains all the information to represent the complete solution of the problem. (We choose to write the Fourier coefficients in the convention of descending order so that the time dependent matrix will agree with that calculated from $\int \Psi_n^*(t) x \Psi_m(t) dx$ where the wave functions are written in the convention of negative exponential of time.)

In classical theory x_0 is continuous but for practical purposes we may always delineate the continuous spectrum into a set of discrete values with infinitesimal increment from one to the next in the sequence. As far as classical theory is concerned we only require the increments to be infinitely small.

It can now be shown that the results of all calculations performed on the Fourier series may be obtained approximately by analogous calculations of the corresponding matrices using the standard rules of addition and multiplications of matrices. The approximation will be better when the discreteness is finer and in the continuous limit the results will be identical. For example, the series of x^2 in Eq. (13–6) may be obtained by multiplying matrix $\mathbf{x}$ with matrix $\mathbf{x}$, the nth row of the product matrix is

$$\frac{1}{4}(\cdots \ 0 \ 0 \ A_n A_{n-1} \ 0 \ A_n(A_{n+1} + A_{n-1}) \ 0 \ A_{n+1} A_n \ 0 \ 0 \ \cdots) .$$

When the discreteness is very fine, A_{n+1}, A_n, and A_{n-1} are nearly the same and the above row thus be replaced by

$$\frac{1}{4}(\cdots \ 0 \ 0 \ A_n^2 \ 0 \ 2A_n^2 \ 0 \ A_n^2 \ 0 \ 0 \ \cdots)$$

which corresponds to the result of Eq. (13–6).

Thus all classical results may be obtained by a corresponding matrix calculation, in particular, the Poisson brackets and the canonical equations of motion. Therefore, in the framework of classical mechanics we may regard the classical equations of motion to be represented by the canonical equation in the matrix form and the solutions may be obtained by matrix calculation. This, of course, is the essence of Heisenberg's matrix mechanics. But Heisenberg's matrices are a collection of empirical numbers having no logical connection with classical mechanics and the conceptual difficulty is to find the rationale to impose the equations of classical mechanics on his matrices. In our presentation the mathematical connection of matrices and matrix calculation on the one hand and classical mechanics on the other is brought out clearly and explicitly.

So far the introduction of matrices into classical mechanics simply represents a novel mathematical substitute without other consequences.

13–3 The commutator of momentum and coordinate. One novel mathematical feature of the matrix system is the non-commutativity of matrix multiplication and we are interested in learning its physical significance. Construct the commutator of $\|\mathbf{p}\|$ and $\|\mathbf{x}\|$ and multiply with $2\pi i$, i.e., the matrix $\|\mathbf{H}\|$ defined in Eq. (13–3). We shall show that $\|\mathbf{H}\|$ is a diagonal matrix. The only possible non-zero elements are $H_{n,n-2}, H_{nn}, H_{n,n+2}$.

$$
\begin{aligned}
H_{n,n-2} &= 2\pi i\{\|\mathbf{p}\| \cdot \|\mathbf{x}\| - \|\mathbf{x}\| \cdot \|\mathbf{p}\|\}_{n,n-2} \\
&= 2\pi i\{p_{n,n-1}x_{n-1,n-2} - x_{n,n-1}p_{n-1,n-2}\} \\
&= \frac{1}{2}\pi i\{i\omega m A_n A_{n-1} - A_n(i\omega m A_{n-1})\} \\
&= 0.
\end{aligned}
\tag{13–9}
$$

Similarly, $H_{n,n+2} = 0$. Thus the matrix is diagonal. The diagonal element of the nth row has the following value

$$
H_{nn} = \pi\omega m A_n(A_{n+1} - A_{n-1}).
\tag{13–10}
$$

Using the following approximation on the assumption that the spacing of the discrete spectrum is very small and well-behaved,

$$
A_n = \frac{1}{2}(A_{n+1} + A_{n-1}),
\tag{13–11}
$$

the above can be changed to

$$
\begin{aligned}
H_{nn} &= \frac{1}{2}\pi\omega m(A_{n+1}^2 - A_{n-1}^2) \\
&= \pi\omega m(A_{n+1}^2 - A_n^2), \quad \text{to the second order}.
\end{aligned}
\tag{13–12}
$$

This represents the difference of a certain quantity $\pi m\omega A^2$ between two adjacent states n and $n + 1$. We next show that this very quantity is nothing but the action integral $\int p\,dx$ in classical mechanics. We evaluate the action integral using Eq. (13–7),

$$\oint_n p\,dx = \oint_n p\frac{dx}{dt}dt = \oint_n (p^2/m)dt$$

$$= \frac{1}{4m}(-\omega^2 m^2 A_n^2)\int_t (e^{2i\omega t} - 2 + e^{-2i\omega t})dt$$

$$= \frac{1}{4}m\omega^2 A_n^2\left(0 + 2\frac{2\pi}{\omega} + 0\right)$$

$$= \pi m\omega A_n^2\,. \tag{13–13}$$

Thus,

$$\oint_{n+1} p\,dx - \oint_n p\,dx = \pi\omega m[A_{n+1}^2 - A_n^2] = H_{nn}\,. \tag{13–14}$$

Note that the diagonal element H_{nn} is non-vanishing because of the discreteness of the states — state n is different from state $n + 1$. In the continuous limit, the difference between two adjacent states becomes vanishingly small and thus $H_{nn} = 0$.

Thereupon the matrices **p** and **x** become commutative as required in the classical limit. Thus the non-commutativity of the matrices is a result of the discreteness of the states. Quantum mechanics is required by empirical evidence to develop a mechanics that deals with discrete states and the non-commutativity of matrices turns out to be just the right mathematical feature that can be used to incorporate the discreteness into mechanics. How the classical continuous spectrum is to be broken up into a discrete sequence desired by experiments of atomic physics is, of course, a principal problem of quantum mechanics and is to be specified by a fundamental assumption designated as the quantum condition. Here we see the opportunity of introducing the quantum condition through the commutator.

The Heisenberg quantum condition of matrix mechanics, Eq. (13–4), calls for the matrix $\|\mathbf{H}\|$ to be a diagonal matrix with all elements H_{nn} equal to to the Planck constant h. Equation (13–4) asserts that the difference of the action integral between two successive states be always h, i.e., the action integral is quantized to the unit of the Planck constant except for an additive constant.

$$\oint p\,dx = nh + \text{undetermined constant}\,. \tag{13–15}$$

By setting the constant equal to zero we have the Sommerfeld quantum condition. The correct value of the constant for the linear harmonic oscillator is $h/2$ which Heisenberg determined by solving the linear harmonic oscillator problem by matrix calculation. Thus Eq. (13–4) carries out the function expected — specifying the rule of quantization — and the result includes the Sommerfeld quantum condition as an approximation as it should.

Incidentally the matrix (13–8) constructed on a classical basis has elements identical with those evaluated from $x_{mn} = \int \psi_m^* x \psi_n dx$ in the high quantum number regions as they should. See Eqs. (10–40, 10–41, 4–59).

From our discussion it is clear that the physical meaning of the non-commutativity of matrices is the quantization of the physical states. It is well known the Heisenberg quantum condition Eq. (13–4) leads to the uncertainty relation between position and momentum (see, for example, pp. 307–309). As a result the uncertainty principle is often thought of as providing the physical meaning of the non-commutativity of position and momentum. This view is misleading. The fact is that the uncertainty principle is the result of the existence of the first quantum state characterized by one unit of the Planck constant h (p. 138) and by itself does not imply quantization beyond the first quantum state. Therefore the uncertainty principle does not embody the full physical contents of the non-commutativity of physical quantities and the view missed the essence of the underlying physical meaning.

References

1. W. Heisenberg, Z. *Physik* **33**, 879 (1925).

Table of Physical Constants

Speed of light in vacuum $\qquad c = 2.997930 \times 10^{10}$ cm/sec

Avogadro's number $\qquad N_0 = 6.02486 \times 10^{23}$/mole

Electronic charge $\qquad e = 4.80286 \times 10^{-10}$ esu

Electron rest mass $\qquad M_e = 9.1083 \times 10^{-28}$ gm

Proton rest mass $\qquad M_p = 1.67239 \times 10^{-24}$ gm

Planck's constant $\qquad h = 6.62517 \times 10^{-27}$ erg·sec

$\qquad \hbar = 1.05443 \times 10^{-27}$ erg·sec

Boltzmann's constant $\qquad k = 1.38044 \times 10^{-16}$ erg/°K

INDEX